Ethno-Indology

Heidelberg Studies in South Asian Rituals

General Editor
Axel Michaels

Volume 4

2006
Harrassowitz Verlag · Wiesbaden

Visualizing Space in Banaras

Images, Maps, and the Practice of Representation

Edited by Martin Gaenszle
and Jörg Gengnagel

2006
Harrassowitz Verlag · Wiesbaden

Cover: Panorama of Banaras (second half of 19th century). Detail with section between Causaṭhī- and Trilocana Ghāṭ. Museum der Kulturen, Basel, IIa 9939.

Bibliografische Information Der Deutschen Bibliothek:
Die Deutsche Bibliothek verzeichnet diese Publikation in der Deutschen Nationalbibliografie; detaillierte bibliografische Daten sind im Internet über http://dnb.ddb.de abrufbar.

Bibliographic information published by Die Deutsche Bibliothek:
Die Deutsche Bibliothek lists this publication in the Deutsche Nationalbibliografie; detailed bibliographic data is available in the internet at http://dnb.ddb.de.e-mail: cip@dbf.ddb.de

For further information about our publishing program consult our website at http://www.harrassowitz.de/verlag

Printed on permanent/durable paper.
Typesetting: Quoc-Bao Do, South Asia Institute Heidelberg
Printing and binding: Hubert & Co., Göttingen
Printed in Germany

ISSN 1860-2053
ISBN 3-447-05187-6 after 1.1.2007: 978-3-447-05187-3

Contents

Social Practice and Everyday Life

MARTIN GAENSZLE AND JÖRG GENGNAGEL

Introduction

The pilgrimage city of Banaras is widely known as an impressively unique and particularly ancient historical place. For many, however, it is above all a universal, cosmic, and in a sense timeless sacred space. Both of these seemingly contrasting images contribute to how the city is experienced by its inhabitants and visitors to it, and there is a great variety of sometimes competing views: Kāśī the Luminous, the ancient Crossing, the City of Death, the place of Hindu-Muslim encounter and syncretism, the cosmopolitan centre of learning, and the like.[1] The present volume deals with the multiple ways this urban site is visualized, imagined and culturally represented by different actors and groups.[2] The major focus will thus be on visual media, which are of special significance for the representation of space. But these cannot be divorced from other forms of expression which are part of the local "life-world" (Ger. *Lebenswelt*). All the contributions look at imaginative constructions–both local and exogenous–of the rich topography of Kāśī and show that such constructions are not static but always embedded in social and cultural practices of representation, often contested and never complete.

Banaras is unique: the city's location on the western bank of the holy Gaṅgā, which flows in a curve to the north and north-east at this point, the elevated riverfront where bathers face the rising morning sun above an empty landscape–these and similar geographical peculiarities contribute to the special character of the town. Historically, Banaras is often praised as one of the oldest continuously settled cities in the world. Whatever the truth, and the archeological research unearthing this complex history is still in its beginning stages, one can say that Kāśī has been an important crossroads for more than 1,500 years: it was at the junction of the east-west route along the Ganges and the north-south route coming from the Kathmandu Valley, and developed into a major trading centre at least as early as the Gupta period. Considering this strategic position on the South Asian subcontinent, it is not surprising that the city has come to be regarded as a centre of Hindu culture though this is a more recent historical development.

Banaras is universal: as Sanskrit eulogies point out again and again, Kāśī is the ultimate place of origins, a place created by Śiva "when there was neither the sphere of the earth nor the creation of water" (*Kāśīkhaṇḍa* 26.28). Kāśī, in this divine perspective, is nothing less than a "microcosm of the universe" where "creation is ... continually replayed" (Parry 1994: 11, 32). It is due to this link with cosmic origins that the city is seen as outside normal time and space, indestructible *(avināśa)* and thus eternal. One popular and vivid image expressive of this particular state of affairs is the one which depicts Kāśī as "propped up on the tip of the trident of Shiva" (*Kāśīkhaṇḍa* 60.61): the city is "above" ordinary ground level. And because of this cosmic dimension of the site, it is not really restricted to a geographical location. Kāśī can be elsewhere: Uttarakāśī in the Himālaya, Dakṣiṇakāśī in South India, or else the Kāśī in one's heart. The city is therefore part of a more complex "imagined landscape" (Eck 1998) in which mythology and topography are intrinsically interlinked.

Banaras, then, exemplifies a certain duality of space, often theorized in terms of concepts of space and place (e.g. Gutschow 1994), in a relation to one another as general to specific. Phenomenological approaches in geography, archeology and anthropology have stressed that place is what is directly experienced, as concrete, specific and foregrounded actuality, whereas space is more general, more abstract: the horizon of understanding, background potentiality (see Tuan 1977, Hirsch & O'Hanlon 1995). While many philosophers have tended to regard space as primary and place as derivative, phenomenologists, on the contrary, stress the primacy of place: "Knowledge of place stems from human experiences, feeling and thought. Space is far more abstract than place. It provides a situational context for places, but derives its meaning from particular places." (Tilley 1994: 15; see also Casey 1996). A clear-cut separation of place and space seems impossible, but a fundamental polarity in human experience remains evident: the specific is always embedded in a larger conceptual frame.

The topic of this book, however, is the cultural representation of space, with space being understood as a general term inclusive of place, that is, the objectification of experience in cultural artefacts and practices. The rich literature of an anthropology of space and place has shown that there is great variety in the ways "sites" are culturally constructed, and that one and the same place may be seen rather differently by different actors. Keith Basso, for example, has pointed out the interconnection of places, place-names and moral narratives among the Western Apache, which makes up a "sense of place", "that close companion of heart and mind, often subdued, yet potentially overwhelming" (1996: 54). Working in a different region but with a similar approach, Steven Feld has explored an "acoustemology of place" (1996: 91), which reminds us that place is experienced through the different senses to varying degrees, and thus in the overall representation one sense may stand out more than another.[3]

Visualization, then, is not the only way to represent space, and one needs to be aware that a certain visualism is deeply rooted in the European tradition. Yet it is certainly true that visual representations of space, in buildings, artefacts, maps, images, icons and the like are found in more or less elaborate forms everywhere. South Asia has a particularly rich heritage which includes scholarly treatises and handbooks with practical advice (e.g. *vastu śāstra* knowledge) and the long tradition of *maṇḍala*-related concepts in both the political and religious sense, which is reflected in indigenous cartography. A place is generally sacralized by the presence of a divine being, local spirits, and goddesses and gods of different orders. Visual representation of space in shrines or other objects of divine presence thus implies some form of interaction with these divinities, be it visiting, "seeing" (*darśana*), offering gifts or receiving blessings. This is an important cultural practice which is embedded in other such practices (bodily movement, ritual, music, speech etc.), and as a social phenomenon it is subject to change. The colonial encounter introduced new forms of visualization, new media and markets, different ways of seeing and treating space, which were variously appropriated, assimilated or rejected. The practice of visual representation has a history of its own.

As spaces can be constructed differently by different actors, there is bound to be conflict over the appropriate "handling" of a place. Brahmans and other high-caste experts of the written tradition tend to organize a complex and disparate reality according to holistic and encompassing patterns (see Gengnagel, this volume), whereas lower castes have their own set of ideas and are generally less concerned with such abstract models (Singh & Singh, this volume). Similarly, regional or ethnic backgrounds lead to diverging views: Maharashtrians attach great importance to Gaṇeśa and celebrate the Gaṇeśa Chaṭ festival in a particularly elaborate manner, while Nepalis revere Bhairava on an almost equal footing with Viśvanātha himself (see Gaenszle, this volume). And last but not least, it should be mentioned that Muslims, who represent about a quarter of the population, have had a significant influence on Vārāṇasī's urban culture, and maintain their own–but not entirely different–vision of cityspace (Visuvalingam & Chalier-Visuvalingam, this volume). Thus there are many perspectives on one and the same location, and though the residents of Banaras have a virtually common ethos, there has always been competition in the appropriation of space and its meaning.

With these general observations in mind, we will now turn to a brief outline of the structure of this book. The first section deals with aspects of the sacred topography of Kāśī. As is well known and has often been described, Banaras is a place with an extreme density of shrines, above all Śiva *liṅgams*, which are grouped in complex *maṇḍala*-ic patterns. Place is inhabited as well as structured by divine forces. In this section, the focus is on the built environment (icons, temples, wells, graves etc.), their interrelationship, and their role in historically changing ritual practice. In the following section, papers discuss the representation of this sacred topography in maps, i.e. in general, two-dimensional schemas, often combined with illustrative images and texts, which bring out the features seen as most significant. While this depiction of sacred space is at least partly a continuation of traditional techniques, the third section, on pictures and images, moves toward a form of representation which is largely a Western, colonial tradition: frontal views, veduta, panoramas and postcard photography. The last section takes a broader look at everyday spaces, more profane sites, little known social spaces and the processes of appropriation, negotiation and contestation which are involved in the construction of meaning. Thus the last section delves into the "unconscious" dimensions of space, which are usually covered over by the "official", idealized self-conscious representations by elites. Spaces are never just there; they are subject to social processes.

Sacred Topography

The section on the sacred topography focuses both on the "classical" textual description of sacred space in Banaras and on the spatial practices of ritual actors in particular performances, such as processions. It will become obvious that conceptions of sacred space are highly varied and incongruous, depending not only on the historical context and the social groups and actors involved (e.g. the pandit and the pilgrim), but also on the interpretive context of one and the same actor. The tradition characterized as *śāstrik* (scriptural) as opposed to *laukik* (popular) is cited as proof whenever something is believed to be authoritative, whether there is indeed textual evidence for it or not; but when it comes to practice even Brahmans often follow local customs which are not sanctioned by the scriptures (Parry 1989). Thus tensions which exist between "high" and "low", or the orthodox and the heterodox, may become visible and be played out among actors in the social arena, but they are also evident on the individual level, and there resolved in ordinary practice.

Hans Bakker's contribution deals with the origin and early developments of the Avimuktakṣetra. He focuses on a visualization of this *kṣetra* as given in the Vārāṇasī-Māhātmya of the *Skandapurāṇa*, tentatively dated to the 6th century. Situated in the ideal centre of this holy field, which extends from the Varāṇa river in the north to the Gaṅgā in the south, lies Madhyameśvara, once a hub for renouncers, who practised the Pāśupata observance (see Bakker Fig. 1). Regarding the localization of the Avimukta-kṣetra, Bakker points out the paradoxical nature of the cremation ground. On the one hand, pilgrims once kept the present Śmaśāna Ghāṭ Maṇikarṇikā on their left-hand side during their circumambulations. On the other, the cremation ground by virtue of its polluting force played an important role in the Pāśupata religion and attracted Pāśupata ascetics to it, thus contributing to the holiness of Banaras. This highlights a constant tension between heterodox, often tantric practices and orthodox ideals, which, as shown in several contributions in the present volume, seems to be of great significance in the construction of the sacred spaces of Banaras.

The paper by Ravi S. and Rana P. B. Singh deals with the cultural, symbolic, mythical and spatial significance of goddesses in Kāśī. Though their number is more than three hundred in the *Purāṇas*, their importance in Śiva's city has often been neglected by scholars. Historically, the veneration of goddesses such as the Yoginīs is linked with ancient folk traditions, which gradually were incorporated into the Brahmanic tradition through the establishment of shrines, temples and temple complexes, and also through the introduction of new images. The present spatial pattern of goddesses in Kāśī is the outcome of a complex acculturation process and the maintenance and continuity of tradition. The physical network of locations evinces an underlying order of patterning, which closely conforms to the symbolic description in mythology.

Annette Wilke's contribution on the Banarsī Navadurgā cycle shows that the concept of the Nine Durgās is localized within the landscape of Banaras, thus lending a local topography and a local bias to the pan-Hindu superstructure. We find a short reference to the Banarsī Navadurgā cycle already in the *Kāśīkhaṇḍa*, which mentions the names of the sites of the nine goddesses, but not their spatial configuration. Nowadays some of the nine temples are among the most famous Devī shrines of the city, while others are hardly known and not visited during other times of the year, or else are confined to non-Banarsī ethnic groups and arouse public communal interest only during Navarātrī. There is a clash between the textual tradition on the one hand

and the actual religious life and diversity of rituals on the other. The traditional Māhātmya known as the *Kāśīkhaṇḍa* and the modern local Hindi Newspaper *Āj* (quoting verses from the *Devīpurāṇa*) both present a uniform picture of the Navadurgā in portraying the nine manifestations as equals. The living tradition, however, presents a rather variegated picture. The textual tradition tends to blur such individuality. The paper suggests that the unifying principles which provide a sense of an unfragmented whole are specific to Banaras: the spatial configuration of the sites in a triangular scheme, which replaced an older circular pilgrimage route. The mental map, according to this interpretation, is the *yoni* of the goddess, which is projected onto the cityspace of Varanasi.

Sacred space is often dense with meaning deriving from different historical layers and different social groups. This is shown in the paper by Elizabeth Chalier-Visuvalingam and Sunthar Visuvalingam which deals with Bhairava, the guardian of Banaras, and more particularly the ritual enactment of the marriage of Lāṭ Bhairõ to the adjacent "maternal" well. This fertility ritual, featuring a procession through the cityspace, used to be celebrated with "lower-caste" Muslim participation, even as the Hindus used to participate in the marriage of the "decapitated" martyr Ghāzī Miyā̃. It is argued that the pillar of Lāṭ Bhairõ, itself an appropriation of an earlier Buddhist icon, has been assimilated to Muslim ritual symbolism in a syncretic manner, and vice versa. Despite a century of more or less peaceful coexistence, the contestation at the sacred ground around the shrine of Lāṭ Bhairõ erupted into the Hindu-Muslim riots of 1809, which event provided the conceptual model for understanding subsequent communalism on the Indian subcontinent. Even today the site of Lāṭ Bhairõ is a paradigmatic case of both syncretic harmony and violent conflict.

Maps

The history of South Asian cartography has long been interpreted based on Western cartographic traditions. Maps of the South Asian subcontinent were assumed to be produced by foreigners, not by South Asians themselves. Maps produced in South Asia proper were neglected as a category in its own right, thereby supporting the assertion made in 1992 by Joseph E. Schwartzberg that "the study of the history of [South Asian] cartography is still in its infancy" (Schwartzberg 1992: 509). The first monograph on South Asian cartography was published by Susan Gole in 1989. Besides a brief introduction to the subject, the book contains many maps that have never previously been reproduced. Then Joseph E. Schwartzberg came out with an *Introduction to South Asian Cartography*, which is partly based on Susan Gole's collections and forms part of *The History of Cartography Project*, headed by the late J. B. Harley and David Woodward (1992). Since then a brief historical overview has been published by P. L. Madan (1997). During the last decades the commonly held view that no independent indigenous traditions of South Asian cartography exist has therefore been called into question (Raj 2003).

The geographical and cartographical construction of British India in the context of colonialism has been studied by M. H. Edney (1997). Aiming at a "deconstruction of the map", J. B. Harley (1988) has greatly influenced and broadened our perspective on cartography in general and non-Western cartography in particular. Within the field of critical human geography, the theory of geographic visualization has shown the need not only to identify accurate topographical maps but to realize the importance of multiple perspectives and multiple maps–in other words, to perceive and study maps as social constructions (Crampton 2001). In spite of this broadened interest in research on cartography, the available data on South Asian cartography remain scarce, and even when cartographic material is available, its study has generally been neglected by the relevant academic disciplines. The visual elements on maps have not been of much artistic interest for the art historian, the geographer seeks in vain for geographical similitude and topographical exactness, while for the Indologist inscribed texts are few in number and lack originality. To study religious maps depicting the sacred space of Banaras, it is necessary to look at both ritual practice, as reflected in processions, and the use of texts with spatial reference.

Taking a more theoretical stance, Axel Michaels demonstrates that in certain religious maps different principles of mapping, which represent different kinds of space, may occur in combination. He compares religious and scientific concepts of space and stresses that religious spaces are omni- and translocal, absolute, more felt than perceived, and represent the idea of an identity of inner and outer space. In the Euclidean concept of space, individual objects are placed in a homogeneous and abstract order defined by three-dimensional coordinates and linked with metric concepts. Space thus becomes equated with the methods of measurement and observation. Michaels shows that in pilgrimage maps there often exist no clear-cut distinctions between religious and scientific concepts of space. Religious maps can to a certain extent integrate scientific topographical representations without difficulty.

The contribution by Jörg Gengnagel presents a case study of a public controversy in the latter half of the 19th century concerning the "correct" performance of the Pañcakrośīyātrā, the circumambulation of Kāśī's holy field. The debate (*śāstrārtha*) is based on religious texts dealing with the sacred topography of Banaras, and most likely influenced the making of lithographic maps. It is shown that the references to authoritative spatial texts, the utilization of surveying techniques and topographical mapping were important tools in the attempt of the local elite to change the route of the Pañcakrośīyātrā. Although the Dharma Sabhā of Banaras, the Mahārāja, the District Magistrate and the famous playwright Bhāratendu Hariścandra were involved in this conflict, the endeavour finally did not succeed. The attempt to reconstruct an ideal practice had to be reconciled with the persistence of pilgrims' everyday pilgrimage practices.

That mapping is not a value-free practice but can serve consequential political agendas is brought out in the paper by Sumathi Ramaswamy dealing with Bharat Mata ("Mother India"), worshipped in Banaras in the form of a map. Ramaswamy examines the complex cultural and political configurations that led to the "enshrining" of the map of India in the Bharat Mata Mandir in 1936 in Banaras. In the context of the "Temple of Mother India", the map of India has become, literally as well as metaphorically, a deity. Ramaswamy argues that the map of India enshrined in the Bharat Mata Mandir metonymically represents the image of Mother India at a time when the latter's visual presence had become a political and cultural embarrassment. In the process, the map, which through a quarter century of ideological work had come to be associated with Mother India, sacralizes the nation's soil, turning it from an abstract territory into a "motherland", an entity worth living and dying for. The article suggests that the map of India, that colonial artefact produced to represent and control Indian space in secular and scientific terms, has been subversively appropriated by India's nationalists, in their quest to create a new "religion", the religion of nationalism. Within this new religion, the map of India occupies the place of a deity, even as the territory of "India" that it represents is re-presented as sacred land.

Images

While all the contributions in this volume deal with images in the larger sense, that is, visual representations of the imagination, this section includes contributions on images in the narrower sense, that is, external images, or pictures, such as paintings, drawings and photographs. These images all show Banaras as an urban landscape, yet there are considerable differences in perspective, the "ways of seeing" they express. For a pilgrim pictures of the city serve not only as souvenirs in the ordinary sense, they also are a physical reflection of the sacred place; they possess agency of their own (Gell 1998). Sellers of postcards and maps (see Fig. 1) often advertise their trade by stressing that putting the images on walls and then visualizing the sacred sites is to have *darśan* similar to that of visiting the real place. Such images do not simply depict a location; rather, they somehow embody its sacred qualities. They show a landscape of a different kind, a place "somewhere else", a heterotopia (Foucault 1986). For the foreign visitors, too, Banaras was such a heterotopia, the oriental city of the Hindus, but seen with eyes trained in their own tradition.

Since early on in the Western encounter with India European travellers have been particularly impressed by the city of Banaras. The unique riverfront as seen from a boat or from the opposite bank has been a favourite vista. Considering that many travellers came up from Calcutta by ship and thus encountered the city in this manner, this is perhaps not surprising. But as will become clear in this section, the frontal view of these early veduta was a peculiar Western form of representation. Soon this perspective and the accompanying techniques were adopted locally, and incorporated into the traditional modes of representation (e.g. maps), and new kinds of images emerged. Thus the colonial situation had a deep impact on the ways the city was perceived, not only by outsiders but also by the inhabitants themselves.

The paper by Niels Gutschow focuses on the genre of panoramas, the unique view of a place which purports to include "everything", that is, all the surroundings as seen in a circular view of 360 degrees. Gutschow traces the history of panoramic views of Banaras from the missionary Joseph Tieffenthaler's early image from across the Ganges (published in 1786) through early colonial paintings and photographs to present-day imageries on pilgrimage maps. It is of particular interest to see that in the course of

Fig. 1: Map sellers at Maṇikarṇikā Ghāṭ (photo by Martin Gaenszle).

the 19th century the idea of an extended horizontal view was adopted by Indian painters and map-makers alike. A new tradition of local panoramic painting developed (probably the painters were trained by Western experts), apparently to serve a growing market in colonial India and outside. At the same time, traditional pilgrimage maps incorporated elements of these panoramas, such as the stylized riverfront with the skyline of temples, roofs and minarets, and depictions of particular buildings–for example, the Town Hall or the new bridge across the river. Thus new ways of seeing are acknowledged, without, however, totally giving up older forms of representation.

Probably no other Indian city appears more often in illustrated Western travel reports than Banaras. The *ghāṭs* are for Banaras what the Taj Mahal is for Agra and the Qutb Minar for Delhi. Joachim Karl Bautze focuses in his paper on the history of Western views of Banaras (paintings and photographs) and shows how artistic representations were strongly reliant on earlier depictions, often directly copying them. The reportedly earliest shoreline view of the city is based on a copper engraving published in 1786 in Berlin by the previously mentioned Joseph Tieffenthaler. Other views, photographs (salt and albumen prints) in particular, were copied again and again with only slight variations. The precise dating of such views is extremely difficult, inasmuch as some may be considerably earlier than the date mentioned on the title page of the publication in which they are contained. What is more important, any non-photographic view fell victim to prevailing Western aesthetics and tastes, so that even a precise date may prove useless for reconstructing how exactly the *ghāṭs* were used and what they looked like between 100 and 220 years ago.

Sandria Freitag in her contribution approaches the local understanding of what it means to belong to Varanasi by examining both images of the city exported and entertained by others, and the activities in which residents themselves engage. As such cultural activities tend to occur repeatedly over time, creating and perpetuating identity of place, the author conceives of them as acts of cultural production. Comparing this production of regional identities in Varanasi, Jaipur and Lucknow, Freitag shows that they have their roots mainly in royal patronage which supported individual artisans, crafts, activities, festivals and the like. Thus the image which residents of a city have of themselves is closely bound up with certain products and practices accorded public recognition and support by the ruling elites. The paper further demonstrates how particular historical circumstances brought about an often syncretic variety of identities and products considered emblematic of the locality. Cultural artefacts are powerful images, metonymically representative of place, but not necessarily uniform and coherent throughout all sections of society.

Social Practice and Everyday Life

The final section covers various representations of everyday space, not so much the conscious, self-reflective and identity-shaping ones, but rather the ordinary, scarcely articulated but nevertheless fundamental imaginings of the lived world, usually less noted. The experience of place and the concomitant visualization of it derives from the use of one's habitat: flats, houses, courtyards, the neighbourhood, the *ghāṭs*, tanks, places of work and leisure, shrines of regular worship and the like. Here cultural representations in the form of images, maps and other artefacts are less elaborate, and visualizations have to be reconstructed on the basis of an intimate understanding of people's habitus and verbal expressions. Therefore the maps in this section are mainly those drafted by the researcher in an attempt to represent what otherwise would remain implicit. How space is organized in everyday practice is a fluid process, dependent on networks of actors (action space), pathways of habitual movement, and the interaction of users who appropriate space in the pursuit of private intentions. Banaras is not one space for all uniformly; rather it is constituted by numerous "mini-spaces", to use a term coined by Freitag in her paper. Above all, these everyday spaces, which may include imaginary boundaries, are never entirely fixed and determined but tend to be subject to negotiations and often contestations.

The paper by Nita Kumar examines the experience of space from the perspective of the child: how does a young person learn to get around–at home, in the neighbourhood, in the city? Kumar's study shows how the very design of a house imposes a certain discipline on its inhabitants, contributing among other things to the socialization of children, their interiorization of a disposition imbued with values. Spatial practices and images are an instrument of education by means of which ideas about gender, generation, the local and the translocal are transmitted. The author presents a lucid phenomenology of spatial experience based on interviews with children in their schools or at home. A special focus of the paper is the construction of the nation's space–which Kumar sees as flawed: images of the nation displayed at school in such forms as photographs of landscapes or historical figures remain pale, distant and little understood. More important for the child are the images of the local milieu encountered on the outside: in *mohallās*, markets, at work and during festivals. In spite of the differences due to class, education and particularly gender, on the level of emotion there is much common ground.

The issue of everyday space is also the topic of Stefan Schütte's paper, which deals with the Dhobīs (washermen) of Banaras. By the very nature of their work, washermen (and -women) have a considerable network of interpersonal relations within the city's space: residences, working places and customers are often spread over a vast area, so that daily activities require a high degree of mobility. Combining a geographical and a social-anthropological approach, Schütte delineates the multiple layers of the Dhobīs' social organisation–which enable them to cope with the fact that this small caste group is dispersed in numerous localities all over the city. There are various groupings for collective action, the most important one of which is called Ṭāṭ, a kind of translocal caste organisation, which not only regulates occupational matters but is also a major reference point for identity–more important than residence. Thus the paper illustrates in an exemplary manner what has been stressed with regard mainly to postmodern conditions: that social space is not necessarily continuous but may be constituted by disjunct places (cf. Appadurai 1996). Spatial coherence is largely a matter of the imagination.

Whereas the construction of "mini-spaces" may be determined by caste, as in the case of the Dhobīs above, it may also be linked with an ethnic or national identity. The paper by Martin Gaenszle (in collaboration with Nutandhar Sharma) focuses on the places which are manifestations of the special significance which Kāśī has had for the people of Nepal. For a long time Himalayan visitors have been coming as pilgrims, students, merchants or refugees. This chapter looks into the ways Nepalis have appropriated particular locations by imposing on them their own visions of the sacred city. Most prominent is the so-called Nepali Mandir at Lalitā Ghāṭ, the well-known pagoda-style temple built in 1843. Today this temple is a contested place: it has become a national symbol of Nepali culture on Indian territory, and thus it may be seen as the result of "Nepalization". Another focus of Nepali identity is the traditional residential area of the Nepali community around Dūdh Vināyak. This area is also known as "*choṭā Nepāl*" ("Little Nepal"), and indeed may be termed the Nepali quarter. Today this neighbourhood is marked by four pilgrimage hostels (*dharmaśālās*), meant especially for Nepali pilgrims and run by an organispation which oversees an expatriate network for the Nepali community in Banaras. Though these two places are quite separate in spatial terms, it is argued in this paper that there have been numerous links, interactions and contestations among actors who have tried to exert control over these places, including their meanings and representations.

The last contribution, by Vasudha Dalmia, leads us into a more ambivalent locality of Banaras. It deals with the long-standing tradition of dancing girls, courtesans and prostitution associated with the quarter of Dalmandi, a *mohallā* right in the centre of the city, near the Chauk. Until the end of the 19th century these women, who were singers and musicians performing in public places on certain occasions, were an integral part of society and held in respect by much of the city's elite. In her reading of Premchand's novel *Sevasadan* (1918), which narrates the story of a wife turned courtesan, Dalmia shows that when this tradition of music and other pleasures was increasingly attacked by reformist activists after the turn of the 20th century the political leadership of the municipality split along various ideological lines, and this led to a heated debate about proper morality and life-styles in the city. Though the reformists eventually won out, the continuing popularity of the novel indicates the significance of the less holy, seamier aspects of life in Banaras.

Space, Place and the Practice of Representation

The perspectives on space in the city of Banaras are not only highly varied and multivocal, they are also often ambiguous or contradictory; marked by tension and sometimes conflict. One is tempted to say that it is precisely this character, a form of liminality, which constitutes the city's continuing appeal and fascination, and ultimately its sacredness. Given that it is a place of transition, a city of both death and the regeneration of life, a place on the threshold to the beyond, located between earth and sky, it is not surprising that it spans a wide range of differences–different emotions, different traditions, different languages, different ethnic cultures–and that people see this as a positive trait. As one resident of the Nepali quarter said: "This is religious cosmopolitanism". And indeed, one could argue that the city represents one of the oldest forms of cosmopolitan culture.

It becomes evident in many of the contributions that a basic tension exists between normative space on the one hand–that is, the idealized, standard and textually sanctioned model of space–and the lived-in place of ordinary experience and everyday activity on the other. Normative space is the subject of exegesis and largely defined by members of the upper castes, in particular the Brahmans. But such norms, even if it is claimed that they exist from time immemorial, continually change, old models are modified and new models are adopted, so that there are layers and layers of meaning which never completely erase the earlier ones. This has often been likened to a palimpsest, which retains a trace of the past.

Although normative space is holistic, always claiming inclusiveness and a consistent order, there is in fact an intractable multitude of life-worlds and mini-spaces which elude this totalizing model. The contributions of this volume present a rich phenomenology of lived-in places defined, for example, by regular worship at local shrines, the use of public spaces in neighbourhoods, movements in translocal action-spaces, processions, festivals and many other activities. Communities are not constituted by spatial boundaries but by spatial practices. And these practices are charged with emotion–for example, feelings of solidarity, longing for the familiar, experience of the divine. Larger spaces abstracted from daily experience–such as the mythical or the national–are visualized only in a mediated manner and require special techniques and strategies of representational practice.

Such visualizations of space as maps, images and photos can be seen as various means of getting a hold on and controlling the multitude of spaces. Power struggles occur over control of the imagination–conflicting claims made by competing groups and individuals. And given the demand and diversity, there is a market for these representations. The ordinary map-seller on the *ghāṭs* knows this and profits from the trade in images of the city used for private consumption. In this he perpetuates a tradition which has absorbed generations of visions and counter-visions, even when depicting them as the timeless truth.

The papers presented in this volume emerged out of an international academic colloquium held at the Internationale Wissenschaftsforum, Heidelberg, in May 2002. We would like to thank the Deutsche Forschungsgemeinschaft (German Research Foundation) for the generous funding. The beginning of the colloquium coincided with the opening of the exhibition "Banaras–Representations of a Sacred City" at the Völkerkunde Museum of the von Portheim Stiftung. The exhibition was made possible by the generous help of the director of the Museum, Margareta Pavaloi, and the German-Indian Society (Heidelberg). We would like to thank Niels Gutschow and Axel Michaels for their constant support while we were preparing this volume. Niels Gutschow has kindly contributed many excellent drawings to this book. Finally, many thanks go to Quoc-Bao Do for his invaluable help in typesetting this volume.

Notes

1 For an introduction into the religious significance and sacred topography of Banaras see Eck (1982), Gutschow & Michaels (1993), Gutschow (2006) and Singh (1993). An extensive bibliography on Banaras by Axel Michaels and Jörg Gengnagel is available on the Internet (*www.sai.uni-heidelberg.de/abt/IND/publikation/bibbanaras/bibbanaras.htm*).

2 It is for this reason that we have not always standardized the spellings of the city's various names: Though we favour the transliteration from common Hindi "Banaras", other connotations are better expressed by the Anglo-Indian "Benares", or the Sanskritic and official "Vārāṇasī".

3 The Bosavi (Papua New Guinea) studied by Feld (1996) have a complex poetics of place-names which mimetically represent the acoustic uniqueness of particular places in the forest through a complex soundscape enacted in songs.

References

Appadurai, Arjun. 1996. *Modernity at Large: Cultural Dimensions of Globalization*. Minneapolis: University of Minnesota Press.

Basso, Keith. 1996. Wisdom Sits in Places: Notes on a Western Apache Landscape. In: Steven Feld & Keith H. Basso (eds.), *Senses of Place*. Santa Fe: School of American Research Press, pp. 53–90.

Casey, Edward S. 1996. How to Get from Space to Place in a Fairly Short Stretch of Time: Phenomenological Prolegomena. In: Steven Feld & Keith H. Basso (eds.), *Senses of Place*. Santa Fe: School of American Research Press, pp. 13–52.

Crampton, Jeremy W. 2001. Maps as Social Constructions: Power, Communication and Visualization. *Progress in Human Geography* 25: 235–252.

Eck, Diana. 1982. *Banaras: City of Light.* Princeton: Princeton University Press.

— 1998. The Imagined Landscape: Patterns in the Construction of Hindu Sacred Geography. *Contributions to Indian Sociology* 32: 165–188.

Edney, Matthew Henry. 1997. *Mapping an Empire: The Geographical Construction of British India, 1765–1843.* Chicago, London: University of Chicago Press.

Feld, Steven. 1996. Waterfalls of Song: An Acoustemology of Place Resounding in Bosavi, Papua New Guinea. In: Steven Feld & Keith H. Basso (eds.), *Senses of Place.* Santa Fe: School of American Research Press, pp. 91–135.

Foucault, Michel. 1986. Of Other Spaces. *Diacritics* 16: 22–27.

Gell, Alfred. 1998. *Art and Agency: An Anthropological Theory.* Oxford: Oxford University Press.

Gole, Susan. 1989. *Indian Maps and Plans: From Earliest Times to the Advent of European Surveys.* New Delhi: Manohar.

Gutschow, Niels. 1994. Vārāṇasī/Benares: The Centre of Hinduism? A Discussion of the Meaning of "Place" and "Space". *Erdkunde* 48: 194–209.

— 2006. *Benares–The Sacred Landscape of Vārāṇasī.* Stuttgart/London: Edition Axel Menges.

Gutschow, Niels & Axel Michaels. 1993. *Benares: Tempel und religiöses Leben in der heiligen Stadt der Hindus.* Köln: DuMont Buchverlag.

Harley, John B. 1988. Maps, Knowledge, and Power. In: Stephen Daniels & Denis Cosgrove (eds.), *The Iconography of Landscape: Essays on the Symbolic Representation, Design and Use of Past Environments.* Cambridge: Cambridge University Press, pp. 277–312.

Harley, John B. & David Woodward. 1992. *Cartography in the Traditional Islamic and South Asian Societies.* (The History of Cartography, Vol. II, 1). Chicago: University of Chicago Press.

Hirsch, Eric, & Michael O'Hanlon. 1995. *The Anthropology of Landscape: Perspectives on Place and Space.* Oxford: Clarendon Press.

Kāśīkhaṇḍaḥ Maharṣivyāsapraṇitaḥ Rāmānandapraṇitayā "Rāmānandī" vyākhyayā; Nārāyaṇapatitripāṭhīpraṇitayā "Nārāyaṇī", Karuṇāpati Tripāṭhī (ed.), 4 vols. Varanasi: Sampurnand Sanskrit University, 1991; 1992; 1996; 1998.

Madan, P. L. 1997. *Indian Cartography. A Historical Perspective.* New Delhi.

Parry, Jonathan. 1989. The Brahmanical Tradition and the Technology of the Intellect. In: Karen Schousboe & Mogens T. Larsen (eds.), *Literacy and Society.* Copenhagen: Akademisk Forlag, pp. 39–71.

— 1994. *Death in Banaras.* Cambridge: Cambridge University Press.

Raj, Kapil. 2003: Circulation and the Emergence of Modern Mapping: Great Britain and Early Colonial India, 1764–1820. In: Claude Markovits, Jacques Pouchepadass & Sanjay Subrahmanyam (eds.), *Society and Circulation: Mobile People and Itinerant Cultures in South Asia 1750–1950.* Delhi: Permanent Black, pp. 23–53.

Schwartzberg, Joseph E. 1992. Introduction to South Asian Cartography. In: Harley & Woodward (eds.), *Cartography in the Traditional Islamic and South Asian Societies.* (The History of Cartography. Vol. II, 1). Chicago: University of Chicago Press, pp. 295–518.

Singh, Rana P. B. 1993 (ed.). *Banāras (Vārāṇasī): Cosmic Order, Sacred City, Hindu Traditions*. Varanasi: Tara Book Agency.

Tilley, Christopher. 1994. *A Phenomenology of Landscape: Places, Paths and Monuments*. Oxford, Providence: Berg.

Tuan, Yi-Fu. 1977. *Space and Place: The Perspective of Experience*. Minneapolis: The University of Minnesota Press.

Sacred Topography

HANS BAKKER

The Avimuktakṣetra in Vārāṇasī: Its Origin and Early Development

Studies in the Skandapurāṇa VI[1]

To judge by the archaeological evidence, the sanctuary called Avimukta or Avimukteśvara was among the earliest ones established in Vārāṇasī.[2] From Gupta times to the "Muslim" period it played a central role in the growth of the holy field around it, which became named after it, Avimuktakṣetra. From the earliest Vārāṇasī-Māhātmya text that we possess, the one contained in the chapters 26 to 31 of the original *Skandapurāṇa* (SP), it seems to emerge that the early size of this *kṣetra* was significantly smaller than the present one.

This Māhātmya, which I tentatively date to the 6th century, describes first the primordial situation, when the holy field was covered by a divine garden called (*udyāna* or *devodyāna*). Śiva, who has taken residence there forever (a *nirukti* of *avimukta*), shows Pārvatī the garden and the *liṅgas* in it, which are either *svayaṃbhu* or have been installed by various gods, twelve altogether. Avimukteśvara is not included within this dozen, as it is conceived of as Śiva's own residence in Vārāṇasī. The twelve primordial *liṅgas* on the other hand are enumerated in such a way that they may be seen as forming a pattern, some sort of circuit around the Avimuktakṣetra, i.e. around the Avimukteśvara sanctuary. This interpretation is based on two assumptions: 1) The locations of the present-day holy places, kindly plotted for me on a map by my friend Niels Gutschow,[3] still correspond roughly to the locations of their namesakes mentioned in the text. 2) The Avimukteśvara temple was originally located at the site of the Jñāna Vāpī Mosque.[4] It seems probable that the Buddhist pilgrim Hsiuen-tsang, who visited the town in the 630s, in his description of the main Maheśvara temple of P'o-lo-ni-sse refers to this sanctuary:

> In the capital there are twenty Dêva temples, the towers and halls of which are of sculptured stone and carved wood. The foliage of trees combine to shade (*the sites*), whilst pure streams of water encircle them. The statue of the Dêva Mahêśvara, made of *teou-shih* (*native copper*), is somewhat less than 100 feet high. Its appearance is grave and majestic, and appears as though really living.[5]

If the sixth-century Hindu pilgrim were to follow the enumeration of *liṅgas* in the *Skandapurāṇa*,[6] he would start in the North (Fig. 1) at Śaileśa, move east to the confluence of the Varaṇā and Gaṅgā, then to the South along the eastern side (i.e. the left bank of the Ganges) to Svarlīna, from where he would make an excursion westwards

Fig. 1: Avimuktakṣetra. The twelve *liṅga* sanctuaries encircling Avimukteśvara according to *Skandapurāṇa* 29.58–59.

to Madhyameśvara, then continuing southwards along Hiraṇyagarbha, Gopreksa and Upaśāntaśiva, again an excursion westwards to Jyeṣṭhasthāna and back, and then, keeping the Avimukteśvara sanctuary on his right hand, to the southern-most shrine of Śukreśvara, from where he would proceed north again along the western side to Vyāghreśa and arrive at Jambukeśvara. From there he might return to Śaileśa to complete the circuit.[7]

The holy space containing these *liṅga* sanctuaries is defined in Adhyāya 167 of the same Purāṇa; it is said to cover the area that stretches from the Varāṇasī/Varaṇā river in the North to the Gaṅgā in the South.[8] In the middle of this field the *tīrtha* Madhyameśvara is said to be situated.[9] Combining both statements we may comprehend what the text means by saying that the holy field lies between the Varāṇasī in the North and the Gaṅgā in the South: the southern point, namely, is where the north-south axis running through Madhyameśvara reaches the western bend in the river Ganges.

Skandapurāṇa 29.52 gives the size of this *kṣetra*: it is said to measure one *yojana*, i.e., it extends one *krośa* in each of the four quarters of the compass. Since no centre is mentioned in this context, this would make sense, if we conceive of the field as resembling a quadrangle of one *krośa* per side, having a perimeter of one *yojana*.

> *krośaṃ krośaṃ caturdikṣu kṣetram etat prakīrtitam* |
> *yojanaṃ viddhi cārvaṅgi mṛtyukāle 'mṛtapradam* ||52||[10]

The above-mentioned north-south axis through Madhyameśvara is c. 3.2 km. The length of the west-east axis, measured from Vyāghreśvara as its westernmost point to the easternmost point co-ordinated to Saṃgameśa, is also about 3.2 km.[11] The text seems to imply that, if one would circumambulate the holy field that encloses all the holy places mentioned in *Skandapurāṇa* 29 (with the possible exception of Kapilāhrada/Vṛṣadhvaja), one would walk one *krośa* in the four directions, a *krośa* to the East, the South, the West, and the North. Madhyameśvara is thought of as its (ideal) centre, lying in between the Varāṇasī and Gaṅgā rivers, though topographically speaking this appears not to be exactly accurate (Fig. 2). This is the oldest, and possibly most original, visualization of the Avimuktakṣetra that our sources allow us to grasp, one which seems to conform by and large with the area of the present-day Oṃkāreśvara and Viśveśvara Khaṇḍas.

The Vārāṇasī-Māhātmya in the *Skandapurāṇa* seems to reflect an historic phenomenon when it describes what happened after these twelve-plus-one holy places were established and pointed out to the Goddess by Śiva.

> Mahādeva looked around in all directions. Then, i.e., after the God of gods (Devadeva) had become fixed (in it), that land all of a sudden became as if it were on fire, while lightning flashed and thunder roared; it became filled with lustres, resembling (the lustre) of the sun and fire in one—as if a million suns were rising.
> Thereupon hundreds of Pāśupata masters (*siddhas*) suddenly arrived—shining whitely, as their bodies were smeared with ashes, lords of the yogins—and also appeared those who followed (Vedic) sacrificial observances (*vaitānikavratāḥ*), and those whose sectarian marks were not visible, blazing in their union with Śiva—and they made obeisances to Maheśvara.[12]

The Māhātmya tells us that the sanctuary of Madhyameśvara is the headquarters of these *siddhas*, who practise the (Pāśupata) observance (*vrata*).[13] Their archetype, according to the Māhātmya, is the saint Jaigīṣavya, who realized a vision of Śiva in his "cave", named the Jaigīṣavyaguhā. The conclusion seems warranted that the Pāśupata order was deeply involved in the development of the holy field of Avimukta, at least from the Gupta times onwards. This is confirmed by Hsiuen-tsang's observations.

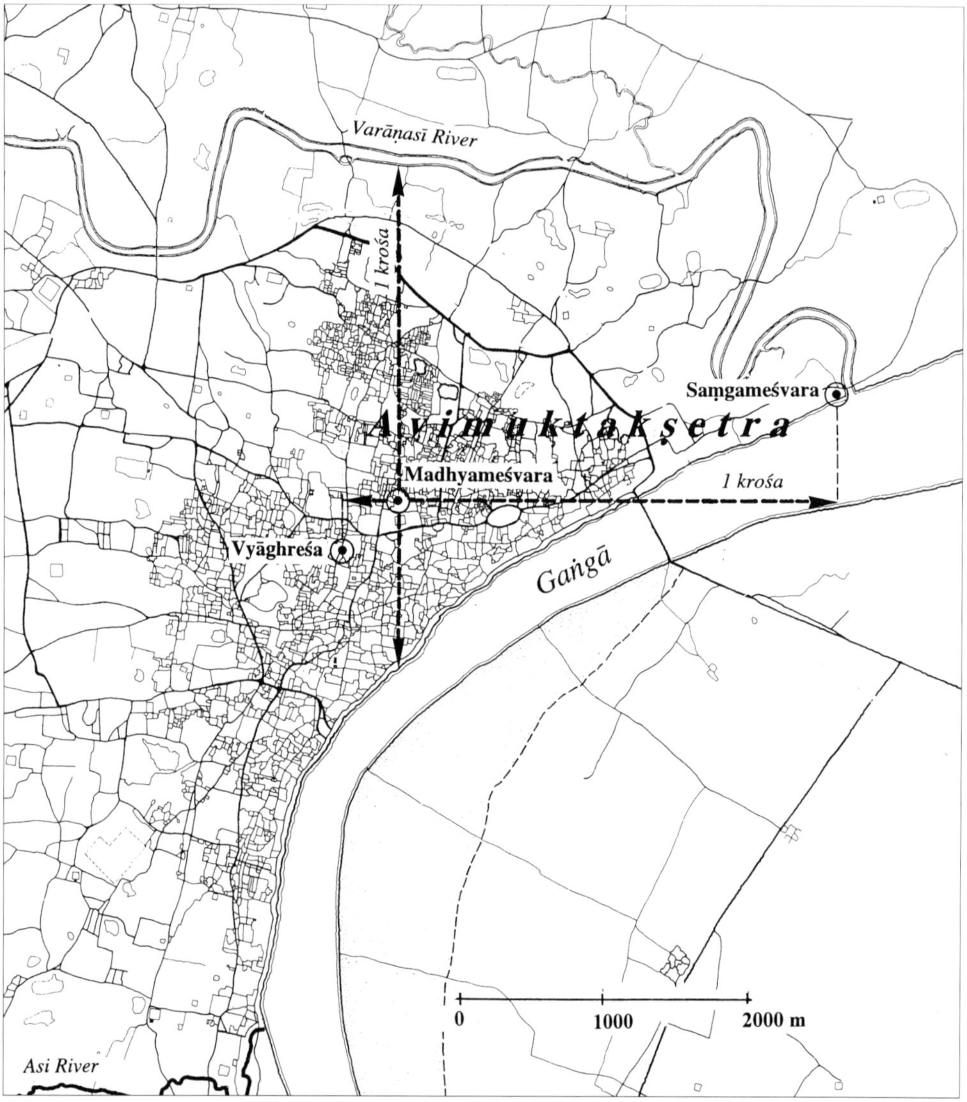

Fig. 2: The north-south and east-west extension of the Avimuktakṣetra according to the S recension of the *Skandapurāṇa*.

> [In this country] there are hundred or so Dêva temples with about 10,000 sectaries. They honour principally Mahêśvara. Some cut their hair off, others tie their hair in a knot, and go naked, without clothes; they cover their bodies with ashes, and by the practice of all sorts of austerities they seek to escape from birth and death. (Beal 1884, II: 44)

Overcoming suffering and death (*duḥkhānta*) was indeed the ultimate goal of the Pāśupata ascetic and, as is well known, he was prepared to go to extremes to realize it. Having forsaken social life, having solicited the status of a social outcast by transgressive behaviour, having withdrawn himself completely from the world in his "cave" (*guhā*), the Pāśupata renouncer finally ended his days on earth in the cremation ground, where he hoped to be saved by God in the hour of death. The cremation ground thus played a very significant role in Pāśupata religion, not as a positive value, though: the renouncer finally settled there precisely because its pollutive force made it the ultimate horror, the last tabu; in this way did he prove that he had transcended the world, the holy and the unholy alike. In this religion the cremation ground should be seen as the final challenge, and this it could be, because it lay outside the temenos, beyond the pale of sanctified ground. It is for this reason, probably, that the cremation ground, the *śmaśāna*, is conspicuous by its absence in the oldest recension of the Vārāṇasī-Māhātmya of the *Skandapurāṇa*. This cremation ground was there, as it was in every Indian city, and in one of the earliest Purāṇas, the *Mārkaṇḍeyapurāṇa*, it is depicted as the unholiest place on earth, the place where king Hariścandra passed his doleful days as a *caṇḍāla*.[14]

If we assume that the present Śmaśāna Ghāṭ, Maṇikarṇikā, still roughly marks the area where this cremation ground always has been, consulting the map of Vārāṇasī (Fig. 3), we notice that this area is situated at the southern fringe of the original Avimuktakṣetra, as it should be—where the north-south axis through Madhyameśvara reaches the holy river, about 400 m east of the Avimukteśvara sanctuary (hypothesized to be located at the site of the Ālamgīr Mosque). This situation is corroborated by the *Mārkaṇḍeyapurāṇa* (8.4), which tells us that king Hariścandra, when he was dwelling in the city of Vārāṇasī and was forced to work in the cremation ground, went in the southern direction, to where the *śmaśāna* of the town was located.[15] Shunned by the laymen, the *laukikas*,[16] who kept it on their left-hand side in their *pradakṣiṇa* around Avimukteśvara, it attracted the Pāśupata *siddha*, and in this way it paradoxically contributed to the holiness of the town.

This paradoxical process becomes discernible in the two later recensions of the original *Skandapurāṇa*, of which it can be shown on text-critical grounds that they go back to one hyparchetype that was known to Lakṣmīdhara. In the *Tīrthavivecanakāṇḍa* (p. 110) three verses are quoted that we find only in these two, so-called R and A recensions, of our text, not in the recension that is transmitted by the old Nepalese manuscripts (S). The translation of the edited verses runs as follows.

> At (i.e., adjacent to) the (*tac*) cremation ground (*śmaśāna*) Avimuktaka, where Rudra was worshipped by the sage Jaigīṣavya, is the [*liṅga*] named Avimukteśvara. Where a Brahman, having performed the *sandhyā* only once, † ... †. One who has seen that town of Vārāṇasī, the cremation ground Avimuktaka, and that (Lord) Avimukteśvara, he shall become a Gaṇapati.[17]

Although the second verse is corrupt beyond repair, we may draw two conclusions from these significant verses. 1) The cremation ground near the Avimukteśvara temple was known as Avimuktaka,[18] and, 2) it had developed into a *tīrtha*, that means into a

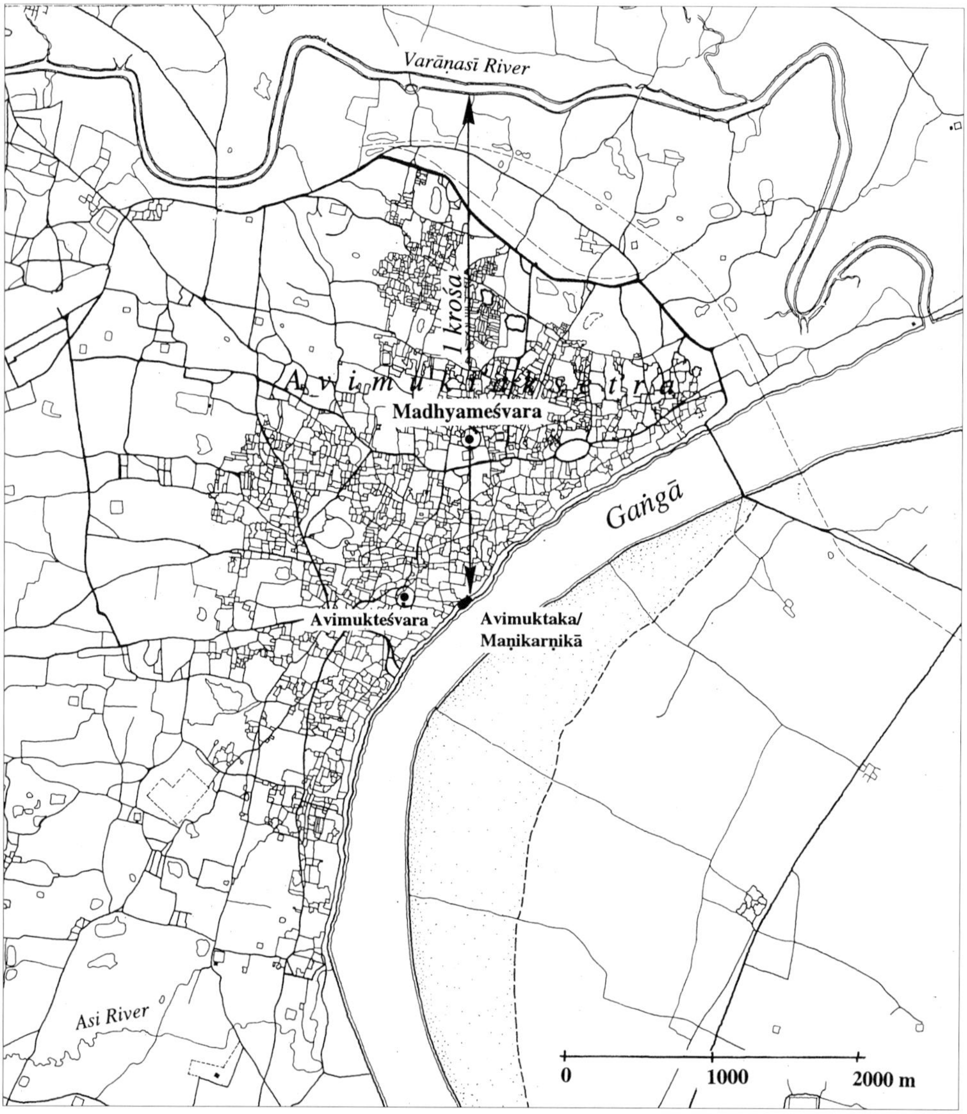

Fig. 3: The position of the Cremation Ground in the Avimuktakṣetra.

place that had become part of the Avimuktakṣetra. Concomitant, it would seem, with this spatial development was the spreading of a new doctrine, which is likewise found, probably for the first time, in these two later recensions of the Māhātmya. In an interpolation after *Skandapurāṇa* 30.50 it is said that men who are dedicated to *japa* and meditation reach their goal in Vārāṇasī:

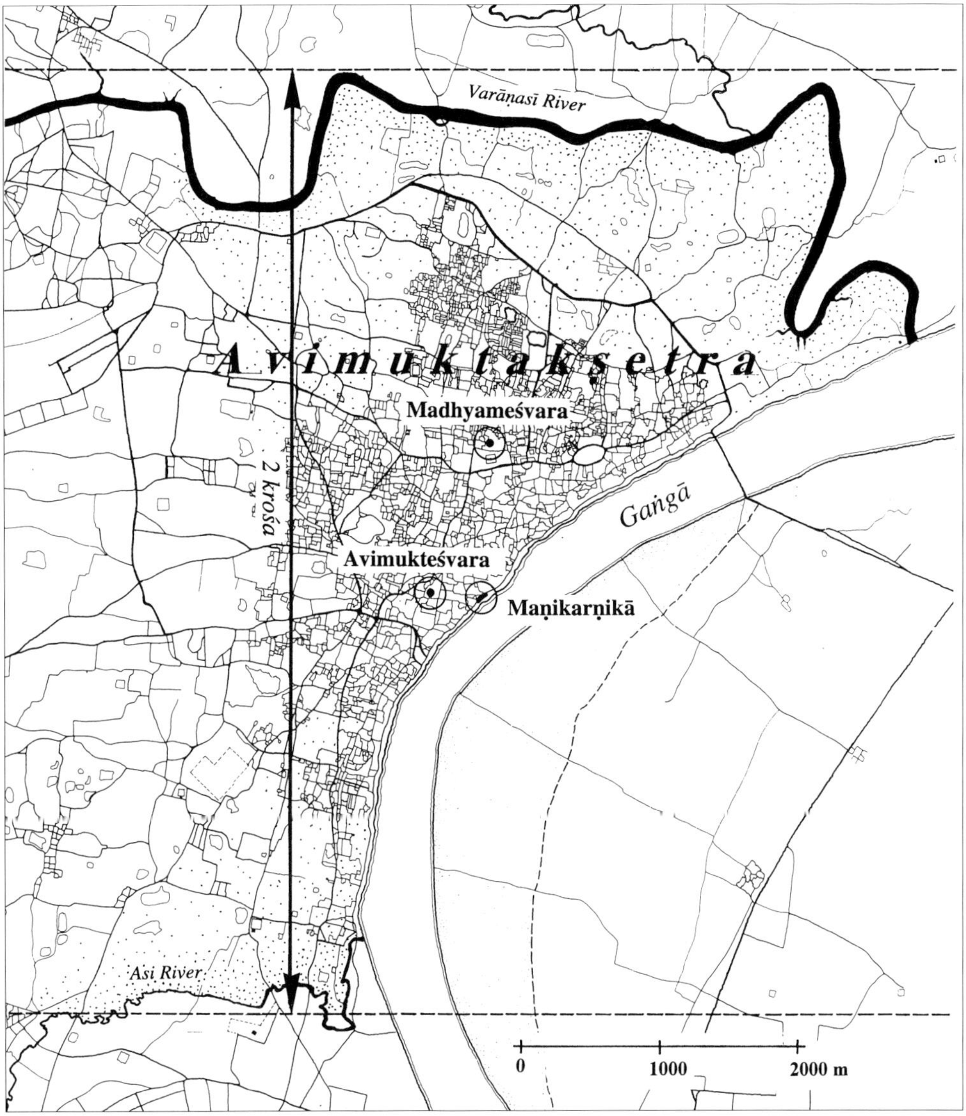

Fig. 4: The Avimuktakṣetra according to the *Brahmapurāṇa* quoted by Lakṣmīdhara (TVK pp.30, 39).

> Here, in the hour of death, Maheśvara himself gives the saving mantra (*tārakaṃ brahma*), thanks to which one becomes imbued with Him.[19]

This verse corresponds closely with a prose passage in the *Jābāla Upaniṣad*. In addition to a significant verbal agreement (given in bold face), this prose passage contains two metrically correct uneven *śloka pādas*,[20] which may be an indication that the Upaniṣad text derives from a metrical exemplar, which might have been slightly different, though, from the verse found in the R and A recensions.[21] The belief of a saving mantra

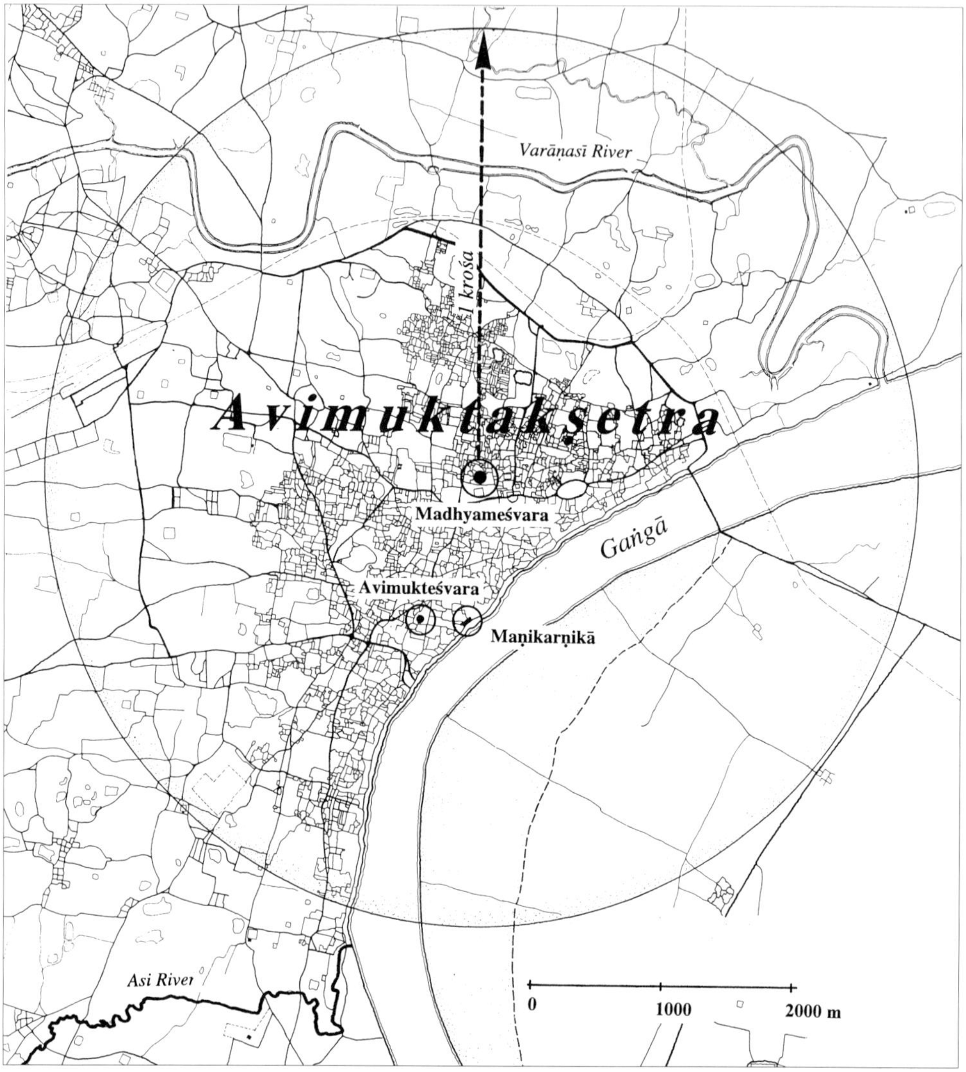

Fig. 5: The Avimuktakṣetra according to the "*Liṅgapurāṇa*" quoted by Lakṣmīdhara (TVK p.40).

in the hour of death is also known to the Vārāṇasī-Māhātmya in the *Matsyapurāṇa*, in which it is called the *karṇajāpa*.[22] In what seems to be an interpolation in the *Matsyapurāṇa*, this tradition is connected to the cremation ground Maṇikarṇī.[23]

Along with its being hallowed, the cremation ground, or rather the adjacent bathing place, received a new name: Maṇikarṇikā. This name, not yet in any of the *Skandapurāṇa* recensions, is first attested in Daṇḍin's *Daśakumāracarita* (7th century?), where the fourth prince, Arthapāla, reports the following.

> O Lord, I, being also on the same mission as my friends here, wandered around the ocean-girded borders of the earth. One day I arrived in the city of the Kāśī, Vārāṇasī. After a bath in the Maṇikarṇikā, of which the water is clear like a slither of crystal, and after having bowed before the Destroyer of Andhaka, Lord Avimukteśvara, I, making a circumambulation, saw a tall man who was girding up his loins with arms strong like iron clubs, whose eyes were red and swollen by incessant weeping.[24]

There might be a connection between this tradition of the "saving mantra", *tārakaṃ brahma*, and the name of this *tīrtha*: *maṇikarṇikā*.[25] The saving mantra is whispered by Śiva in the ear of the dying and, one assumes, should be repeated by him, hence the name *karṇajāpa*. The word "*maṇikarṇa*", taken as a Bahuvrīhi compound, could be taken to mean "whose ear is, i.e., possesses a jewel", which would be apposite to the dying man who has received the *karṇajāpa* or *tārakaṃ brahma*. The feminine derivative stem *maṇikarṇikā* refers to the bathing place at the river (*nadī*) where this soteriological event takes place. It became naturally interpreted as a goddess.[26]

It would thus seem that by the 7th century the holy field of Avimukta began to extend southwards. The pollutive place of the dead was enveloped within sacred space and transformed eventually into one of India's most holy places.[27] The growth of the *kṣetra* can be read in the descriptions of it collected from various Purāṇas by Lakṣmīdhara and collated in his *Tīrthavivecanakāṇḍa* (TVK pp. 39f.).

In what seems to be a conscious rewriting of *Skandapurāṇa* 29.52, the so-called "*Liṅgapurāṇa*" quoted by Lakṣmīdhara says that the *kṣetra* extends one *krośa* in four directions *from Madhyameśvara*,[28] which means that, if we conceive of it as a perfect circle, the surface of the holy field has increased by the factor π; it would have a diameter of two *krośas*, and a mathematical circumference of 6.28 *krośas* (Fig. 5).[29] Whereas in *Skandapurāṇa* 29.52 the mentioning of one *yojana*, being the sum of four *krośas*, is significant in that it may correspond with the perimeter of the field, in the rewritten "*Liṅgapurāṇa*", which clearly intends to describe a circle, its occurrence seems to be merely a vestige of the original, having no longer any other meaning than saying that four *krośas* amount to one *yojana*. Geographically this modified description could be taken to refer to a holy field extending from the north bank of the Varaṇā river in the North to the Asi river in the South.

The new historic situation is acknowledged also in the *Brahmapurāṇa* quoted by Lakṣmīdhara, which explicitly situates the holy field between the Varaṇā and Asi rivers (Fig. 4),[30] and describes the extent of the *kṣetra* as "measuring (*pramāṇa*) five *krośas*", which may refer to a perimeter of 5 *krośas*.[31] This may be seen as an adjustment to the historic reality, viz. a holy field that conforms to an (approximate) ellipse situated between the two boundary rivers. The alleged circuit of five *krośas* may have given rise to the idea of the Pañcakrośī Yātrā.

The *Matsyapurāṇa* agrees with the "*Liṅgapurāṇa*" in that the north-south extension of the holy field is said to be half a *yojana*, that is two *krośas*,[32] but it expands the west-east width to unrealistic dimensions by declaring it to be two or two-and-a-half *yojanas*.[33]

Finally, the idea of an Avimukta(kṣetra) situated between the Varaṇā and Nāśī (sic) is found in the *Jābāla Upaniṣad* 2. This text gives its location as: "The juncture (*saṃdhi*) of the eyebrows and the nose".[34] It remains as yet uncertain whether this conception, devoid of any geographical correspondence, came first and became reified eventually, when the holy *kṣetra* became defined as being situated between two rivers, the Varaṇā and Asi, or whether the allegorical interpretation of the *Jābāla Upaniṣad* reflected an historic reality. The latter possibility appears more probable. This would entail a considerably later date, by at least 900 years, for the *Jābāla Upaniṣad* than the 300 BC that has been proposed by Sprockhoff.[35] Such a later date seems also suggested by the passage concerning the "saving mantra" (*tārakaṃ brahma*), a semi-metrical periscope that this Upaniṣad has in common with the RA recension of the original *Skandapurāṇa* (see above).

From the evidence presented above we may conclude that by the 11th century the Avimuktakṣetra was conceived as lying between two rivers: the Varaṇā in the North and the Asi in the South. This vast expansion of holy space, a process that had gained momentum in the seventh century, accommodated an increasing number of sanctuaries and bathing places. The dozen or so *tīrthas* mentioned in the *Skandapurāṇa* and the twenty Dêva temples seen by Hsiuen-tsang had exploded to 285 in the "*Liṅgapurāṇa*" quoted by Lakṣmīdhara. Within this labyrinth the cremation ground took central stage. It gave Vārāṇasī its unique texture of holiness.

Notes

1 For Studies in the *Skandapurāṇa* I, II, III, IV, and V see Bakker 1996a, Yokochi 1999, Bakker 2000, Yokochi 2000, and Bisschop 2002.

2 Bakker 1996b: 33, 46 n. 6. "Avimuktakeśvara" is attested in a pre-Angkorian inscription (i.e. before AD 800): Bhattacharya 1961: 52.

3 I would like to express my sincere thanks to Niels Gutschow for providing me with accurate maps, which helped significantly with the interpretation of the texts. These maps were drawn by Niels Gutschow on the basis of a tracing of the 64-inch-to-1-mile map surveyed in 1928–29.

4 "*Liṅgapurāṇa*" quoted by Lakṣmīdhara in his *Tīrthavivecanakāṇḍa* (TVK) p. 109: *avimuktaṃ sadā liṅgaṃ yo 'tra drakṣyati mānavaḥ* | *na tasya punarāvṛttiḥ kalpakoṭiśatair api* || *devasya dakṣiṇe bhāge vāpī tiṣṭhati śobhanā* | *tasyās tathodakaṃ pītvā punarjanma na vidyate* ||. Bakker 1996b: 42f. Gutschow & Michaels 1993: 64f.

5 Transl. Samuel Beal (1884), II, 45. "*the sites*" evidently is an addition of Beal. The translation raises some questions. I have asked my colleague Dr. Harimoto to look at the Chinese text (Taisho ed.) and he kindly came with the following observations. "First of all, there is a slight possibility that the '100 feet' part might not be referring to the image of the 'Dêva Mahêśvara'. Even if it does, the text is ambiguous, it remains doubtful whether it refers to the height of the image. The character to which '100 feet' is predicated simply means 'measurement' (often used to render the Sanskrit *pramāṇa* or *māna*), a character which is used by Hsiuen-tsang to say that something has a certain size. He does use the character when he says certain Buddha images are of life size, as in 'The image measures the same as the body of the Tathāgata', which is equivalent to saying 'the size of the image is equal to that of the Tathāgata.' However, he does not use the character in

combination with numeral measurement without specifying what, such as height, weight, length, or circumference. The character is not used by him specifically in connection with the measurement of the height of statues at other sites, such as the Bamiyan sculptures. His description of the latter images as measuring 140–150 ft. (41–44 m) and 100 ft. (30 m) in height seems fairly accurate, since they were actually 55 m and 38 m high. The measurement of '100 feet' may refer to the size or circumference of the shrine in which the image stood, if it is not a textual corruption. The Chinese pilgrim's specification of the material, 'native copper', may also be compared with his description of the Bamiyan Buddhas. One of these too is said to be made of copper, while in fact it was of stone. The word Beal translates as 'native copper' is generally used in the sense of 'brass,' but Hsiuen-tsang seems to use the word to describe anything made of copper, pure or an amalgam like bronze. We may conjecture that the Bamiyan stone sculpture was covered by a metal coating, or something that resembled it. The Dêva Mahêśvara image may likewise have had a coating or, if it had an ordinary size, may have been a bronze. The translation of the 'trees' and 'streams' part seems to be quite wrong. The following translation may be proposed: 'The trees give shades to one another. The clean streams form a network'". The latter may refer to the drainage system of rivulets and ponds that connected both main rivers along the boundaries of the city (cf. Eck 1982: 50).

6 SP 29.58–59 (≈ SP_{Bh} 29.64cd–66cd): *śaileśaṃ saṃgameśaṃ ca svarlīnaṃ madhyameśvaram | hiraṇyagarbham īśānaṃ gopreksaṃ savṛṣadhvajam ||58|| upaśāntaśivaṃ caiva jyeṣṭhasthāna-nivāsinam | śukreśvaraṃ ca vikhyātaṃ vyāghreśaṃ jambukeśvaram | dṛṣṭvā na jāyate martyaḥ saṃsāre duḥkhasāgare* ||59||.

7 Although it may be interpreted as resembling one, the text does nowhere explicitly refer to the actual practice of this *yātrā* (a reference to the *yātrā* practice in general is made in SP 26.37). In Deva's and Devī's stroll along the *liṅgas* no topographical pattern is discernable (SP 29.8–57). In interpreting these two verses (SP 29.58–59) as we do, we try to uncover the underlying structure or idea that is supposed to have been in the mind of the composer when he wrote these verses. This idea appears for the greater part to have been informed by a geographic image of the holy field, though sometimes mythological associations seem to prevail. The latter might be true in the problematic case of the mentioning of Vṛṣadhvaja after Gopreksa. The sanctuary known as Vṛṣadhvaja belongs to the oldest stratum of Śaiva holy places in Vārāṇasī, being the only *tīrtha* mentioned in the critical text of the *Mahābhārata* (MBh 3.82.69). To date it is situated on the northern bank of the Varāṇasī river, on "mound 2" of Rajghat, a site that "after a settlement of about 600 years (*i.e.*, period I [800–200 BC]) ... was abandoned for a considerably long time and was reoccupied only during the late medieval times" (Singh 1985: 3). Why Vṛṣadhvaja is mentioned after Gopreksa in the "circuit" remains unclear; mythological association may have got the upper hand, since Vṛṣadhvaja/Kapilāhrada and Gopreksa are intrinsically related and described together in the Māhātmya (SP 29.8–19). It is conceivable, though, that these two holy places were adjacent to one another at the time our text was composed. After the abandonment of "mound 2", the original sanctuaries may have (temporarily) been eclipsed by sanctuaries in the city itself near Gopreksa.

8 SP_S 167.149 (S recension, edit. by P. Bisschop): *gaṅgā dakṣiṇatas tasminn uttareṇa varāṇasī | tat tayor madhyataḥ kṣetram avimuktaṃ pinākinaḥ* ||149|| (≈ SP_{Bh} 167.161). In a source that may be 500 years earlier, Aśvaghoṣa's *Saundarananda*, it is said that the city (Kāśi) is encircled by the Varāṇasī: *avabudhya caiva paramārtham ajaram anukampayā vibhuḥ | nityam amṛtam upadarśayituṃ sa varāṇasīparikarām ayāt purīm* ||3.10||.

9 SP 29.44–45 (≈ SP_{Bh} 29.48cd–50): *idam anyan mahat kṣetraṃ nivāsaṃ yogināṃ param | kṣetramadhye ca yatrāhaṃ svayaṃ bhūtvā samāsthitaḥ | madhyameśvara ity evaṃ khyātaḥ*

sarvasurāsuraiḥ || 44 || *siddhānāṃ sthānam etad dhi madīyavratacāriṇām* | *yogināṃ mokṣalipsūnāṃ janmamṛtyujitātmanām* | *dṛṣṭvedaṃ madhyameśānaṃ janma prati na śocati* || 45 ||.

10 SP 29.52 (≈ SP_{Bh} 29.58). This verse is part of a passage that suffered badly in transmission. The first hemistich is preserved only in the oldest Nepalese MS S_1 (and the R and A MSS). After this hemistich, this Nepalese MS has also a lacuna, so that the second hemistich is only known from the R and A MSS and the testimonium of Lakṣmīdhara (TVK p. 40). With regard to the measures of *krośa* and *yojana*, compare *Mārkaṇḍeyapurāṇa* 46.40: *krośo dhanuḥ sahasre dvau gavyūtis taccaturguṇam* | *proktaṃ ca yojanaṃ prājñaiḥ saṃkhyānārtham idam param* ||40||. 1 *krośa* = 2000 *dhanus*, 4 *krośa* = 1 *gavyūti* or *yojana* (cf. *Vāyupurāṇa* 101.126). Kane 1930–62, III pp. 145f., n. 185, IV pp. 589, n. 1337. Hayashi (1995: 116) notes that mathematical works usually give the relations "2000 *dhanus* = 1 *krośa*" and "4 *krośas* = 1 *yojana*". Cf. *Arthaśāstra* 2.20.25–26, where the *yojana* is said to comprise 4 *gorutas* (synonymous with *krośa*) of 2000 *dhanus*. A variant reading, however, gives 1000 *dhanus* to one *goruta* (cf. Barnett 1913: 217f.). This, again, is partly in line with the Bakhshali MS, which counts 1000 *dhanus* in one *krośa*, but—and this seems not to be attested in the variants of the *Arthaśāstra* passage—8 *krośas* in one *yojana* (Hayashi 1995: 114).

11 It thus appears that the length of the *krośa* used in Vārāṇasī at the time our text was composed is equivalent to about 3.2 km, or somewhat more, if we allow for some space on both sides of the outermost *tīrthas*. However, we should take heed of Basham's words, saying that "distances as given in texts are thus very unreliable" (Basham 1954: 506). Basham (ibid.) gives as the "commonest" equivalence 4 *krośas* (3.6 km) in one *yojana* (14.5 km).

12 Paraphrase of SP 29.60–63 (≈ SP_{Bh} 29.67–71ab). In the first stage of the Pāśupata practice (*sādhana*) the ascetics adopt the signs of their order, notably the application of ashes (*Pāśupatasūtra* 1.2). In the second and following stages, according to Kauṇḍinya, they drop these outward marks, i.e. they become *avyaktaliṅgin* (*Pāśupatasūtra* 3.1) and go around like madmen etc. Representatives of both groups seem to be introduced here. In addition, a third group seems to be mentioned, viz. that of the orthodox Vedic Brahmans who are devoted to (Vedic) fire sacrifices (*vaitāna*).

13 SP 29.44–45 (≈ SP_{Bh} 29.48cd–50) quoted above n. 9.

14 See the description in MārkP 8.109–130.

15 MārkP 8.108f.: *iti pratisamādiṣṭo jagāma śavamandiram* | *diśaṃ tu dakṣiṇāṃ yatra vārāṇasyāṃ sthitaṃ tadā* ||108|| *śmaśānaṃ ghorasaṃnādaṃ śivāśatasamākulam* |.

16 Cf. what is said to Hariścandra by his wife: *śmaśānavad varjanīyo naraḥ satyabahiṣkṛtaḥ* (MārkP 8.17).

17 This translation is based on the text edited by P. Bisschop (2002: 233), who notes: "These verses can be identified as SP_{RA} 167.4.5–7. Neither the manuscripts of R and A nor the TVK provide us with a satisfactory text, but with the external evidence of the TVK we are in a better position to account for the variant readings and partly restore the common hyparchetype": *avimukteśvaraṃ nāma tac chmaśāne 'vimuktake*| *jaigīṣavyeṇa muninā yatra rudro 'bhirādhitaḥ* ||5|| *yatra sandhyām upāsitvā brāhmaṇaḥ sakṛd eva tu* | † *sandhyām upāstavān so hi ṣaṭpadākhyaḥ samāḥ samāḥ* † ||6|| *purīṃ vārāṇasīṃ tāṃ ca śmaśānaṃ cāvimuktakam* | *avimukteśvaraṃ taṃ ca dṛṣṭvā gaṇapatir bhavet* ||7|| (see SP IIA Appendix p. 280f.). *Kūrmapurāṇa* 1.29.27, like the *Matsyapurāṇa* (see below n. 28), seems to identify Avimukta and the cremation ground (*śmaśāna*). The standard *nirukti* of *avimukta* ("never left (by Śiva)") may reflect the historic origin of this name. The Avimukteśvara sanctuary may have been closely connected with the nearby cremation ground from the days of its foundation. The name seems to express the belief that those who die and are

cremated there will reach Śiva, since He will never forsake them there in the hour of death. This belief developed eventually into the doctrine of the saving mantra.

18 See also KūP 1.29.59, quoted below in n. 21.

19 SP$_{RA}$: *atrotkramaṇakāle tu svayam eva maheśvaraḥ* | *dadāti tārakaṃ brahma yenāsau tanmayo bhavet* || (SP$_{Bh}$ 30^2.52cd–53ab).

20 Beginning with *prāṇe°*... and *yenāsāv*..., see below.

21 The prose text runs: **atra** *hi jantoḥ prāṇeṣ***ūtkrama***māṇeṣu rudraḥ* **tārakaṃ brahma** *vyācaṣṭe*, **yenāsāv** *amṛtī bhūtvā mokṣī* **bhav***ati* (JābU 1). Cf. the verse in KūP 1.29.59: *y***atra** *sākṣān mahādevo dehānte svayam īśvaraḥ* | **vyācaṣṭe tārakaṃ brahma** *tatraiva hy avimuktakam* ||59||. The *Rāmottaratāpanīya Upaniṣad* (RUTU 1.5) and the *Agastyasaṃhitā* (AgS 7.31) accommodated this tradition to the Vaiṣṇava, i.e. Ramaite, faith (Bakker 1986 I, 76).

22 MtP 182.23cd–24ab, quoted in TVK p. 17; also in the "*Liṅgapurāṇa*" quoted in TVK p. 35. In both passages this mantra is called *karṇajāpa*, which is explained by Vācaspatimiśra, who quotes Lakṣmīdhara, as *tārakabrahmopadeśaḥ* (TCM p. 347). A text-critical analysis of the Māhātmyas of the *Skanda-* and *Matsyapurāṇas* shows that this portion of the *Matsyapurāṇa*, at least the text as we have it, has been borrowed from the SP (see SP IIA p. 8).

23 MtP 182.24cd, not quoted in TVK p. 17; cf. SkP *Kāśīkhaṇḍa* 1.32.114–116. Bakker 1996b: 49 n. 20.

24 *Daśakumāracarita*, Ucchvāsa 4, first two sentences: *deva so 'ham apy ebhir eva suhṛdbhir ekakarmormimālinemibhūmivalayaṃ paribhramann upāsaraṃ kadācit kāśīpurīṃ vārāṇasīṃ* | *upaspṛśya maṇibhaṅganirmalāmbhasi maṇikarṇikāyām avimukteśvaraṃ bhagavantam andhaka-mathanam abhipraṇamya pradakṣiṇaṃ paribhraman puruṣam ekam āyāmavantam āyasapārigha-pīvarābhyāṃ bhujābhyām ābadhyamānaparikaram aviratаruditocchūnatāmradṛṣṭim adrākṣam* |.

25 Cf. "*Liṅgapurāṇa*" quoted in TVK p. 103, 121. *maṇikarṇī* in MtP 182.24.

26 E.g. SkP *Kāśīkhaṇḍa* 2.61.86.

27 The *Vārāṇasīmāhātmya* in the *Matsyapurāṇa* (MtP 182–185) is virtually equating the cremation ground (*śmaśāna*) and Avimukta. Śiva's Palace is said to hover above it: *śmaśānasaṃsthitaṃ veśma divyam antarhitaṃ ca yat* ||6|| *bhurloke naiva saṃyuktam antarikṣe śivālayam* | *ayuktās tu na paśyanti yuktāḥ paśyanti cetasā* ||7||.

28 TVK p. 40: *asmād* (i.e., Madhyameśvara) *ārabhya liṅgāt tu krośaṃ krośaṃ caturṣv api* | *yojanaṃ viddhi tat kṣetraṃ mṛtyukāle 'mṛtapradam* ||. The *Tristhalīsetu* p. 103 suggests that the SP too contains a similar verse, when it, after an explicit quotation from the *Skandapurāṇa* (SP 29.52, presumably borrowed from the TVK), continues with *tathā* introducing the following verse that is not found in any of our MSS, but in the *Kāśīkhaṇḍa* (2.97.151) instead: *muktikṣetrapramāṇaṃ ca krośaṃ krośaṃ ca sarvataḥ* | *ārabhya liṅgād asmāc ca puṇyadān madhyameśvarāt* ||.

29 Cf. SkP *Kāśīkhaṇḍa* 2.97.151. The quoted "*Liṅgapurāṇa*" gives yet another description, which takes its starting-point from Kṛttivāsas: *kṛttivāsasam ārabhya krośaṃ krośaṃ caturdiśam* | *yojanaṃ tatra tat kṣetraṃ gaṇai rudraiś ca saṃvṛtam* ||. The relationship between both descriptions remains opague.

30 TVK pp. 30 and 39: *varaṇā cāpy asiś caiva dve nadyau suravallabhe* | *antarāle tayoḥ kṣetraṃ dharaṇyāṃ na viśet kvacit* ||.

31 TVK p. 39f.: *pañcakrośapramāṇaṃ tu kṣetraṃ dattaṃ mayā tava* | *yatra codaṅmukhī gaṅgā* † *prācī vāpi sarasvatī* † || (cf. SkP *Kāśīkhaṇḍa* 2.55.44). What the reference to the Sarasvatī implies in this context is not clear to me. In the hemistich that follows, the Gaṅgā is said to run to the North over a distance of two *yojanas* (*udaṅmukhī yojane dve gacchate jāhnavī nadī* |), which is incompatible with the preceding measures, unless one assumes that it refers to the whole stretch in which the Ganges runs northwards.

32 MtP 183.61cd and MtP 184.51ab, quoted both, with minor variants and corruptions, in TVK p. 39: *ardhayojanavistīrṇaṃ dakṣiṇottarataḥ smṛtam* |.

33 MtP 183.61ab and MtP 184.50cd (TVK p. 39). In view of the corrupt state of both passages, both in the MtP as published and the TVK, this measurement may also eventually turn out to be due to a textual corruption. Cf. the discussion in Sukul 1977: 49–51.

34 *ghrāṇa* = *nāsā*/*nāsikā*. JābU 2.

35 Sprockhoff 1976: 95–109. Cf. Oberlies 2000, 181 n. 28. This early date is mainly based on an observation made by Kane 1930–62 (II, 421 n. 1001) to the effect that *Baudhāyanadharmasūtra* 2.10.2 and 18 "probably have in view the *Jābāla Upaniṣad* 4 or some other similar Upaniṣad passage". The latter might be the case. What the *Baudhāyanadharmasūtra* and the *Jābāla Upaniṣad* have in common is that both envisage the possibility that one enters *saṃnyāsa* directly after the stage of *brahmacarya*. One could find this view also in 20th century texts of which no one will maintain that the author of the *Baudhāyanadharmasūtra* had these texts in mind. A fixed *terminus ad quem* for 'the tradition (*śruti*) of the Jābālas' is Śaṃkara's commentary on *Brahmasūtra* 1.2.32.

References

Texts

Agastyasaṃhitā
Agastyasaṃhitā. 1–23. Introduction, Critical Edition and Annotated Translation by Roelf Barkhuis. Groningen 1995. 2 vols. (thesis).

Arthaśāstra
The Kauṭilīya Arthaśāstra, edit. and transl., with a study, by R.P. Kangle. Bombay 1960–1965. 3 vols. University of Bombay studies, Sanskrit, Prakrit and Pali, No. 1–3.

Aśvaghoṣa
The Saundarananda of Aśvaghoṣa. Crit. edit. and transl. with notes by E. H. Johnston. Delhi 1975 (1st ed. Lahore 1928).

Baudhâyanadharmasûtra
The Baudhâyana-Dharmasûtra, [edit. by] Eugen Hultzsch. Leipzig 1922. Abhandlungen für die Kunde des Morgenlandes Bd. 16, No. 2.

Daṇḍin
Daśakumāracarita of Daṇḍin. Text with Sanskrit Commentary, Various Readings, a Literal English Translation, Explanatory and Critical Notes, and an Exhaustive Introduction by M.R. Kale. Fourth Edition, Delhi 1966.

Jābāla Upaniṣad (JābU)
The Jābāla Upaniṣad. In: S. Radhakrishnan. *The principal Upaniṣads edit. with Introduction, Text. Translation and Notes*. London 1953. pp. 893–900.

Kūrmapurāṇa (KūP)
The Kūrmapurāṇa, crit. edit. by Anand Swarup Gupta. Varanasi 1971. All-India Kashiraj Trust.

Lakṣmīdhara (TVK)
Bhaṭṭaśrīlakṣmīdharaviracite Kṛtyakalpatarau Aṣṭamo bhāgaḥ. Tīrthavivecanakāṇḍam, edit. by K.V. Rangaswami Aiyangar. Baroda 1942. Gaekwad's Oriental Series XCVIII.

Mahābhārata (MBh)
The Mahābhārata. For the first time critically edited by V. S. Sukthankar and others. Poona 1927–1959. 19 vols.

Mārkaṇḍeyapurāṇa (MārkP)
The Mārkaṇḍeyamahāpurāṇam edit. (with a Hindi translation) by R. N. Śarmā. Venkateshvara Steam Press, Bombay 1910. Reprinted with an *Ślokānukramaṇī* by N. S. Siṃha. Nag Publishers, Delhi 1983.

Matsyapurāṇam (MtP)
Śrīmad-Dvaipāyanamuni-praṇītaṃ Matsyapurāṇam, etat pustakam ānandāśramasthapaṇḍitaiḥ saṃśodhitam. [Poona] 1981. Ananda Ashrama Sanskrit Series 54. [Reprint of the ed. of H. N. Apte AD 1907].

Pāśupatasūtra (PS).
Pāśupatasūtra. With the Pañcārthabhāṣya of Kauṇḍinya, edit. by R. Ananthakrishna Sastri. Trivandrum 1940. Trivandrum Sanskrit Series CXLIII.

Rāmottaratāpanīya Upaniṣad (RUTU).
Die Râma-Tâpanîya-Upanishad. Von A. Weber. Berlin 1864. Abhandlungen der königlichen Akademie der Wissenschaften zu Berlin. Philologische und historische Abhandlungen.

Śaṃkara
The Brahmasūtra Śaṅkara Bhāṣya. With the commentaries Bhāmatī, Kalpataru and Parimala, and with an alphabetical index of quotations. Edit. with notes etc. by Anantakṛṣṇa Śāstrī. Bombay Saka 1860 (AD 1917).

Vācaspatimiśra (TCM)
Tīrthacintāmaṇi of Vācaspatimiśra, edit. by Kamalakṛṣṇa Smṛtitīrtha. Calcutta 1910–12. Bibliotheca Indica 195.

Vāyupurāṇam (VāP)
Mahāmuniśrīmad-Vyāsa-praṇītaṃ Vāyupurāṇam, etat pustakam ānandāśramasthapaṇḍitaiḥ saṃśodhitam. [Poona] 1983. Ananda Ashrama Sanskrit Series 49. [Reprint of the edition by H. N. Apte AD 1905].

Skandapurāṇa
(SkP) *Śrī-Skāndamahāpurāṇam, [edit. by] Kṣemarāja Śrīkṛṣṇadāsa.* Venkateshara Steam Press, Bombay, V.S. 1967 [AD 1910]. 7 vols. [Reprinted by Nag Publishers, Delhi 1986 (2nd ed. 1995). 7 vols. Vol. 8 *Ślokānukramaṇī* by Nāgaśaraṇa Siṃha. Nag Publishers, Delhi 1990].

(SP I) *The Skandapurāṇa, Volume I. Adhyāyas 1–25. Critically Edited with Prolegomena and English Synopsis by R. Adriaensen, H. T. Bakker & H. Isaacson.* Groningen 1998. Supplement to Groningen Oriental Studies.

(SP IIA) *The Skandapurāṇa, Volume IIA. Adhyāyas 26–31.14. The Vārāṇasī Cycle. Critical Edition with an Introduction, English Synopsis & Philological and Historical Commentary by H. T. Bakker & H. Isaacson.* Groningen 2004.

(SP_{Bh}) *Skandapurāṇasya Ambikākāṇḍaḥ, saṃpādakaḥ Kṛṣṇaprasāda Bhaṭṭarāī.* Kathmandu 1998. Mahendraratnagranthamālā 2.

(SP_S) Draft edition of *Skandapurāṇa* 167 in the Nepalese (S) recension (see Bisschop *forthcoming*).

(SP_{RA}) Draft edition of *Skandapurāṇa* 167 in the *Revākhaṇḍa* (R) and *Ambikākhaṇḍa* (A) recensions (see Bisschop *forthcoming*).

Secondary Literature

Bakker, Hans T. 1986. *Ayodhyā. Pt.I The History of Ayodhyā from the 7th century BC to the middle of the 18th century. Its development into a sacred centre with special reference to the Ayodhyāmāhātmya and the worship of Rāma according to the Agastyasaṃhitā. Pt.II Ayodhyāmāhātmya. Introduction, Edition, and Annotation. Pt.III Appendices, Concordances, Bibliography, Indexes, and Maps*. Groningen. Groningen Oriental Studies vol. I.

— 1993. Early Mythology relating to Vārāṇasī. In: Rana P.B. Singh (ed.), *Banāras (Vārāṇasī). Cosmic Order, Sacred City, Hindu Traditions. Festschrift to Prof. R.L. Singh*. Varanasi, pp. 21–28.

— 1996a. Pārvatī's Svayaṃvara. Studies in the *Skandapurāṇa* I. *Wiener Zeitschrift für die Kunde Südasiens* 40: 5–43.

— 1996b. Construction and Reconstruction of Sacred Space in Vārāṇasī. *Numen* 43: 32–55.

— 2000. Somaśarman, Somavaṃśa and Somasiddhānta. A Pāśupata tradition in seventh-century Dakṣiṇa Kosala. Studies in the *Skandapurāṇa* III. In: Ryutaro Tsuchida & Albrecht Wezler (eds.), *Harānandalaharī. Volume in honour of Professor Minoru Hara on his Seventieth Birthday*. Reinbek, pp. 1–19.

Barnett, L. D. 1913. *Antiquities of India: an account of the history and culture of ancient Hindustan*. London.

Basham, A. L. 1954. *The Wonder that was India. A Survey of the Culture of the Indian Sub-Continent before the Coming of the Muslims*. London.

Beal, Samuel. 1884. see Hsiuen-tsang.

Bhattacharya, Kamaleswar. 1961. Les religions brahmaniques dans l'ancient Cambodge d'après l'épigraphie et l'iconographie. EFEO, Paris. Thèse principale.

Bisschop, Peter. 2002. On a quotation of the *Skandapurāṇa* in the *Tīrthavivecanakāṇḍa* of Lakṣmīdhara's *Kṛtyakalpataru*. Studies in the *Skandapurāṇa* V, *Indo-Iranian Journal* 45: 231–243.

— *forthcoming*. Early Śaivism and the *Skandapurāṇa*. Sects and Centres. Groningen 2004 (thesis).

Eck, Diana L. 1982. *Banaras City of Light*. Princeton.

Gutschow, Niels & Axel Michaels. 1993. *Benares. Tempel und religiöses Leben in der heiligen Stadt der Hindus*. Köln.

Hayashi, Takao. 1995. *The Bakhshālī Manuscript: an ancient Indian mathematical treatise. [Edit. and transl. from the Sanskrit and commentary] by Takao Hayashi*. Groningen. Groningen Oriental Studies Vol. 11.

Hsiuen-tsang. *Si-Yu-Ki. Buddhist Records of the Western World. Transl. from the Chinese of Hiuen Tsiang (A.D. 629) by Samuel Beal*. London 1884 (reprinted Delhi 1969).

Kane, P. V. 1930–62. *History of Dharmaśāstra (Ancient and Mediaeval Religious and Civil Law in India)*. Poona. 5 vols. Government Oriental Series, Class B, No. 6.

Oberlies, Thomas. 2000. Kriegslisten und ungeziemendes Benehmen: die Askesepraktiken der Pāśupatas. In: Ryutaro Tsuchida & Albrecht Wezler (eds.), *Harānandalaharī. Volume in honour of Professor Minoru Hara on his Seventieth Birthday*. Reinbek, pp. 175–191.

Sukul, K. N. 1977. *Vārāṇasī Vaibhav*. Patna.

Sprockhoff, Joachim Friedrich. 1976. *Saṃnyāsa. Quellenstudien zur Askese im Hinduismus. I Untersuchungen über die Saṃnyāsa-Upaniṣads*. Wiesbaden (Abhandlungen für die Kunde des Morgenlandes Band XLII, 1).

Yokochi, Yuko. 1999. Mahiṣāsuramardinī Myth and Icon. Studies in the *Skandapurāṇa* II, *Studies in the History of Indian Thought* 11: 65–103.

— 2000. The story of the seven brahmans in the *Harivaṃśa*. Studies in the *Skandapurāṇa* IV. In: Ryutaro Tsuchida & Albrecht Wezler (eds.), *Harānandalaharī. Volume in honour of Professor Minoru Hara on his Seventieth Birthday*. Reinbek, pp. 525–552.

RAVI S. SINGH AND RANA P. B. SINGH

Goddesses in Kāśī (Vārāṇasī): Spatial Patterns and Symbolic Orders

Introduction

In the major eulogic Sanskrit text of Kāśī, the *Kāśīkhaṇḍa* (most of whose parts are dated to the 14th century CE), the virtuous King Divodāsa made Kāśī the capital of his kingdom after Śiva and all the Devas had left the city (cf. KKh 43). Eventually Śiva and his wife Pārvatī, who lived in the mountains, no longer wanted to live away from their homely abode, Kāśī. But it was difficult to attack such a flawless king. Śiva realised: only when King Divodāsa would fall, he and other gods would get resort in the ordered and holy city of Kāśī. So in order to create a crisis in the city and dislodge Divodāsa from Kāśī Śiva sent sixty-four Yoginīs who entered the city. Being thrilled by the cosmic order and sacred power of the city, they settled down there. Similarly, other gods, planetary divinities, goddesses, demi-gods, assisting divine beings, and divine vehicle-animals in their different forms, all went there and settled down. Only after a cycle of time was completed, Śiva and Pārvatī returned home on Viṣṇu's appeal and took up residence again in Kāśī. Eventually, Divodāsa himself worshipped Śiva by installing a *liṅga*, which was later on known after him as *Divodāseśvara*. In the course of time the varieties and different groups of goddesses and gods came to support Śiva and to enjoy the sacred power of the city.[1]

This myth tells us something about the historical growth of goddesses, their male partners, and their multiplicity. It indicates that the Yoginīs are the oldest female divinities, originally non-Brahmanic folk deities, and that the other deities came later in the process of cultural transformation. Through a fusion of "high" and "low" and with the supreme integration of *prakṛti* (feminine power of nature) and *puruṣa* (masculine power of nature) an equilibrium is maintained. Further, this myth also talks about the processes and stages of non-equilibrium and equilibrium and their final merger resulting in mutual convergence.

Metaphorically speaking, through the intense power of deeper consciousness (*icchā-śakti*) "cosmic man", *puruṣa*, generated the energy of creation, *prakṛti*, i.e. feminine power. The *prakṛti* gets settled down at different places, giving them strong energy: *sthāna-śakti*. The inherent power possessed by these places is awakened by the power of words and sounds, *vāk-śakti*, and by the power of rituals, *kriyā-śakti*. The above powers (*śaktis*) of consciousness, sound, and actions together took form at the power places; thus places turn into more powerful sacred places, constituting a sacredscape. In the course of time the varieties, distinctions and uniqueness of these sacredscapes are mythologised and a glorifying literature is developed. Consequently there emerged a series of hierophanies in different groups. All such groups independently create their own systems of articulation, which in other contexts are in a state of non-equilibrium while with respect to interdependency they make up a complex system in a state of equilibrium. This way the opposite poles converge toward an ordering mode of a complex network system, which is in itself a mosaic of several independent but mutually cohesive sub-systems. This is reflected in the varieties and groupings of goddesses in Kāśī/Vārāṇasī.

In the Rāj Ghāṭ excavations, some remains of goddess worship are found, like ring stones, the circular discs which seem to have been associated with goddesses since the Indus Valley civilisation (Banerjea 1974: 168–173, 489). On one of such ring stones from the Mauryan period in the 4th or 3rd century BCE, there are relief carvings of full-hipped *devīs* (female divinities) with honey-suckle between them (Eck 1982: 75). In Kāśī, one finds a great variety of goddesses, which are attested since the 4th–3rd century BCE. Kāśī, or Vārāṇasī, the abode of Śiva, possesses different forms of the goddess, which are oriented according to their complexity, variety, cosmogony and sacred manifestation. These forms can be interpreted in the context of a multiplicity of layers, orderings, locations, hierophanies, and so on. The *Kāśīkhaṇḍa* (KKh), which constitutes the fourth canto of the *Skandapurāṇa*, is the main source of mythologies related to the goddesses in Kāśī. The following lists and categories are prepared on the basis of the *Kāśīkhaṇḍa*, and are further compared with results from field studies. According to the *Kāśīkhaṇḍa* there were 324 forms (identified by names) of goddesses in Kāśī. They are categorised into several groups, like 96 Śaktis, 8 Kṣetra-rakṣikas, 64 Yoginīs, 9 Durgās, 12 Gaurīs, 12 Mātṛkās, 9 Caṇḍis, 41 Kṣetradevīs, 12 Devīs on the Pañcakrośī route, 12 Independent Devīs, and 42 Loka ("folk") Devīs. Of course, out of the 324, only 144 are presently existing in Kāśī, and of these 36 are repeated in two or more groups. As a result at present only 96 are enumerated. For most of them worship is prescribed independently, and also the sacred times and orderings are fixed for pilgrimage. All the goddesses are associated with their male partner, always some form of Śiva, who, of course, is sometimes not mentioned explicitly. The *Kāśīkhaṇḍa* describes 324 main forms of Śiva, who together with the 324 forms of the goddesses, runs and regulates the rhythm of the cosmic cycle in the universe by integrating the power of male and female. By this integration the territoriality of the cosmic equilibrium is maintained. This paper deals with the spatiality, cosmic geometry and related myths and metaphors, and finally projects the idea of a self-organised system in order to explain the complex distribution of goddesses-shrines.

The Cosmic Circuit and Protecting Goddesses

There are five main sacred territories, marked by the respective travel routes, which symbolise the ecological cosmology and wholeness of Kāśī. These routes also symbolise the five *kośas* ("sheaths") and are analogous to the five gross elements of the organism according to Hindu mythology. The outer sacred circuit is identical to a circle whose centre lies at Madhyameśvara and the radial point at Dehalīvināyaka, covering a distance of 5 *krośa*, i.e. 11 miles/17.6 km. The circuit is also described as an aura of cosmic light that illumines the world. It is an example of the archetypal nature of spatial transposition, called "hierophany" by Eliade (1959: 12).

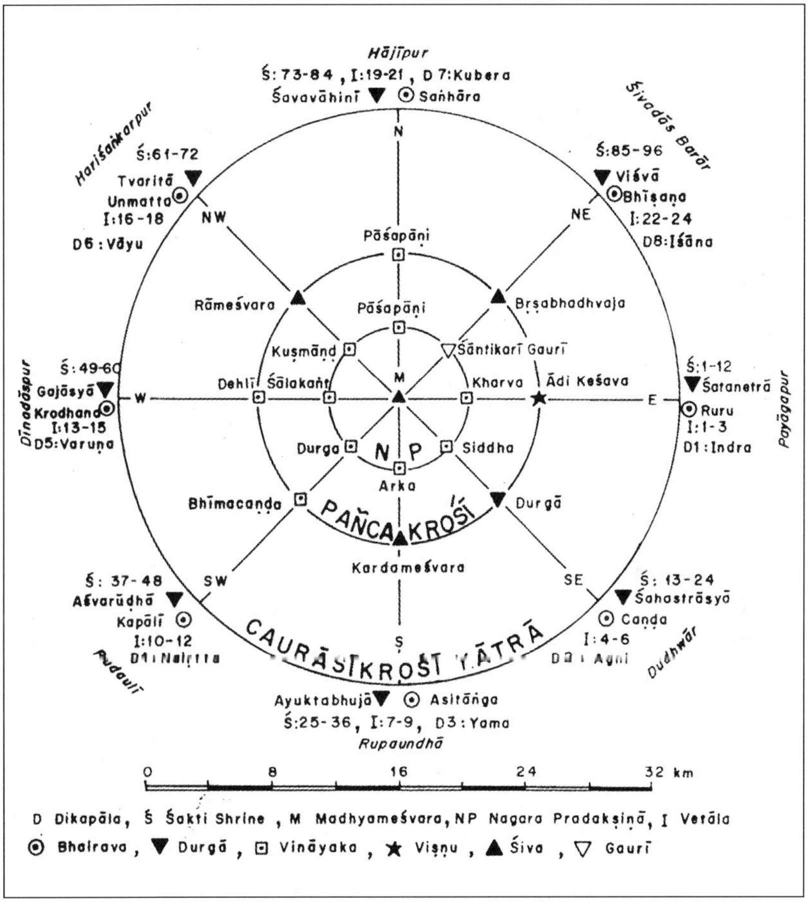

Fig. 1: Kāśī Maṇḍala: Sacred Yātrās and directional deities (drawn by the authors).

The outer circuit, known as Caurāsīkrośīyātrā route is divided into eight directions (*aṣṭa-dik*), and protected by the directional deities (Fig. 1). Each of these directions is regulated and controlled by goddesses of the annual cycle (i.e. 12 zodiacs/12 months), called Śaktis ("power goddesses"), and thus their number reaches 96 (i.e. 8 x 12). In a cognitive pilgrimage map prepared by Kailāsanātha Sukul in CE 1876 the location of shrines and deities along the sacred route is depicted as a *maṇḍala* (Michaels Fig.1, this volume). The myth recounts that the 96 Śaktis had helped the goddess Durgā in killing the demon's army. At the end of the war Durgā had cited their names (KKh 72.3–13).

The act of these Śaktis in each direction is supervised by Kṣetradevī ("territorial goddess"), a form of the Great Goddess Durgā, and further assisted by a form of Bhairava (her male partner), a directional deity (*dikpāla*) and three Vetālas ("demi-divinity assistant"). In this way the sacrality of time (twelve months) and the temporality of space (eight directions) finally merge into a cosmic state controlled and maintained by the Śaktis. At the centre resides Saubhāgyagaurī who supervises the articulation and ordering (KKh 72.91).

The numerical symbolism related to the number of divinities on this route can also be interpreted in a cosmogonic frame. There are 96 Śaktis (i.e. 8 directions x 12 months), 8 Kṣetradevīs, 8 Bhairavas, 8 Dikpālas and 24 Vetālas; thus the total number reaches to 144 (see Appendix, Table 1). The number 144 (12 x 12) refers to the cosmogony where the macrocosmos (12 zodiacs) leads to the microcosmos (12 months). Thus by performing an earthly journey one can experience the cosmological occurrences. The number 144 is interpreted as the product of 9 Hindu planets (*macrocosmos*) multiplied by 8 directions (*mesocosmos*) and 2 parts of the day (day and night, *microcosmos*). According to the *Kāśīkhaṇḍa* (72.101–102) the 96 Śaktis protect the territory and help the devotees to get relief from any sort of difficulty. It is difficult to infer whether their worship had been a part of tradition. In contemporary times, there is no trace of their independent worship. Among the 96 the twelve are spatially manifested at different places in the main city, mostly as auxiliary deities in the main temples. And, some of them received the power of great goddesses by benefiting from identical names and thus capturing the glories of *purāṇic* mythologies, e.g. Trailokya-Sundarī, Tripura-Bhairavī, and Kāmākṣā. The rest have lost their identity in the city area; of course many of them are existent in the form of ruins, or folk goddesses along the Caurāsīkrośī route.

The Kṣetrarakṣikadevīs

The *Kāśīkhaṇḍa* (72.88) describes the Nine Durgās as protecting deities of the outer territory of Kāśī. Durgā had also created a subordinate power representing herself who protects the inner area of the cosmic territory (*Kṣetrarakṣikadevīs*); they are eight, each one in one direction and an additive at the centre named Saubhāgyagaurī (KKh 72.90–91; cf. Fig. 2). Their spatial arrangement is comparable to a cone in which the base is an octagonal wheel showing the eight directions, and the apex is the controlling centre. Worship of the Kṣetradevīs together with Durgās is obligatory. By this process one gets protection from the sides, the outer boundary and inner space. This hierophanic symbol of spatial manifestation representing 96 forms of goddesses (Śaktis) is an example of the exteriorisation of space, an example of expansion (*vistara*). At the other end, the manifestation of the 9 Kṣetradevīs shows an interiorisation of space, an example of shrinking (*saṃkoca*). In an independent condition, both situations show a state of non-equilibrium, which results into further retreat, causing them finally to converge in a state of equilibrium. This is close to the perspective of Tantra ("to weave"), which metaphorically explains creation by the interaction of two opposite forces. This metaphor can easily be explained in the above context of arranging territorial goddesses who protect nature's realm of space-time and the human realm of deeper consciousness. In another perspective, the integration of the two sub-states mentioned above represents the meeting of transcendence and immanence, whereby one understands reality (cf. Singh 2000: 67).

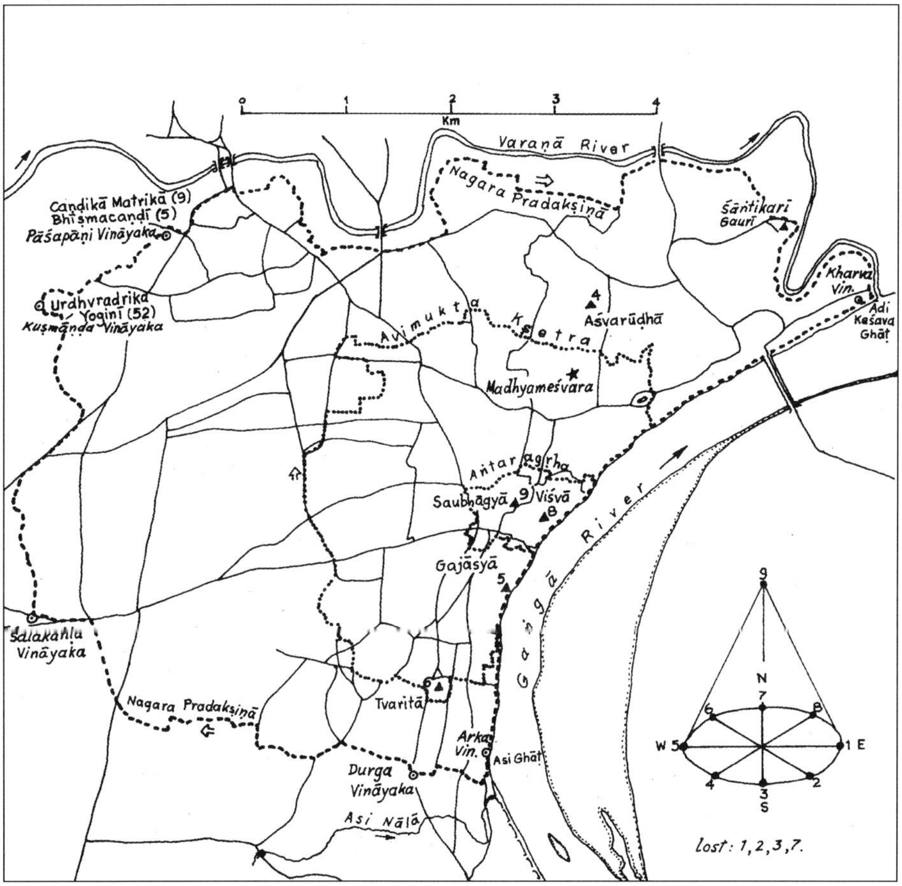

Fig. 2: Kṣetrarakṣikadevīs (drawn by the authors).

The Sixty-Four Yoginīs

The story of Divodāsa and the supremacy of Śiva recount how the 64 Yoginīs were sent at first to disturb the cosmic order and to overrule the city. This story suggests that Yoginīs are "surely among India's oldest deities, and they have always been linked to place—to some particular hill or grove, some particular village or cross-road, to some particular river or cave. These goddesses unquestionably occupied their 'benches' (*pīṭhas*) in the Forest of Bliss long before that mythical time when Śiva chose the city to be his home" (Eck 1982: 157–158). During the 10th–11th centuries, the worship of Yoginīs was very popular in the religious arena. It is possible that some portions of the *Kāśīkhaṇḍa* may have been composed in the late 11th century. Out of 64 Yoginīs' shrines 60 are assumed to be in the Rāṇā Mahāl at Causaṭhī Ghāṭ (cf. KKh 45.34–41; also see Dehejia 1986: 215–216). Presently only 13 Yoginī images exist and are mostly attached to other temples (Fig. 3).

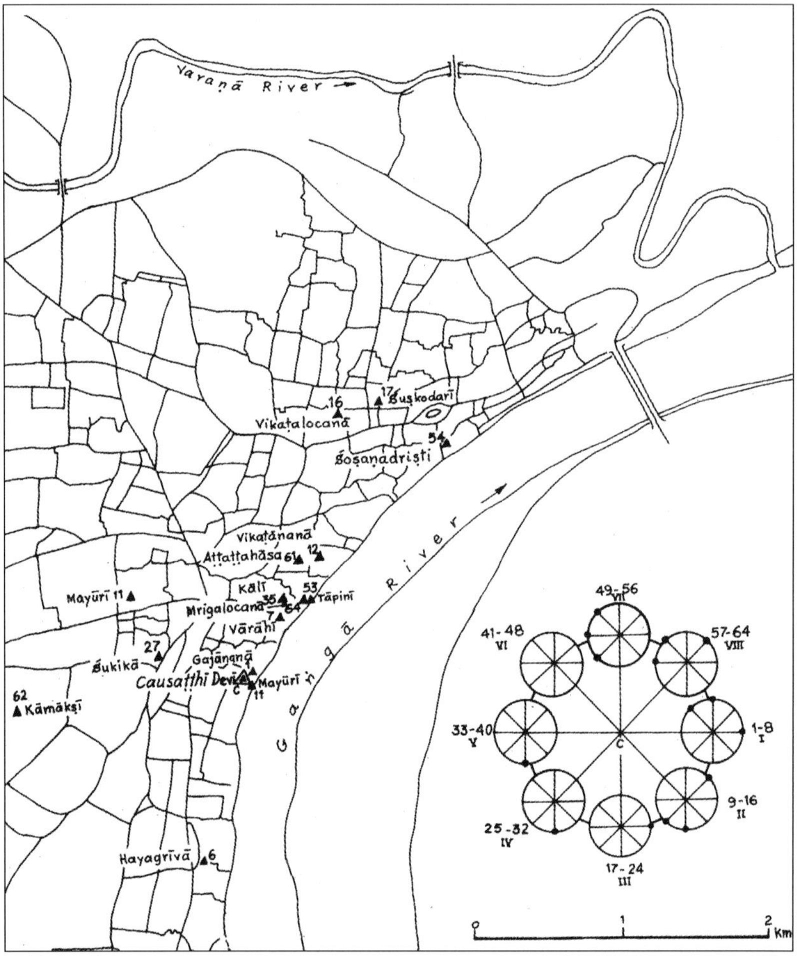

Fig. 3: The Yoginīs (drawn by the authors).

The symbolic number 64 is a product of the 8 directions and the replication of each one in the *aṣṭa-maṇḍala/aṣṭaka-cakra* (eight points on the cosmic circuit). This theoretic or esoteric interpretation is explained in the *Agnipurāṇa* (52, 146) that prescribes that one goddess must preside over each group of 8 goddesses. The primordial number of eight *mātṛkās* (mother goddesses) symbolises the condensation (*saṃkoca*) of the cosmic rhythm, and the number 64, resulting in the emergence of the *mātṛcakra* or *mātṛmaṇḍala*, represents its expansion (*pūrṇa-vikāsa*). This pattern is clear by the geometrical arrangement of the 64 Yoginīs. Thus, the number eight and sixty-four are two sides of reality, referring to the two phases of creation. Following the framework of the *Agnipurāṇa*, the placement of 64 Yoginīs would be arranged into an eight-petal (*lotus*) *Yantra* where each petal is the place of eight forms of Yoginī. Each of the petals is ascribed to a specific direction; and each direction is under the control of one of the eight forms of Mātṛkās: Brahmī (east), Maheśvarī (southeast), Kaumārī (south), Vaiṣṇavī (southwest), Vārāhī (west), Aindrī (northwest), Cāmuṇḍā (north), and Narasiṃhī (northeast). Although at one time there might have been individual Yoginī temples, at present they are worshipped together in a single temple, Causaṭhī Devī, close to the same named *ghāṭ*. The identities of the Yoginīs and their characteristics are not very clear, however most of them are place-specific, or task-specific. Many of these Yoginīs ("sorceress") are allied to Yakṣīs ("tree-spirits"), Mātṛkās ("mothers"), Vetālīs ("vampires"), Śaktis ("power goddesses"), or Gaurīs ("white goddesses"). Association of these goddesses with place and archetypal connotations that all of the goddesses are forms of the Great Goddess show the mythology to be a sacred geography.

The Yoginīyātrā (pilgrimage) at Causaṭhī Ghāṭ is performed on the first day of the dark-half of Caitra (March-April), and for all the Yoginīs pilgrimage is performed on the 14th of the dark-half of Aśvina (September-October). Many pilgrims follow the traditional time of the journey, i.e. dark-half 12th of Caitra, and the majority of them pays visit only to the Causaṭhī Devī in Rāṇā Mahāl. On the 11th of the light fortnight of Phālguna (February-March), an initiation rite of a festival is celebrated with wet colour and colour powders in the temple of Causaṭhī Devī.

The Nine Durgās

Though there are many versions of Durgā's origin myth, She is commonly accepted as a most powerful warrior goddess who appeared on earth under many names for the destruction of demons. She obtained the name Durgā because She slew a demon named Durga (cf. KKh 71; 72.81). According to the *Devīmāhātmya* (11.38–51; 12.32) Durgā is narrated as cosmic force incarnate in many forms to kill the demons and maintain the cosmic order. Her origin in the backdrop of cosmic crisis highlights her as cosmic mother who intervenes on behalf of her devotees. By contextual inference Durgā means "who destroys misfortune"; this supreme power is the spirit of the trinity of forces of creation, preservation, and destruction. Durgā is described in nine forms, parallel to the eight directions and the centre as She protects the cosmic territory (cf. Fig. 4). The number 9 may also be compared with the number of planets in Hindu mythology, which directly influence the earthly happenings and human realm. The spatio-geometrical arrangement of the 9 Durgās is like the three layers of spire crossing at the cardinal points, each of which is taken care of by one of the forms of Durgā.

Pilgrims and devotees especially visit these nine forms of Durgā during the nine nights (*navarātrī*) of spring and autumn, sequentially one of the Durgās on each of the nine days (see Wilke, in this volume). Each form is described in short.

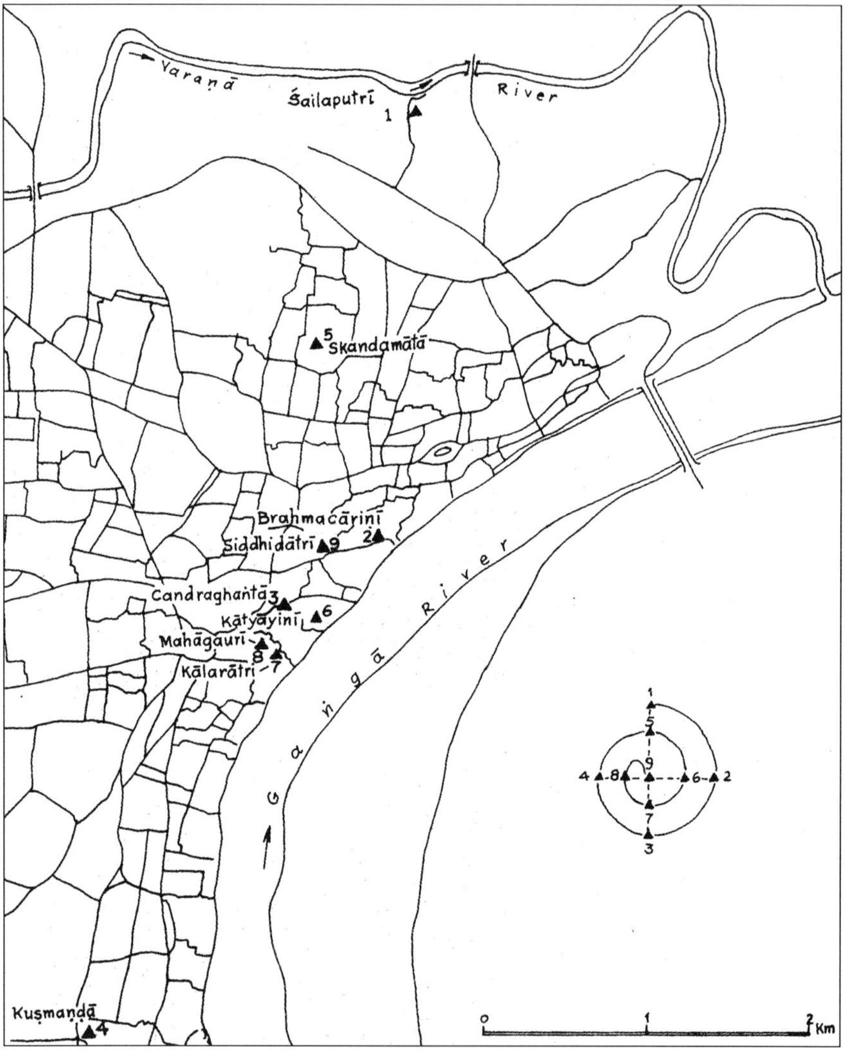

Fig. 4: The Durgās (drawn by the authors).

1. *Śailaputrī.* As the name identifies she is the daughter of the Himālaya Mountains. In her earlier incarnations she was the wife of Śiva and committed suicide in repentance for negligence of her husband. She rides a bull.

2. *Brahmacāriṇī.* This form reflects her great devotion to tapas which is reflected as an aura around her. Having a rosary in one hand and a sacred water pot in the other she symbolises the nature of meditative power and supreme bliss. This form is associated with the myth of Pārvatī, Śiva's wife, who as virgin girl performed a great strenuous *tapas* to have Śiva as her husband. She is also known as Umā.

3. *Candraghaṇṭā.* Having a shaped-moon on the forehead her face radiates like gold. She has three eyes and ten hands. She shows the gesture of being ready for war. She wears a bell-necklace whose sound creates fear among demons and sinners.

4. *Kūṣmāṇḍā.* Her name refers to "pumpkin" that symbolises the cosmic egg in the cosmos and ten rays from her body radiate the world. She has eight hands and rides a lion. She is popularly perceived as the protector of the holy territory on the south as one of the fierce goddess guardians (Eck 1982: 165). The Durgā at Durgākuṇḍa is known as Kūṣmaṇḍā. This Durgā is "particularly pleased with sacrificial offerings (*bali*), which, though not stated explicitly, are most likely construed as blood sacrifices (*mahābali*)" (Rodrigues 1993: 190). The term *bali* is used almost synonymously with blood sacrifice in the later Purāṇas (cf. Kooij 1972: 21, 52–54). Still today during the autumn festival of nine nights, on the early morning of the eighth day a few cases of goat sacrifice are in evidence. The sacrificial altar possesses signs of goat sacrifice.

5. *Skandamātā.* Being the mother of Skanda, the first son of Śiva, she is known as Skandamātā. She represents the flame in the pyre having Skanda in her lap. She has three eyes and four hands, and also known as Vāgīśvarī. Her face "consists of a compound of eight kinds of metal, which is of a pale hue, and highly burnished. She wears on her head a large crown, surmounted with balls, like the coronets of the nobility" (Sherring 1868: 90).

6. *Kātyāyanī.* Pleased by the great *tapas* of sage Kātyāyana, who completed the final version of Pānini's book on grammar *Aṣṭadhyāyī*, the goddess was born as his daughter. That is how the name is Kātyāyanī. In the Brāj Maṇḍala she is the patron deity whose stories are described in the mythologies of the region. She is portrayed as three-eyed and eight-armed.

7. *Kālarātri.* Having a dark complexion, dishevelled hairs, and a rosary shining like an electric light, she rides a donkey. From her three eyes electric currents flow and with her nasal breath come out flames of fire. She provides fearlessness to devotees and grants a boon of "auspiciousness" (*śubha*); that is how she is also known as "Śubhaṅkarī". She is one of Durgā's energy manifestations (*śakti*) who served as a messenger to tell the demon Durgāma to desist. And finally she burned the demon's attendants with her breath and prepared Durgā for war (Rodrigues 1993: 191).

8. *Mahāgaurī.* She is described as an innocent girl of eight years and her garments, ornaments and complexion—all are white. She rides a bull and appears very quiet. When Gaurī was in deep *tapas* her body got covered with dust and mud, later on cleaned by Śiva who used holy water of the Gaṅgā river. Afterward she became more fair and beautiful—Mahāgaurī. Mahāgaurī is recognised as Annapūrṇā; and She is Kāśī's queen, reigning alongside the king, Viśveśvara Śiva. She is also eulogised as Bhavānī, the feminine energy of Lord Śiva (see KKh 61.123–38). The myth says that "For happiness, there is no home like Kāśī; there is no father like Vishveśvara; there is

no mother like Bhavānī, destroyer of rebirth; and there is no household like the people of Kāśī, who are Viṣṇu's incarnate!" (Eck 1982: 160–161).

9. *Siddhidātrī*. As giver (*dātrī*) of eight-fold mystical powers (*siddhi*) she is considered as the supreme. At one stage even Lord Śiva worshipped her and she granted him all the mystical powers. That is how half of Śiva's body became female (androgynous), called Ardhanarīśvara. Sages, gods, and noble souls always surround her—all worshipping her to receive her kind blessing.

Durgā Pūjā in Vārāṇasī

During the period of *Aśvina Navarātrī* (light-half 1–9), i.e. 26th September to 5th October 2003, interviews with 250 pilgrims and devotees were conducted at Asi Ghāṭ, Daśāśvamedha Ghāṭ and near the Durgā temple at Durgākuṇḍa with a view to understanding their background and various characteristics. Out of 250 persons there were 162 males and 88 females. As to caste, 43% were Brahmans, followed by Kayastha (13%) and Rajput (10%); this trend clearly indicates dominance of higher castes. With respect to age, the group of 20–30 years made up 47%, and that of 30–40 years 27% of the respondents. It is notable that the autumn season is famous for many celebrations, especially the installation of Durgā images and related festivities, therefore many young people from the nearby areas came to take part in the festival and also the accompanying recreation and entertainment. Around 54% of the respondents belonged to the countryside of the Vārāṇasī district, and the rest came from the bordering districts of the two states, Uttar Pradesh and Bihar (less than 100 km). For moving from one place to another in the city area, more than half of the people used bicycle rickshaws. A little over 60% of the devotees were schoolteachers, and 15% were engaged in their own small business. Almost all the devotees took a holy dip in the Gaṅgā River, and half of them recorded their earlier frequency of bathing here as around ten times. 80% of people paid visit to and performed rituals at three goddess temples, viz. Durgā, Annapūrṇā and Kāmākṣadevī. Concerning the merit received about 60% explained that it is a religious act which a Hindu has to perform as rite of passage. And 15% explained their motive in terms of self-purification and blessings from the mother goddess.

About ten per cent of the respondents (mostly young) accepted that they came for recreation and entertainment and at secondary level they also took part in religious activities. The active participation of youngsters in the Durgā festival shows the tendency of using this festival as a mark of identity. Many of the low caste people have also formed their committees and installed Durgā images (made of clay) as part of the celebration. This is a sign of using the traditions and symbols of higher castes as a means for upward mobility. With the increasing impact of this tendency, now the people of higher castes and wealthier classes are improving and adding more decorations to their Durgā-pavilions, and they also make the celebrations grander and bigger in terms of show, expenditure, decorations, stage performance, and distribution of *prasāda* (offered-food items to divinities).

The Gaurīs ("White Goddess")

One of the forms of the Great Goddess is Gaurī, of whom there are nine, as is the case also in Vārāṇasī (KKh 100.68–72). After passage of time seven other Gaurīs, viz. Viśvabhujā, Śāntikārī (LP in KKT 45; KKh 67.17), Ambikā (KKh 67.19–20), Pārvatī (KKh 33.128), Virūpākṣī (KKh 70.36), Vijayabhairavī and Trilokasundarī (TS: 221, 225) and their shrines were added in the list, thus reaching a total of sixteen (Fig. 5).

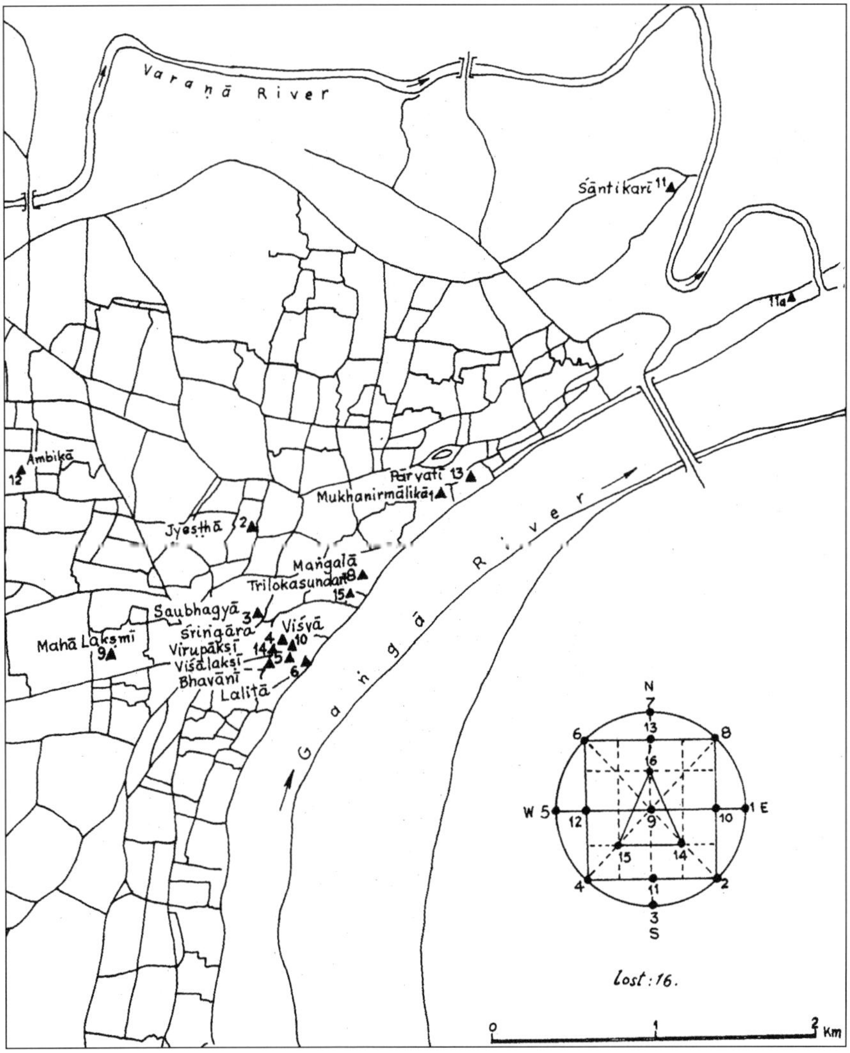

Fig. 5: The Gaurīs (drawn by the authors).

One of them, namely Vijayabhairavī, does not exist at present. Except for Śāntikārī, the rest of the Gaurīs are located within the Avimuktakṣetra. As with the Mātṛkās and Durgās, the number of the Gaurīs too was eight at first, and later on an integrating power point was added, thus totalling nine, which can be represented as a square. At two successive stages a triangle with three images and finally a covering circle with four images were superimposed. The three basic geometric designs in the arrangement of the Gaurīs may be interpreted as the three parts of a *cosmogram*: the base of a *square* (symbol of "water") that denominates 9 points (planets); the covering circle ("light") marking 4 points of cardinality; and the *triangle* ("wind") symbolising the 3 mythic realms. The total number reaches 16. And their multiplication (9 x 4 x 3) comes to 108—in itself a numerical cosmogram. This is comparable to the integration of macrocosmos (9 planets), mesocosmos (4 directions), and microcosmos (3 qualities of human consciousness) (cf. Singh 1994a: 189–190).

It is notable that these Gaurīs do not have any direct interrelationship among themselves except for being forms of the Great Goddess. Mythology recounts that the Gaurīs possess mystic power to grant boons and bliss, especially for fulfilment of all kinds of wishes and a glorious life (KKh 100.72–73). Her pilgrimage is performed during the nine-day period of the light half of Caitra (March–April). The initiation rite starts with the installation of the Gaurīs' images made of clay and cow dung in the house at a sacred place, and on the next day, which marks the first day of the Hindus' lunar calendar, the worship and pilgrimage starts. By glancing every day at one of the forms in a sequence the worship is completed on the ninth day but the concluding ritual is performed on the tenth day, marked by immersing all the ritual items, including the sacred pots, water, flowers, incense, etc. in the Gaṅgā river. On this day the great festival of Rāmanaumī ("the 9th of Rāma", the birthday of Lord Rāma), is celebrated. In the nine-day period, the eighth day marks the zenith when Annapūrṇā is worshipped; formerly this day was known for the worship of Bhavānī.

Two of these Gaurīs, i.e. Bhavānī and Viśālākṣī, are historically very popular. On the basis of mythology it is believed that as late as CE 1585, Bhavānī was the most prominent goddess of Vārāṇasī. In the course of time her glory was captured by Annapūrṇā (*Kāśīrahasya* 20.102), and in this process Bhavānī merged with Annapūrṇā. Annapūrṇā ("She of Plenteous Food") is a completely gracious and luminous goddess, the "Mother of the Three Worlds". Today, she is revered as the cosmic mother in Kāśī. She is also perceived as the basic spirit among all the forms of Durgā (KKh 61.30); in the list of nine Durgās she represents Mahāgaurī. Being the consort of Viśvanātha, the patron deity of Kāśī, she is the city's governing goddess (Eck 1982: 160–161). The great festivals associated with the goddess relate her to the production of cereal grains (*anna*). Annakūṭa (the "Food Mountain") festival falls on the first day of the waxing fortnight of the autumn month of Kārttika (October–November). During the second festival, celebrated in Mārgaśīrṣa (November–December), Annapūrṇā is decorated with green sprouts of paddy, and the same is given as blessed-food (Eck 1982: 163–164).

Viśālākṣī (the "wide-eyed" goddess), though counted among the Gaurīs in Vārāṇasī, is the only goddess in the city included in the list of 51 and 108 *Śakti Pīṭhas* ("sacred spots of Goddess"). It is said that Sati's eyes/ear ring had fallen here (Sircar 1973: 27

and 36). Her mention in the *Kāśīkhaṇḍa* at several places (70.13–14, 79.77, 100.70) is a clear indication of her significance.

Fig. 6: The Mātṛkās (drawn by the authors).

The Mātṛkās

The reference to Mātṛkās ("Mothers") can be found as early as the *Mahābhārata* (13.215.16, 18), however the best known mythological account is given in the third episode of the *Devīmāhātmya* (8.12–21) which includes the story about killing the demon brothers Śumbha and Niśumbha. The *Devīmāhātmya* refers only to the Seven Mātṛkās. Along with Kālī (also known as Caṇḍikā), Śivadūtī (another form) and the Great Devī herself, the Mātṛkās killed the demons (Kinsley 1986: 156). At the end the demons' spirit came to Kāśī and worshipped Śiva while installing the *liṅgas*, later known by their names as Śumbheśvara and Niśumbheśvara. The malevolent nature of the Mātṛkās seems clear in several myths, and they are perceived as a dangerous group of goddesses (Kinsley 1986: 159). They are always related to a male deity of the Hindu pantheon, and took form to serve in the maintenance of the cosmic equilibrium and order. The *Varāhapurāṇa* (17.33–37) describes the Mātṛkās as symbols of human emotions: Brāhmī is the symbol of pride, Maheśvarī/Mahendrī of anger, Kaumārī of attachment, Vaiṣṇavī of greed, Vārāhī of envy, Aindrī of jealousy, Cāmuṇḍā of depravity, and Yogiśvarī/Narasiṃhī of lust (cf. Bhattacharyya 1974: 102). These Mātṛkās are also described as the feminine deities together with their male partners controlling the eight directions, sequentially from right to left, i.e. from the east to north. In the Yoginīyantra, they are represented as the controllers of the eight directions (cf. *Agnipurāṇa* 146).

The earlier number seven of Mātṛkās was a symbol of "seven planets seen with naked eyes". This number in the course of time reached up to nine, also denoting the common cosmo-vision of eight directions of the cosmic circumference and additionally the centre; this is also comparable to the nine planets in Hindu mythology. After the passage of time the number increased from nine to twelve, parallel to the 12 zodiacs or months, as in case of Vārāṇasī (Fig. 6). In the latter case some of the important goddesses of the other groups are given the status of Mātṛkās. Among the twelve, Raudrī has disappeared. The rest are worshipped in association with other pilgrimages and groups.

The Mātṛkās are said to protect their devotees against all difficulties and fulfil their wishes with motherly affection (cf. KKh 97.45–46). The *Kāśīkhaṇḍa* (83.33) eulogises their glories. Some of the Mātṛkās, e.g. Narasiṃhī is described separately (KKh 70.31). Vikaṭā is also identified as Kātyāyanī, Pañcamudrā (KKh 83.37–38; 97.40–41), and Saṅkaṭā Devī (cf. Sukul 1977: 6–97). Saṅkaṭā (the "Goddess of Dangers") vanquishes dangers for her devotees. She is worshipped on Fridays (Eck 1982: 168–170). Worship of Saṅkaṭā is very popular among women devotees. She is said to fulfil all wishes of the devotees and provides security in difficult moments of life.

The Ten Caṇḍīs

The "fierce, passionate" derivative of the Great Goddess Durgā is called Caṇḍī. She is eulogised in three ways: (a) as the daughter of Himālaya, (b) as the wife of Śiva, and (c) horrific form of Durgā (Coburn 1988: 96). Of course, Caṇḍī is called Caṇḍikā in the group of Mātṛkās, She is also mentioned as an independent goddess who like others took ten forms to control the eight directions and also the upward and the downward. Therefore she is represented in ten forms (cf. Fig. 7).

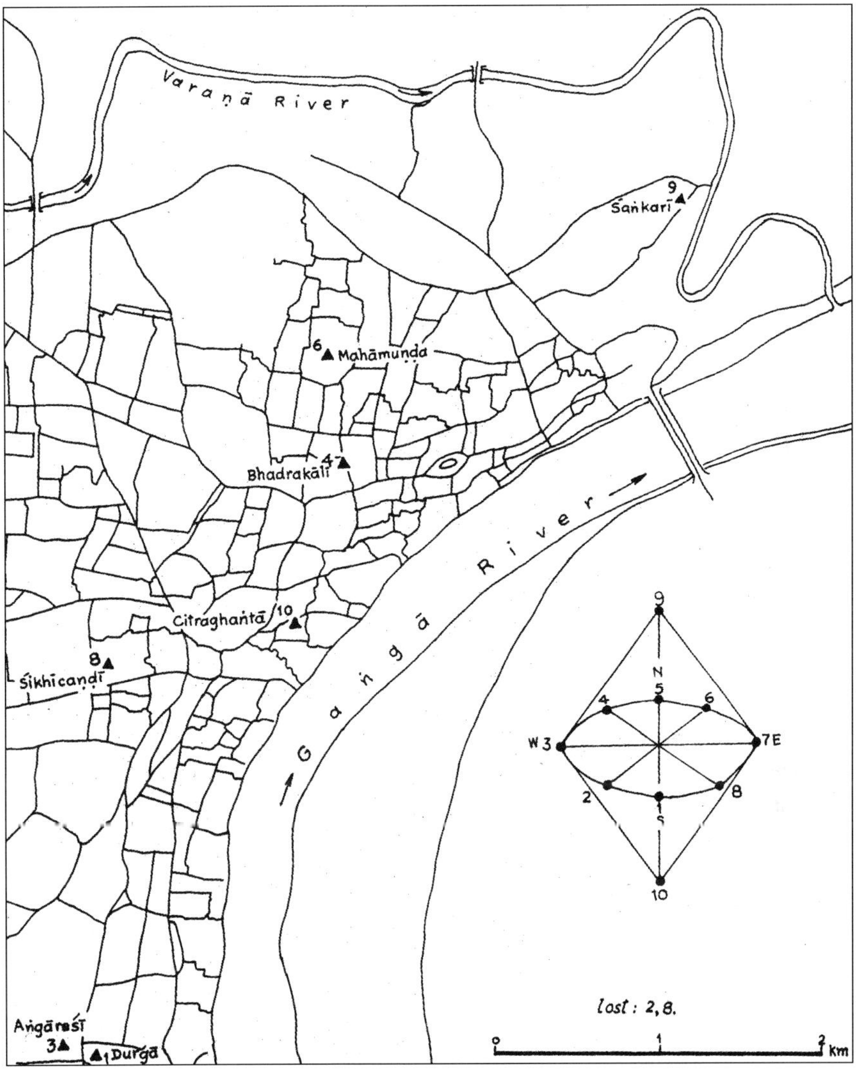

Fig. 7: The Caṇḍīs (drawn by the authors).

In the *Liṅgapurāṇa* the ten Caṇḍīs are mentioned, while in other texts there appear some variation of names and their association with place and direction (cf. Sukul 1977: 84–85). Among the ten, Uttareśvarī and Adhaḥkeśī have now disappeared. Moreover, one of them has got her name changed, like Śikhīcaṇḍī is known as Śikhīkaṇṭhī. This Yātrā is to be performed in the same period of time as the Durgās.

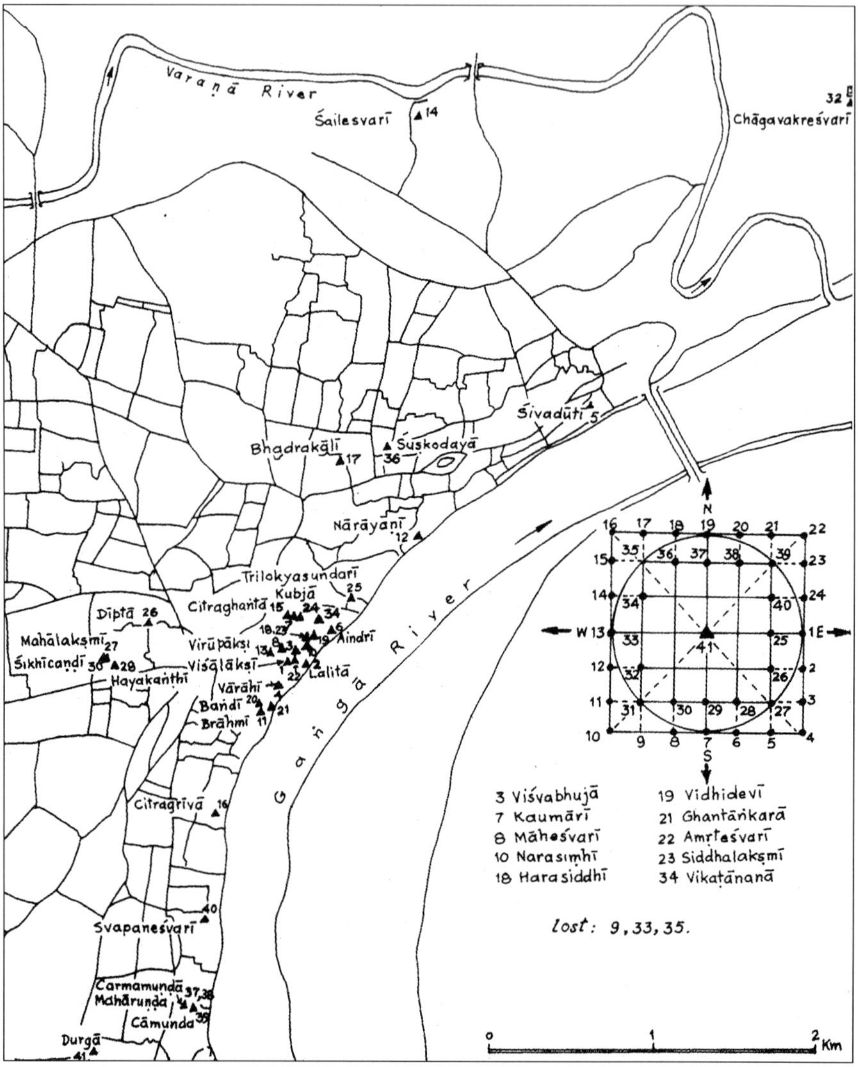

Fig. 8: The Kṣetradevīs (drawn by the authors).

The Kṣetradevīs

The *Kāśīkhaṇḍa* (70.10–97) describes the list of Kṣetradevīs that includes all Durgās, Gaurīs (of course, sometimes-variant names), and many other goddesses of Kāśī in addition to Caṇḍīs who in different contexts and from different directions and different places protect the territory of Kāśī. Kṣetradevīs, in fact, represent the combination of all the goddesses, thus the total number reached 41 (Appendix, Table 2; Fig. 8). By this combination the numerical symbolism of the inner portion of the Śrī Yantra is represented; of course the total number of triangles in the inner part of the Śrī Yantra comes to 45, based on the crossing and superimposition of nine triangles. Performance of pilgrimage to their shrines is prescribed on each of the ninth or eighth day of the waxing or waning of the lunar month. However, this pilgrimage is not undertaken these days, and most of the goddesses of this group are worshipped together with other goddesses.

The *Kāśīkhaṇḍa* (70) describes the spatial and divine characteristics of the Kṣetradevīs. Viśālākṣī is described as adjacent to Viśalakuṇḍa ("water pool"), which was once connected to the Gaṅgā River. However, at present this water pool does no longer exist. This sacred place is eulogised as *Mahāpīṭha* ("the great seat"), which possesses the strong power to liberate the soul from transmigration (KKh 70.16). The myth that her worship helps to get conception attracts a large mass of newly married ladies (KKh 70.15). The other prominent Kṣetradevī has been Mahālakṣmī, located in the neighbourhood named after her, i.e. Lakṣmīkuṇḍa. In the month of Āśvina (September–October), the bright fortnight is especially auspicious for goddess worship.

The Daśamahāvidyās

The Mahāvidyās (the goddesses of "Great Knowledge") are a rather late addition to the list of goddesses, which happened around the turn of 14th century C.E. They are considered as "all one", and "every one of them is strongly associated with Durgā in her role as demon slayer" (Kinsley 1997: 5). According to the *Vāmanapurāṇa* (30.3–9) the Great Goddess produced goddesses from different parts of her body. In the *Devīmāhātmya* (11.38–50) it is said that She "often announces to her petitioners that she assumes different forms at different times in order to maintain cosmic stability" (Kinsley 1986: 161). Yet the general people hardly know them as a group. Some of them, like Kālī, Tārā and Kamala are well recognised, probably due to their unique appearance and independently glorified mythologies. Kālī and Tārā are also counted as prominent divinities in the list of Yoginīs. They are very closely associated with Śiva according to their origin myths. Therefore, it is no wonder to find shrines of all of them, except Mātaṅgī, in the city of Śiva, Kāśī (Fig. 9). However, there is little textual reference to some of them. Most of them are worshipped locally and there is no indication of an organised and patterned pilgrimage to the Daśamahāvidyās. In the course of time three other goddesses have been counted under the Mahāvidyās, viz. Rājārājeśvarī, Lalitā and Jvālāmukhī, thus the total number reached thirteen. Many of them are described in the list of the Kṣetradevīs (cf. KKh 70) and the Mahā Śaktis (cf. KKh 72). The festive days and times of worship of the Mahāvidyās are prescribed according to the seasonal conditions and the basic qualities of the deities. The annual auspicious dates are particularly used for celebrations, special decorations (*śṛṅgāra*) and musical performances.

1. *Kālī*. Among the Mahāvidyās, Kālī's position is prominent. Usually she is considered the first in most of the lists. The complexion of Kālī is black. She stands on a corpse. Her appearance is frightening, and she is depicted as laughing. She has four arms. She holds a cleaver in one hand, a severed head in another, making the sign of *abhaya* ("fearlessness") by another, and the fourth hand is in the gesture of granting

boon(s). She is naked except the girdle of severed hands. The cremation ground or the battlefield is her abode. Her hair is dishevelled. In the Tantric tradition the corpse on which she is standing is identified as Śiva who after falling in deep trance had stopped the rhythm of his body. There are two such images of Kālī, known as "Śava-Śiva Kālī": the image in Devnāthpura shows Śiva lying on a corpse that lies on a platform of five heads while Kālī is standing on Śiva's body. There are images of a jackal and a lion on either side of the platform. Another Śava-Śiva Kālī image is close to the Prahlād Ghāṭ, where Kālī is in standing pose. A similar image can also be seen in Maṇikarṇikā Lane, close to the cremation *ghāṭ*.

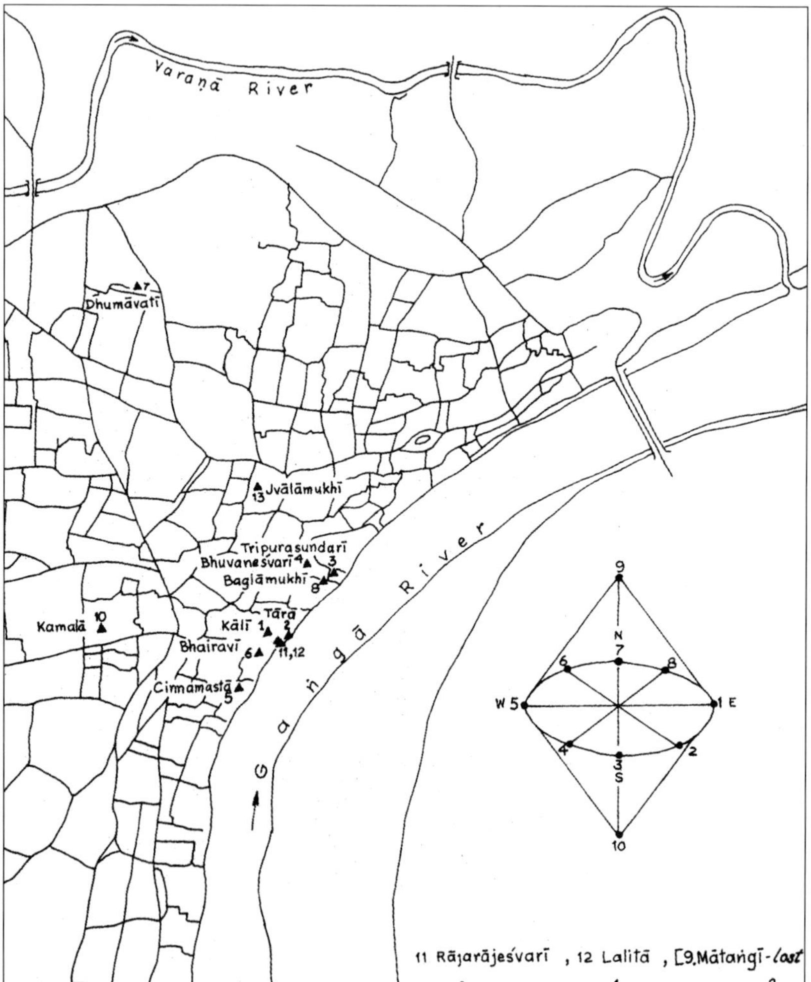

Fig. 9: The Mahāvidyās (drawn by the authors).

2. *Tārā*. Generally Tārā is depicted as similar in appearance to Kālī. Her complexion is black, she is pot-bellied, wears tiger skin, and has four arms. She holds a chopper, a severed head, scissors, and makes the sign of granting boon(s) with the fourth. Her left foot is set on a corpse (sometimes on Śiva). She is often depicted as standing amidst a cremation fire. There are four images of Tārā in Vārāṇasī city, viz. (i) in Bengali Tola, built by Rāṇī Bhavānī around 1776; (ii) close to the stairways to Rājārājeśvarī Ghāṭ, in the turret (*maṛhī*); (iii) close to the gate of Kālikā, in Kālikā Galī; and (iv) in Brahmanāl, built by queen Padmāvatī in 1855.

3. *Ṣoḍaśī*. Also known as Tripurasundarī, she is a sixteen years old young and beautiful girl. Her appearance is like youthful sun. She has three eyes and four arms. She holds a noose, goad, bow, and arrows.

4. *Bhuvaneśvarī*. She is said to be worshipped by the whole world and is of bright complexion. She wears a moon as crown. Her breasts are large. She smiles pleasantly and has three eyes. She is four armed and holds a goad, noose, by the third one grants wish(es), and makes the sign of assurance by the fourth.

5. *Chinnamastā*. She is shown standing in an aggressive posture with one leg put forward. She holds a cleaver in one hand and her own severed head in the other. She is naked and drinks the blood gushing (through the middle stream) from her headless body. She is standing (or sitting) on a copulating couple of Rati ("goddess of sexual desire") and Kāma ("god of sexual desire"). To her right is Varṇinī, possessed by the quality of *rājasa* (passion), who holds a sword and a skull-cup. She happily drinks blood gushing from the Devī's severed neck. Dākiṇī, who is on the left side, also drinks blood from the Devī's headless body. She possesses by the quality of *tāmasa* (ignorance) and enjoys the dissolution of the world.

6. *Bhairavī*. Her complexion is like the glow of thousands of rising suns. She wears a garland of skulls and covers her body with skin of demons she has killed. Her feet and breasts are blood smeared. She is four-armed and holds a rosary, a book, makes the sign of fearlessness, and of granting boons. She is three-eyed and sits on a lotus. The crown she wears is gem-studded and beautified by a crescent-shaped moon. Her smile is pleasant and superb.

7. *Dhūmāvatī*. She is tall with a stern and unsmiling face. Her complexion is pale and her hair dishevelled. She wears dirty clothes. She is the "smoke-faced" typical widow. Her carrier is a chariot with a flag on which there are prints of the emblematic crow. She has long pendulous breasts, only a few teeth, and lifeless eyes. She holds a winnowing basket in one hand and makes the sign of fearlessness, by the other. Her nose is large and crooked.

8. *Valga/Bagalāmukhī*. She is sitting on a jewel-studded golden throne in the ocean of nectar. Her complexion is bright golden and she wears yellow clothes. She holds a club in her right hand with which she is about to beat an enemy. She is pulling his tongue with her left hand.

9. *Mātaṅgī*. She is dark-complexioned and has three lotus-like eyes. The moon adorns her forehead. She is seated on a jewelled throne. She is clothed in beautiful garments and a garland of flowers is around her neck. She holds a sword, goad, noose and club in her four hands.

10. *Kamalā*. Her complexion is golden. She is sitting on a lotus in an area surrounded by snow-clad mountains. She is flanked by elephants, usually either two or four, who pour water on her. She is four-armed and holds a lotus and a disc in two hands, makes the sign of boon-granting, and by the fourth hand assures fearlessness.

Independent Goddesses

There are many images and shrines of the goddesses, which are not grouped in a particular pattern, however many of them are already given a seat or traditionally accepted as part of different pilgrimages and their routes. The cosmic territory and the holy city are both perceived as goddesses. The various images are quite similar to each other, like other secondary goddesses. However on the neighbourhood level special celebrations are held every year. The Kāśī- and Vārāṇasīdevīs are eulogised in the *Kāśīkhaṇḍa* (97.8–9). According to Sukul (1977: 98) "perhaps these images are of the same deity, installed at three locations due to some perception of threat". Or they are to reflect their local area as unique and representative of the whole territory. This process of manifestation has been very common in the medieval period.

Śītalā is known as the one who wields influence over fever diseases, both causing and curing them, especially smallpox. She is regarded as terrible, though also seen as gracious and popularly called *Mātā* ("Mother"). She is considered to be a representative of the village tradition of goddesses, who after the passage of time was accepted in the written tradition. Śītalā is the standard name for the smallpox goddess worshipped throughout India under a variety of names, e.g. Mahāmāi and Mātā in north India, and Mariyammān and Ammā in South India (Bakker 1983: 12). Her shrines are often located at crossroads, at former cremation grounds, or beside a nīma tree (margosa, *Azadirachta indica A. Juss*). Her vehicle is the donkey. The most prominent Śītalā temple is at the Daśāśvamedha Ghāṭ, after which a portion of the *ghāṭ* is called Śītalā Ghāṭ. There is also another shrine of Śītalā near Gaia Ghāṭ, further downstream; here too the adjacent *ghāṭ* is called Śītalā Ghāṭ. In fact, originally this was the shrine of the "Snake Goddess", Nāgeśvarī Devī, who in the course of time was named Śītalā (see Eck 1982: 231). The 9th day of the waning fortnight (dark half) is the popular day for her worship. Of course, people also visit her on a daily basis together with other deities.

Saṅkaṭā Devī, which means "Goddess of dangers", is perceived by the city dwellers as the one who vanquishes dangers for her devotees. In the *purāṇic* stories (e.g. KKh 83.33) she is described as one of the Mātṛkās who, together with the nine heavenly mother-goddesses, is said to have been the foster mother of a young prince of Kāśī who was born under an unlucky star and left to their care (cf. KKh 10.100–101; 83.26). It is believed that this goddess has no husband. Friday is the common day for her worship (cf. Eck 1982: 169–170). The shrine of Saṅkaṭā Devī lies in the temple compound of Siddheśvarī, to which is attached an ancient well known as Candra Kūpa ("well of the moon"). On the day of full moon in Caitra (March–April), a large mass of devout Hindus pays visit to this shrine.

According to a *purāṇic* story the Great Goddess, after killing the demon, took rest and settled on the hill of the Vindhyā. That is how she was given the name of

Vindhyāvāsinī ("Dweller of the Vindhyā Mountains"). Her original shrine is on the right bank of the Gaṅgā river in the adjacent district of Mirzāpur, about 83 km southeast of Vārāṇasī. In the process of spatial manifestation her image has been installed at six places in Vārāṇasī. Some scholars may prefer to call her a regional goddess (cf. Kinsley 1987: 99). However, she is already included in the *purāṇic* list of 108 main goddesses, and the whole territory of Vindhyācala together with two other forms of goddesses, viz. Kālī and Aṣṭabhuja, is an example of the complex network of a landscape cosmogram (cf. Singh 1997). Another form of the goddess, named Kālarātri, who had killed the demon Durg and got the epithet of "Durgā" after killing the demon's army, visited Vindhyāvāsinī and informed her about the misdeeds and sins committed by the demon (KKh 71.61). This myth indicates that Vindhyāvāsinī was one of the elder forms of the Goddess. The 3rd day of the waning fortnight (dark-half) of Bhādrapada (August–September) is assumed to be her birthday, therefore on this day special annual celebration and musical performances take place at the shrine.

Conclusion: Spatiality, Complexity and Self-Organised Systems

It is notable that all the inner four circuits are intact even today and always regulated through the pilgrimages. However, the outer circuit is subject to a symbolic and abbreviated representation of space. According to *purāṇic* myth the circuit covers exactly a circular circumference of 84 *krośa* (i.e. 184.8 miles, or 297.4 km), but in reality if the radius is only 5 *krośa* (i.e. 11 miles, or 17.6 km), then the circumference will be 110.6 km (i.e. 31.2 *krośa*, or 68.7 miles). The number "84" symbolically represents the integration of the 12 signs of the zodiac (divisions of time) and 7 sheaths of the body (*kośas*) or 7 layers of the atmosphere (division of space), thus 12 x 7 = 84.

Each route and each of the images on the routes is valuable and meaningful for the pilgrim, each providing different benefits, of course also having some general function. The shrines are neither regularly spaced nor randomly scattered, but their sequential placement seems to obey a fundamental organisation that is deep, hidden, and yet natural (Malville 1999: 6).

The spatial patterning of goddess shrines reflects a system of complexity, which resembles the geometrical order of a self-organised structure, known as *fractal* (see Bak 1996: 19). It is an example of the endless inclusion of patterns within patterns, replicating self-similar structures. No matter how much its tiniest element is magnified it still contains the essence of the entire universe. The mythologies support this characterisation when glorifying each form of goddess as specific, general and finally universal, i.e. an archetype. The psycho-spiritual design, or cosmogram, of each group of goddesses has the dual characteristics of distinctiveness and analogy.

By plotting all the 96 goddess-sites on a map (Fig. 10), the idea of opposite poles and corresponding opposite forces becomes obvious. The spot near Annapūrṇā Devī is the epicentre of spirituo-magnetic force from where power gets diffused towards the periphery. This is an interplay between *exteriorisation* (diffusion) and *interiorisation* (contraction), by which the system of self-similarity and scale invariance is maintained. With respect to the four sacred territories (*kṣetras*) and the increase of the mean radius of circuit, the distribution of the 96 goddess-sites (cf. Appendix, Table 3) shows a tendency of decrease from the inner to outer. Conversely, by grouping the distances in descending order the concentration of shrines shows a unique tendency, which on a double logarithmic graph shows a straight line.

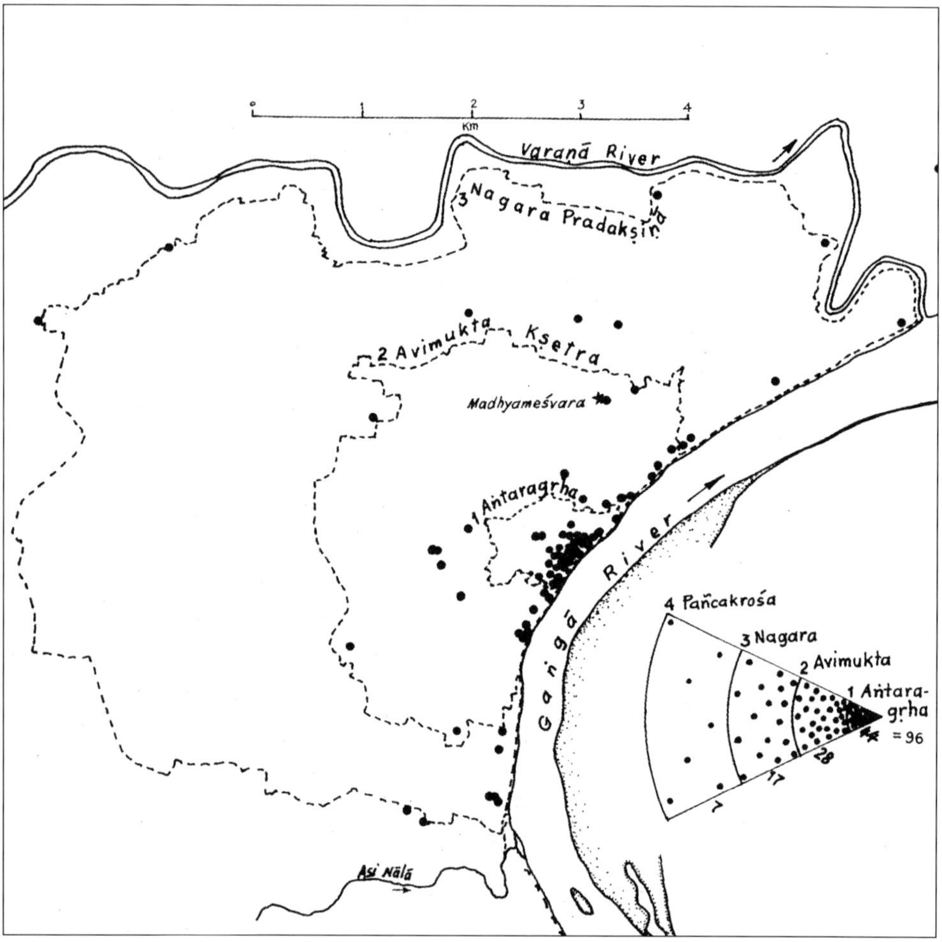

Fig. 10: Distribution of Goddess temples and shrines (drawn by the authors).

This type of distribution tendency is called "power law", since it shows that some quantity, or number of objects or events (N) can be expressed with respect to the proportion to size, or some power of another quantity (s) raised to a negative power (α), $N = s^{-\alpha}$ (see Bak 1996: 27; Malville 1999: 6). In such a situation, like in fractals, there are many more structures than the large ones. The diffusion ratio decreases from inner to outer and its reciprocity shows correspondingly the increase of distortion (see Appendix, Table 3). Taking the diffusion ratio as a representative of power intensity, one can easily predict that from core to periphery the power decreases, and on the other side the total power collapses into the nucleus.

This patterning is in close correspondence with a *self-organising system*. A change in scale does not change the significance or efficacy of the pilgrims' movement (see Malville 1999). These sacred places "reveal themselves", and because of the special features within the complex of landscape, tradition and belief, a site may become an "inexhaustible source of power and sacredness" and may operate according to "laws of its own dialectic" (cf. Malville & Singh 1997: 18). Eliade (1958: 369) suggests that if the choice of sacred places were left entirely to human design and intention there could be no explanation for their permanence and continuity.

Epilogue

The Goddess is viewed as omnipresent and omniscient in the sacredscape of Vārāṇasī. The city-dwellers believe that "without the power and energy of Śakti, Śiva (...) is a *śava*, a corpse. It is she who embodies the vibrance of both life and death" (Eck 1982: 174). Śiva and Śakti, both are represented independently in 324 forms (i.e. = 108 x 3) whose integration maintains the rhythm of the cosmic cycle in the universe. The aim of getting in close contact with the divine spirit, the acceptance and recognition of folk traditions by the Great Tradition, and the respectability of the feminine divine in the symbolic expression of identity, all these factors have shaped the human quest to establish goddess-images in various forms and at different places. Local mythologies (*sthalapurāṇas*) have been superimposed to justify their historicity and religious merit, and after time they became part of the living tradition. In the chain of "existence-maintenance-continuity" the worship of goddesses is flourishing (cf. Singh 2000: 96).

The spatial patterning of goddess shrines can be explained in the context of a *fractal and self-organising system* in which the complexity of disequilibrium itself forms an order and finally results in equilibrium. This system is always re-created and rejuvenated by the ritual processes and the sacred performances by the devotees, while of course the earlier system always persists and serves as base and pre-condition. The remark by Cambel (1993: 4) that "complex systems are dynamic and not in equilibrium; they are like a journey, not a destination, and they may pursue a moving target" aptly describes this complex system of goddesses in Vārāṇasī.

Notes

1 The story of King Divodāsa goes back to the Vedas and in the course of time was modified in the *purāṇic* literature to fit the local traditions and the rulers. In the Vedas a king named Divodāsa is mentioned in association with Indra's destruction of a hundred cities (cf. Ṛgveda Saṃhitā, 4.30.20) but it is difficult to accept the same figure as king of Kāśī/Vārāṇasī (Eck 1982: 387). In the *Mahābhārata* (12.31.26–28) Divodāsa is described who had fortified the city of Vārāṇasī and also taken away the sacrificial fire of his enemies. This act resulted in conflict with Śiva who wished to dwell there (12.97.20; see Shulman 1980: 77).

References

Texts

Agnipurāṇa. Anandashrama Sanskrit Series No. 41. Poona: Anandashrama Mudranalaya, 1957.

Ṛg Veda Saṃhita. Ed. Shripad Sharma. Svadhyayamandala, Anudha, 1940.

Kāśīkhaṇḍa (of *Skandapurāṇa*). Gurumandala Granthamalaya No.XX vol.IV, Calcutta, 1961.

Kāśīrahasya (an appendix to the Brahmavaivarta Purāṇa), 1957. Gurumandala Granthamalaya, No. XIV, vol. III, Calcutta.

Kṛtyakalpataru, by Lakṣmīdhara, 1942. Ed. K.V. Rangaswamy Aiyangar. Gaekwad's Oriental Series Vol. XCVIII. Oriental Institute, Baroda.

Devībhāgavatapurāṇa. Banaras: Pandita Pustakalaya, 1969.

Devīmāhātmya. Transl. & ed. by V. S. Agrawala. Varanasi: All-India Kashiraj Trust,1963. [Text with English].

Mahābhārata. Critical ed. by V. S. Sukthankar (& others). 19 vols. Bhandarkar Oriental Research Institute, Baroda, 1933–1959.

Liṅgapurāṇa. Trans. and ed. by J. L. Shastri, 2 vols. Motilal Banarasidass, Delhi, 1973.

Vāmanapurāṇa. Ed. A. S. Gupta. Critical Ed. All-India Kashiraj Trust, Varanasi, 1975.

Vārāhapurāṇa. Ed. A. S. Gupta. Critical Ed. All-India Kashiraj Trust, Varanasi, 1981. 2 vols.

Secondary Literature

Bak, Per. 1996. *How Nature Works. The Science of Self-Organized Criticality*. New York: Copernicus.

Banerjea, J. N. 1974. *The Development of Hindu Iconography*. 3rd ed. New Delhi: Munshiram Manoharlal. 1st ed. 1956.

Bhattacharyya, Narendra Nath. 1974. *History of Śākta Religion*. New Delhi: Munshiram Manoharlal.

Bhattacharyya, Raghunath. 1986. Lokmātā Mahārānī Bhavānī. In: R. D. Singh (ed.), *Bengal and Varanasi*. Calcutta: Bibliographical Society of India, pp. 1–18.

Cambel, A.B. 1993. *Applied Chaos Theory: A Paradigm for Complexity*. Boston: Academic Press.

Coburn, Thomas B. 1988. *Devī Māhātmya*. Delhi: Motilal Banarasidass.

Dehejia, Vidya. 1986. *Yoginī Cult and Temples. A Tantric Tradition*. New Delhi: National Museum.

Eck, Diana L. 1982. *Banaras. City of Light*. New York: Alfred Knopf.

Eliade, Mircea. 1959. *The Sacred and the Profane*. Princeton: Princeton University Press.

— 1958. *Patterns in Comparative Religion*. London: Sheed & Ward.

Kinsley, David. 1986. *Hindu Goddesses*. Delhi: Motilal Banarasidass Publ.

— 1997. *Tantric Vision of the Divine Feminine*. Berkeley: University of California Press.

Kooij, K. R. Van. 1972. *Worship of the Goddess according to the Kālikā Purāṇa*. Leiden: E. J. Brill.

Malville, J. McKim. 1999. Complexity and Self-organization in Pilgrimage Systems. Paper presented at the Conference on *"Pilgrimage and Complexity"*, at Indira Gandhi National Centre for the Art, New Delhi, January 5–9; 9 pp.

Malville, J. McKim & Rana P. B. Singh. 1997. Pilgrimage Cosmologies at Chitrakut and Vijayanagara: Emergent Geometries and Self-organization. Paper presented at the 26th *Annual Conference on South Asia,* Panel on *"Cosmology and Complexity of Pilgrimage"*, October 19, University of Wisconsin, Madison. 22 pp. + 12 figs.

Rodrigues, Hillary P. 1993. Some Puranic Myths of the Durgākuṇḍ Mandir in Varanasi, *Purāṇa* (Varanasi) 35, 2: 185–201.

Shulman, David. 1980. *Tamil Temple Myths: Sacrifice and Divine Marriage in South Indian Śaiva Tradition*. Princeton: Princeton University Press.

Singh, Rana P. B. 1987. The Pilgrimage Mandala of Varanasi/Kashi: A Study in Sacred Geography. *National Geographical Journal of India* 33, 4: 493–524.

— 1993. (ed.), *Banaras (Varanasi). Cosmic Order, Sacred City, Hindu Traditions.* Varanasi: Tara Book Agency.

— 1993a. Cosmos, Theos, Anthropos: an inner vision of Sacred ecology in Hinduism. In: Rana P.B. Singh (ed.), *Environmental Ethics.* Varanasi: National Geographical Soc. of India, Pub. 40, pp. 113–130.

— 1997. Sacredscape, Cosmic territory, and Faithscape: Goddess Territory of Vindhyachal, India. *National Geographical Journal of India* 43, 3: 237–263.

Singh, Ravi S. 2000. *Goddesses in India: A Study in the Geography of Sacred Places.* Unpub. Ph.D. Dissertation, Dept. of Geography, J.P. University, Chapra (Bihar), 211 pp. (39 figs., 25 tables).

Sirkar, D. C. 1973. *The Śākta Pīṭhas.* Delhi: Motilal Banarasidass Publ.

Sukul, Kubernath. 1977. *Vārāṇasī Vaibhava.* Patna: Rastrabhasha Parishad.

Zimmer, Heinrich. 1992. *Myths and Symbols in Indian Art and Civilization.* Princeton: Princeton University Press.

Appendix

Table 1. Kāśī: Goddesses and associated deities on the cosmic circuit

EAST	SOUTH	WEST	NORTH
Śaktis	**Śaktis**	**Śaktis**	**Śaktis**
(1–12): only five exist	(25–36): # all lost	(49–60): # all lost	(73–84): only one exists
1. Trailokya-Vijayā	25. Vidyujjihvā	49. Bhāsī	73. Daityatāpanī
2. Tārā*	26. Śivā	50. Garutmatī	74. Staṃbhanī
3. Kṣamā	27. Ravā	51. Padmāvatī	75. Mohanī
4. Trailokya-Sundarī*	28. Trinetrā	52. Padmakeśī	76. Māyā
5. Tripurā	29. Trivaktrā	53. Padmāsyā	77. Bahumāyā
6. Trijaganmātā	30. Tripādā	54. Padmavāsinī	78. Balā*
7. Bhīmā*	31. Sarvamaṅgalā	55. Akṣarā	79. Utkaṭā
8. Tripura-Bhairavī*	32. Huṃkāraheti	56. Tryakṣarā	80. Uccāṭanī
9. Kāmākhyā*	33. Tāleśī	57. Tantu	81. Maholkāsyā
10. Kamalākṣi	34. Sarpāsyā	58. Praṇaveśī	82. Danujendra
11. Dhṛiti	35. Sarva-Sundarī	59. Svarā	83. Kṣayaṃkarī
12. Tripuratāpanī	36. Siddhi	60. Ātmikā	84. Kṣemakarī
Kṣetradevī 1: Śatanetrā	**Kṣetradevī** 3: Ayutabhujā*	**Kṣetradevī** 5: Gajāsyā	**Kṣetradevī** 7: Śavavāhīnī
Bhairava 1: Ruru	**Bhairava** 3: Asitāṅga	**Bhairava** 5: Krodhana	**Bhairava** 7: Saṃhāra
Dikpāla 1: Indra	**Dikpāla** 3: Yama	**Dikpāla** 7: Varuṇa	**Dikpāla** 7: Kubera
Vetālas 1–3: Vidyujjihvā, Lalajjihvā, Krūrāsya	**Vetālas** 7–9: Vakramukha, Vakranāsika, Jaṃbhaka	**Vetālas** 13–15: Gartanetra, Mahānetra, Tucchanetra	**Vetālas** 19–21: Kharvagrīva, Mahāhanu, Mahānāsa

SOUTH-EAST	SOUTH-WEST	NORTH-WEST	NORTH-EAST
Śaktis	**Śaktis**	**Śaktis**	**Śaktis**
(13–24): # all lost	(37–48): only one exists	(61–72): only one exists	(85–96): only four exist
13. Jayā	37. Buddhi	61. Trivargā	85. Siddhikarī
14. Jayantī	38. Svadhā	62. Garvarahitā	86. Chinnamastā*
15. Vijayā	39. Svāhā	63. Ajapā	87. Śubhānanā
16. Jaleśī	40. Mahānidrā	64. Japahāriṇī	88. Śakaṁbharī
17. Aparājitā	41. Śarāśanā	65. Japasiddhi	89. Mokṣalakṣmi*
18. Śaṅkhinī	42. Pāśapāṇī *	66. Tapaḥsiddhi	90. Trivargaphaladāyinī
19. Gajavaktrā	43. Kharamukhī	67. Yogasiddhi	91. Vārtālī
20. Mahiṣaghnī	44. Vajratārā	68. Parā	92. Jaṁbhalī
21. Raṇapriyā	45. Ṣaḍānanā	69. Amṛtā*	93. Klinnā
22. Śubhā	46. Mayūravadanā	70. Maitrīkṛt	94. Aśvārūḍhā*
23. Nandā	47. Kākī	71. Mitranetrā	95. Sureśvarī
24. Koṭarākṣī	48. Śukī	72. Rakṣoghnī	96. Jvālāmukhī*
Kṣetradevī 2: Sahasrāsyā	**Kṣetradevī** 4: Aśvārūḍhā	**Kṣetradevī** 6: Tvaritā	**Kṣetradevī** 8: Viśvā
Bhairava 2: Caṇḍa	**Bhairava** 4: Kapālī	**Bhairava** 6: Unmatta	**Bhairava** 8: Bhīṣaṇā
Dikpāla 2: Agni	**Dikpāla** 4: Nairṛtta/ Rudra)	**Dikpāla** 6: Vāyu	**Dikpāla** 8: Īṣanā
Vetālas 4–6: Krūralocana, Ugra, Vikaṭadaṃṣṭra	**Vetālas** 10–12: Jṛṃbhaṇamukha, Jvālānetra, Vṛkodara	**Vetālas** 16–18: Antramandana, Jvalatkeśa, Kaṃbuśira	**Vetālas** 22–24: Laṃbakarṇa, Karṇaprāvaraṇa, Anasa

Sources: in the *Kāśīkhaṇḍa*: 96 Śaktis (KKh 72.3–13); 8 Kṣetradevīs (KKh 72.90–91); 8 Bhairavas (KKh 72.93); 24 Vetālas (KKh 72.97–99). #: Lost / no trace; *: Existing (location given). See Fig. 1.

Table 2. Vārāṇasī: The Kṣetradevīs

	Name	Meaning	Location in Varanasi
1	Viśālākṣī	Wide-eyed G.	Mīr Ghāṭ, near Dharmakūpa, D 3/85
2	Lalitāgaurī	The Amorous G.	Lalitā Ghāṭ, in Madhi, D 1/67
3	Viśvabhujā	World's covering hands	Viśvabahuka, Dharmakūpa, D 2/15
4	Vārāhī	Boar-Form Viṣṇu's Power	Mān Mandir, Tripurabhairavī, D 16/84
5	Śivadūtī	Śiva's messenger	Pañcāgni Akhāḍa, Prahlād Ghāṭ, A 11/30 (attached to Svarlīneśvara temple)
6	Aindrī	Indra's Power G.	Near Maṇikarṇikā Ghāṭ, c. CK 9/2
7	Kaumārī	Skanda's Power	Kashmirimal, Kalaśakūpa, CK 7/102
8	Maheśvarī	Śiva's Power	In the court of Viśvanātha, CK 35/19
9	Vṛṣayaṇavatī	Amorous Desire G.	# Lost
10	Narasiṃhī	Man-lion Viṣṇu's G.	Maṇikarṇikā Galī, Sureka Bl, CK 33/43
11	Brāhmī	Brahma's Power	Brahmeśvara, Khalispura, D 33/66
12	Nārāyaṇī	Viṣṇu's Power	Gaurī Śaṅkar T., Lāl Ghāṭ, K 4/24
13	Virūpākṣī	Virūpāksha's G.	Viśvanātha T. compound, CK 35/19
14	Śaileśvarī	Mountain's Daughter	Maḍhia Ghāṭ at Varaṇā's bank, A 40/11
15	Citraghaṇṭā	Wonderfull Bells	Lakkhi Chabutara, CK 23/34
16	Citragrīvā	Wonderfull Neck	Near Kṣemeśvara Ghāṭ, B 14/118
17	Bhadrakālī	Noble Black G.	# Lost (Madhyameśvara: K53/63)
18	Harasiddhi	Śiva's Power	Back of Siddhi V., Maṇikarṇikā,CK 9/9
19	Vidhidevī	Law Maintainer	Vindhyavāsinī c. Saṅkaṭā T., CK 2/33
20	Nigaḍabhañji (Bandī)	Remover of Iron-chain	Daśāśvamedha, D 17/100
21	Ghaṇṭāṅkarā	Bell-sound Maker	Śūlataṅkeśvara, D 17/111
22	Amṛteśvarī	Nectar giver G.	Amṛteśvara, Nīlakaṇṭha, CK 33/28
23	Siddhalakṣmī	Perfected G.of Wealth	Jagadhātri, E of Siddha Vin., CK 8/1
24	Kubjā	Hump-backed G.	Prapītamaheśvara, CK 7/92
25	Trilokyasundarī	Most Beautiful G.	Pītamaheśvara, Śitalā, CK 60/92
26	Dīptā/ Dīptāśakya	Radiant One	Near Sambāditya, Surāj Kuṇḍ, D 51/90
27	Mahālakṣmī	The Great G. of Wealth	Lakṣmīkuṇḍ Temple, D 52/40
28	Hayakaṇṭhī	Horse-headed G.	Kālīmaṭh, Lakṣmīkuṇḍa, backside
29	Kaurmīśakti	Tortoise-form Viṣṇu's Power	Bhīṣmacaṇḍī, Pāśapāṇi II, Sadarbazar
30	Śikhīcaṇḍī	Tall-hair G.	Mahālakṣmī T., backside, D 52/40
31	Bhīmacaṇḍī	The Strongest Caṇḍī	On Pañcakrośī route, Bhīṣmacaṇḍī
32	Chāgavakreśvarī	She-goat Face G.	S wall of Vṛṣabhadhvaja, Kapildhārā
33	Tālajaṅgheśvarī	Full-buttocks G.	# Lost
34	Vikaṭānanā	Toughest One	Scindhia Ghāṭ, Ātmavīreśvara, CK 7/158
35	Yamadaṃṣṭrā	Destroyer of Yama's curse	# Lost
36	Śuṣkodarī	Dry-belly G.	Vṛddhakāla, Daranagar, K 46/23
37	Carmamuṇḍā	Leather-headed G.	Lolārka Kuṇḍa, Munnar Panda house, B 2/62
38	Mahāruṇḍā	Mutilated G.	Lolārka Kuṇḍa, Munnar Panda house, B 2/62
39	Cāmuṇḍā	Killer of Caṇḍa & Muṇḍa	Near Lolārka Kuṇḍa
40	Svapneśvarī	G. of Good Dreams	Hanumān Temple, Śivālā, B 3/150
41	Durgā	Inaccessible One	Durgā Kuṇḍa, B 27/1

Source: KKh 70.10–97. G = Goddess. Out of 41, four are lost: No. 9, 17, 33, and 35.

Table 3. Kāśī / Vārāṇasī: Nature of distribution of goddess shrines

Sacred Territory, Kṣetra	Mean radius of circuit from centre, km *	No. of Goddess shrines	Diffusion ratio: from in to out (2/ 44 base)	Reciprocity of diffusion ratio	Power ratio of radius (1/ 1.27 base)	Inter-circuit Power ratio increase, times
(inner to outer)	*1*	*2*	*3*	*4*	*5*	*6*
1. Antargṛha	1.27	44	1.00	1.00	1.00	1.00
2. Avimukta	1.97	28	0.64	1.56	1.55	1.55
3. Nagarapra-dakṣinā	3.95	17	0.38	2.63	3.11	2.00
4. Pañcakrośī	14.08	7	0.16	6.25	11.07	3.56
5. Caurāsīkrośī	17.60	96	2.18	0.46	13.85	1.25

Source: the distribution maps; * refers to Route length/ 2 π.

ANNETTE WILKE

The Banarsī Navadurgā Cycle and its Spatial Orientation

Arriving in Varanasi[1] in autumn during the great festival of the goddess Durgā (*śārada*-Navarātri), the temporary pavilions (*paṇḍāl*) enshrining fanciful clay representations of the demon-slaying Durgā cannot but catch the eye of even a casual tourist. This kind of Navarātri celebration is rather common in urban North India, particularly in Calcutta. My paper, however, is devoted to another form of celebration which is more quiet and may likely escape a foreigner's attention.[2] It takes place in nine goddess temples of the city, and it is certainly more specific to Varanasi, since it localizes the concept of the Navadurgā ("Nine Durgās") within the cityscape (Fig. 1). On each of the nine festival days a particular representation of the Durgā of the day, starting with Śailaputrī, is worshipped in a different shrine. The full list of names—1. Śailaputrī, 2. Brahmacāriṇī, 3. Candraghaṇṭā (Citraghaṇṭā)[3], 4. Kūṣmāṇḍā, 5. Skandamātā, 6. Kātyāyanī, 7. Kālarātri, 8. Mahāgaurī, and 9. Siddhidātrī[4]—is well known from the *kavaca* section of popular *Durgāsaptaśatī* editions for ritual use, too. But the Banarsi celebration translates this temporal sequence into space, and thus lends a local topography to the Navadurgā concept by identifying each of the nine manifestations with a Goddess of Kāśī. There are already references to the *śārada-navarātrayātrā* in the *Kāśīkhanda* (KKh) of the *Skandapurāṇa*, mentioning four of the Devīs still worshipped today (KKh 70.23, 37–42; 72.83–89). Most of the *Kāśīkhaṇḍa* dates back to the 14th century,[5] and from the Gāhadavāla inscriptions we know that Navarātri was one of the most popular festivals already in the 12th century Varanasi (Eck 1982: 174). Until today it is very much in the awareness of the Banarsi population and perpetuated by the media. On each of the nine festival days the local Hindi newspaper *Āj* announces on its front page the Durgā of the day, presenting her name, her picture, her location, and her *dhyāna-śloka*.[6] Thus the local Navarātri celebration is still very much alive, though there may have been severe breaks in the flow of tradition. It is noteworthy that the present Navadurgā do not appear yet as a *group* in the *Kāśīkhaṇḍa*.[7] The set of names is in fact not local, but pan-Hindu. Yet, it has become the "Cycle of Kāśī Goddesses", as Diana Eck puts it.[8] Archaeological and textual evidence indicate that there have been several stages of re-interpretation and re-invention of tradition which led to this cycle. Except for Kātyāyanī whose (Mahiṣāsuramardinī-)image can be dated to the 9th/10th century,[9] and maybe Bāgeśvarī, the original Skandamātā, none of the *mūrtis* seems to be older than two hundred years. The same holds true for the temples, even though the sacred site itself (like the one of Śailaputrī) might date back to the time of the *Kāśīkhaṇḍa*. In the *Kāśīkhaṇḍa*, however, we find other goddess cycles related to Navarātri and some insights on a different spatial orientation in former times. My main concern does not

Fig. 1: Location of the Navadurgā (drawing by Niels Gutschow).

primarily lie in a historical reconstruction, but in the projection of the pan-Hindu Navadurgā scheme onto the local topography as we find it today. I would like to stress two arguments: a) the strong local bias toward the pan-Hindu superstructure, and b) the replacement of an older circular scheme by a triangular spatial pattern.

Pan-Hindu Scheme and Living Tradition

The Navadurgā list starting with Śailaputrī itself is not Varanasi specific, but well known all over India being found in North and South Indian *Durgāsaptaśatī* editions for ritual use.[10] It was known in North India as early as the medieval period (Mishra 1973: 23).[11] The same list is also found in the *Tattvanidhi,* a Southern Śilpa text (Nagaswamy 1982: 134). Thus it can be termed pan-Hindu. Besides this list, there exist a large number of Navadurgā catalogues in which the names of the nine manifestations vary considerably depending on diverse textual and regional traditions and different theologies.[12] Compared with such other catalogues, the set starting with Śailaputrī seems to have gained the widest distribution apart from lists including the (Sapta-/Aṣṭa-)Mātṛkās which are much closer to the Durgā myth. The latter kind is found in the *Kāśīkhaṇḍa* even twice. *Kāśīkhaṇḍa* 70.26–36 presents the Mātṛkās who assisted Durgā in her battle as part of a group of nine precisely localized goddesses.[13] This group is placed in a Navarātri context (70.22–25; 37–41), whereas the second set (KKh 83.33)[14] appears in a *vrata* chapter glorifying Śiva-Vīreśvara (83.54f., 106–124) in whose temple today's sixth Durgā Kātyāyanī is worshipped. Unlike Banarsi scholars, Awasthi (1978: 167) understands the fierce second group of nine as Navadurgā.

Even though the full list of the Śailaputrī group is not found in the *Kāśīkhaṇḍa*, four of the present Durgās appear already in a Navarātri context: Śaileśvarī (today's first Durgā Śailaputrī) and Citraghaṇṭā (the third, also called Candraghaṇṭā) are loosely "paired" by their successive verses (KKh 70.37–38ff.), whereas the "southern" Durgā at Durgākuṇḍa (today's fourth Durgā Kūṣmāṇḍā) and Kālarātri (the seventh) appear in the same verse (KKh 72.88) and are coupled by the Durgā myth. The two pairs, however, are completely unconnected. We cannot conclude for certain that the complete Navadurgā cycle was unknown by the time of the *Kāśīkhaṇḍa*.[15] Whereas we can deduce from the highly diverse Navarātri passages that there may have been alternative ways to celebrate the festival and possibly more than one "Navadurgā cycle".

Names often function as narrative theologies. Much more than the Mātṛkā lists, the Śailaputrī list provides a sense of an integrated whole, the mere names being suggestive of both mild and fierce aspects, spouse goddess and independent Great Goddess features. The "Mountain's Daughter" (1. Śailaputrī) can be read as a synonym of Pārvatī who performs as virgin or "Chaste" (2. Brahmacāriṇī) austerities to win Śiva as her husband. As Śiva's creative Śakti, the goddess is the "Wonderous Bell" (3. Citraghaṇṭhā, or Candraghaṇṭhā, "Moon Bell") who initiates creation[16] and the cosmic mother whose womb contains the universe like a "Pumpkin Gourd" (4. Kūṣmāṇḍā) contains numerous seeds.[17] Being Śiva's mythical spouse and the "Mother of Skanda" (5. Skandamātā), she has a heroic demon-slaying son, but as independent goddess, she herself defeats the forces of evil and thus encompasses also dark and destructive sides: (6.) Kātyāyanī is an epithet of Durgā-Mahiṣāsuramardinī ("the Destroyer of the Demon Mahiṣa"), whereas (7.) Kālarātri, "Black Night", is an epithet of Kālī, signifying the most bloodthirsty aspect of the war faring goddess, as well as the cosmic night of dissolution and the destruction of ignorance (Coburn 1988: 186–188). Her ferocious aspects being tamed, however, she is "the Great Golden (or: Fair) One" (8. Mahāgaurī) and Lord Śiva's mild and gracious wife.[18] As "Bestower of Success and Accomplishment" (9. Siddhidātrī), she finally represents (Śiva-)Ardhanarīśvara[19] or the supreme divinity, incorporating and transcending the gender categories.

The pan-Hindu popularity of the Śailaputrī list may be due to the fact that it indicates a Śaiva orientation, a strong brahmanical bias, and a balanced concept regarding the most distinctive goddess features. Since it implies Durgā's identification with Śiva's spouse Pārvatī, its adoption in Varanasi, the holy city of Śiva-Viśvanātha, is quite meaningful.[20] Since names are open systems, there is no limit to find ever new meaning in them. Traditional Pandits are masters in such hermeneutics. Acharya Vagish Shastri, a well-known Sanskrit Pandit in contemporary Varanasi, explains the names esoterically as an inner journey through the mental world: penance (Śailaputrī) and celibacy (Brahmacāriṇī) lead to the mystical knowledge of hearing inner sounds (Citraghaṇṭhā). A further step on the spiritual path is winning over the delusion of the world by absorbing it (Kūṣmāṇḍā). Seeing in everything the same, one is no more bound and lives as a *jīvanmukta* (Skandamātā), becomes a Ṛṣi like Kātyayana (Kātyāyanī), and conquers death (Kālarātri). From the *tamas* (Kālarātri) one moves to *sattva* (Mahāgaurī) and reaches *mokṣa,* the final perfection (Siddhidātrī). Mystic topographies of real places have a long history.[21]

While such kind of hermeneutics may have been decisive in favouring the present Navadurgā list, the devotees visiting the nine temples are generally of a less speculative type. They rather seek Durgā's help for daily life and its problems, or maybe just a diversion from the daily routine. Contrasting the pan-Hindu Navadurgā scheme, the living Kāśī tradition presents a variegated picture, each of the Durgās having a very distinct profile and local colouring. Mahāgaurī for instance, is known in Varanasi as Annapūrṇā, the one "Full of Food and Nourishment", residing next to the golden temple of her husband Śiva-Viśvanātha, the Lord of Kāśī. Like Kūṣmāṇḍā's, her temple belongs to the largest and best frequented temples in the city, while other Navadūrgā sites are hardly known. The nine temples differ not only considerably in terms of popularity and size, but also regarding their neighbourhood, owners, priests, *mūrtis*, rituals, and opening hours. Each must be described as a social and ritual space of its own. In the following, I restrict my discussion to some of the more striking features.

The Diversity of Cultural Spaces

Śailaputrī, the First

Out of all the Durgās, Śailaputrī, whose remote shrine at the northern border of the city (Maḍhīya Ghāṭ, No. 40/11) is not much frequented on ordinary days, attracts the largest crowd.[22] Thousands of devotees, coming from all stratas of society, many from the surrounding rural area, take the pain to reach Śailaputrī before sunrise and are willing to wait one or two hours for *darśan*. The space around the site serves well for a *mela.* Moving towards the temple grounds the street gets more and more densely packed with booths where devotional paraphernalia, toys, and jewellery are sold. Ascetics display their penance, snake charmers attract children, beggars and lepers beg for alms.

The spacious temple ground includes a Śītalā shrine which is passed first. On the first day a small girl accepts the offerings. The temple is run by several male Brahman and non-Brahman *pūjārīs*. Before the head *pūjārī* took charge in 1952, a lady *pūjārī* was doing service. Today's main priest is said to be a Tāntrika, consuming meat and wine.

Śailaputrī's site is permeated by an atmosphere of "folk religion". Between the main sanctum and the *havankuṇḍ* there is a platform with several conic *samādhi* stones. They are believed to enshrine the bodies of holy men to whom the goddess appeared, and to be imbued with the spiritual power of sages who had lived in the temple grounds. Many women, especially barren ones, first worship the *samādhis* in order to get a son before worshipping Śaileśvarī in the main sanctum.[23] According to my informants,[24] only on the first day of *āśvina* and *caitra* Navarātri does Śailaputrī wear a mask and a dress. Two male priests pass the devotee's offerings to her: coconuts, flowers (preferably red Hibiscus) and red cloth. Thereafter, many women turn to the Nīm tree in the back of the shrine which is considered the original dwelling place of Śailaputrī. Without priestly assistance, they worship the tree and the aniconic *mūrtis* on its root with red cloth, strings, vermillion, food and coins, seeking to obtain children or a husband or to secure health and well-being of husband and family.

Brahmacāriṇī, the Second

Brahmacāriṇī's small temple, close to Durgā Ghāṭ, is hard to find and better reached by boat (Fig. 2). The neighbourhood is dominated by Banarsis, Maharashtrians and Gujaratis. In close vicinity to the temple, there is the "Kashi Maharashtra Bhavan" serving as lodge for Maharashtrians who come for pilgrimage or choose to die in the holy city. Out of all Durgās, Brahmacāriṇī best demonstrates how neighbourhood and ethnicity are an important factor. The site has a strong Maharashtrian influence. The Brahman *pūjārī*[25] claims that it was established by his priestly Maharashtrian ancestors about three hundred years (five generations) ago and still belongs to his family. He inherited his office from his father and grandfather, and is a full time priest and Śrīvidyā initiate. It is the only Navadurgā site in which the highly esoteric *śrīcakrapūjā* takes place (each Friday behind closed temple doors).[26]

Due to donations from Maharashtrians, Gujaratis and Nepalis who live close by or come for pilgrimage, the temple is fairly wealthy, owning jewellery and six masks, one of them made of silver over coated with gold. The goddess wears this precious mask only on the second day of Navarātri, says the priest. But in fact, the masks worn in 1994 and 1995 were not identical. In 1995 "dharmic pictures" of Brahmacāriṇī and a small pamphlet called *Śrī Brahmacāriṇī Mantrapuṣpāñjali* were sold.

The original black stone *mūrti* (with pasted eyes and beautifully dressed) can be seen without mask only on *aṣṭamī* of the autumn and spring Navarātris. On autumn *aṣṭamī*, several additional rituals take place in the evening/night apart from *śṛṅgār*. *Garbha* dance is performed by Gujarati girls and young women from the neighbourhood. They are colourfully dressed as *gopīs* and dance with their sticks around a little

Fig. 2: Brahmacāriṇī, the Second Durgā (Brahmacāriṇīmandir). Navarātri 1995 (photo by A. Wilke).

boy dressed as Kṛṣṇa.[27] In front of the temple *ṣoḍaśopacāra* is offered on a decorated *kalaśa* representing the goddess, followed by *havana* with a litany of 1008 names.[28] In the neighbouring Maharashtra Bhavan about 16 (Maharashtrian and Gujarati) women dance in front of a rice representation of Mahālakṣmī.[29] The ladies implore the goddess (or their *iṣṭadevatā)* to inhabit them. While dancing, they blow into earthen *kalaśas* which have been consecrated in the temple. Possessed by the Devī some continue their dance in the temple, some act as oracles for the watching crowd of ladies. Obviously, the Brahmacāriṇī temple fulfils different spiritual demands. Its attraction seems to be largely confined to a special ethnic clientele.

Candraghaṇṭā, the Third

The third Durgā resides in an even smaller, but free standing shrine which belongs to a private owner.[30] Candraghaṇṭā is known in Varanasi also as Citraghaṇṭā. Out of all Durgās she demonstrates best historical shifts and local colouring. A goddess called Citraghaṇṭā was important in the 11th century as we know from a Pāla inscription (1026 AD) which mentions that the king of Gauḍa installed hundreds of temples of Īśāna (Śiva) and Citraghaṇṭā in Kāśī (Verma 1971–84: 199). Today she has lost her previous glory. Citraghaṇṭā's tiny shrine is rather insignificant and does not attract a large crowd even on the third day of Navarātri. There are two representations of the goddess, each within a row of other goddesses. Both icons reveal the special Banarsi colouring of the pan-Hindu scheme and convey the message: I am a Navadurgā shrine, since the names of the Devīs are inscribed on the pedestals. Citraghaṇṭā's image in the front row is flanked by Annapūrṇā and Kūṣmāṇḍādurgā, the most prominent goddesses of today. In the two back rows we find the whole Navadurgā group with Candraghaṇṭā. Two of the Navadurgās differ from the "pan-Hindu scheme", namely the old local Banarsi Skandamātā Vāgeśvarī, and Annapūrṇā, the Banarsi Mahāgaurī. While the name Citraghaṇṭā (goddess in the centre of the three goddesses in the front row) coincides with the *Kāśīkhaṇḍa,* the name Candraghaṇṭā (goddess belonging to the Navadurgā in the back rows) coincides with the name of the third Durgā given by the *Āj* and the *Durgāsaptaśati kavaca.* This may be taken as a visible token that the shrine seems to have preserved the memory of two distinct goddesses who were identified for the phonetic similarity of their names. All present images are made of black stone, covered with rich brocade dresses and of the same height (ca. 30 cm). Only due to her central position, the rich golden frame and her mask Citraghaṇṭā can be determined as the main icon.

Kūṣmāṇḍā, the Fourth

This Durgā resides in the most southern Navadurgā site, located at the river Asi which formed the natural southern border of the city by the 11th century.[31] Her beautiful pink sandstone temple is known as Durgākuṇḍmandir because of its vast water tank (*kuṇḍa*). It is one of the largest and most prominent temples of Varanasi,[32] exceeding in popularity all other Devī shrines of the city and attracting a large number of visitors throughout the year. It belongs to a wealthy Mahant. While the prestigious office of the Mahant is hereditary, the *pūjārīs* come from different families.

The Durgā(kuṇḍ)mandir is the only Navadurgā temple where animal sacrifice is sanctioned nowadays. Great animal sacrifices *(mahābali)* belong, according to the *Kāśīkhaṇḍa,* intrinsically to the worship of Durgā and even some generations ago such sacrifices were conducted on a large scale in the Durgāmandir. Today only occasionally a goat sacrifice takes place, generally on *aṣṭamī,* the eighth day of Navarātri, sometimes also on the fourth, Kūṣmāṇḍā's day. The initiative, however, is not taken by the temple officials but by private persons.[33] Their motivation is to seek Durgā's help for a specific undertaking.

Durgādevī has no *mūrti* but a "self-manifested" (*svayaṃbhu*) silver mask to represent her.[34] From the mask is draped a tinselled red cloth[35] which makes her look like other Devīs. On the fourth day of Navarātri, however, she literally gets embodied, being represented by a metal figure dressed in a red Sārī and bedecked with ornaments and garlands.

Skandamātā, the Fifth

The Skandamātā temple stands out for its Brahman priestess and its goddess Bāgeśvarī whose *darśan* can only be obtained on the fifth day of *āśvina* Navarātri. The free standing, flat-roofed, two-storied temple is located in Jaitpurā, north-west of the city, where many Muslims live. It shelters two goddesses, a fact which is hardly known to anyone and can only be ascertained during the fifth day. On this day the smaller sanctum on the ground floor is open. Otherwise it is closed for the rest of the year. It shelters a small, black stone image of Bāgeśvarī whom some of my informants call "Vāgīśvarī" (about 20–30 cm in height).[36] According to Kedarnath Vyas (1987: 153) she is the original Durgā of the fifth day, but the lady priest, the *Āj*, and the devotees consider the main Skandamātā image on the first floor as fifth Durgā. Some even miss the *darśan* of Bāgeśvarī. Although the shrine is closed for lay people throughout the year, Bāgeśvarī receives *pūjā* every day by the officiating priestess. On the fifth day of Navarātri, the icon is adorned with flowers and a huge, silverish looking metal mask which covers her body.

In the main sanctum on the first floor Skandamātā resides in an elevated *garbhagṛha*. Her iconography (described by the head priestess) matches the Śilpaśāstras and the drawing presented by the *Āj:* the (nude) Skandamātā sits on a lion (of foam), carries a lotus in each of her two hands and in her lap reclines her son Skanda.[37] The priestess is proud to be allowed to see the goddess naked which is forbidden to male priests. In fact, many of the male *pūjārīs* in charge of the other Durgā temples pretended that they are not able to describe the iconography of the respective *mūrti* because they are not supposed to see the goddess nude. Lay devotees can rarely detect the iconographic traits themselves because the images are always draped in a Sārī and covered with flower garlands. This is especially true for Navarātri where the images vanish completely behind the many flower garlands and the masks. The black stone icon of Skandamātā with wide pasted eyes and thick layers of vermillion on lips and forehead is adorned with a metal mask every day for two hours after the morning bath. She owns only one mask made of five metals. The head priestess herself had it made as well as a mask for Skanda.

This Brahman priestess inherited her office because her parents had no son. In the mid-nineties she was already in her eighties and stated that after her death her son will take over. In her personal accounts she oscillated constantly between a dominant male and dominant female priest perspective, some legendary and some actual description, e.g.: first she says that the priestly office is hereditary and it always falls to the son, but if there is no son, a daughter or daughter-in-law can succeed. Her predecessor happened to be her mother-in-law. But then again she presents a legendary explanation where female succession is stressed as a special trait of this temple. She says that her grandmother came by foot from Jhansi to this place which was by that time covered by jungle full of lions and panthers. The grandmother, feeling uneasy, asked herself: "What am I going to do here?" In a dream the goddess appeared and ordered: "You stay here and do service for me". The place got thus sanctified and Skandamātā being installed was considered a *svayaṃdevī* ("selfmade, selfborn", not man-made *mūrti*). Of course, Jaitpura was hardly a place of wild beasts two or three generations ago, but the story of origin functions as a special form of cultural memory, as a reconstruction of the past to create an imaginary space in which the female element dominates and lends authority to the present. According to the story the *pūjārīṇī's* own grandmother started the *pūjā,* but then there was an initial account of a male ancestor who initiated the worship and later her own son will carry on. This kind of multiple discourses I have often acknowledged in my discussions, especially with regard to the iconography.

Kātyāyanī, the Sixth

The sixth Durgā does not have a temple of her own. She is worshipped in a partly damaged Mahiṣāsuramardinī image carved on stone plate and situated in a dark side niche of the (Ātma-)Vīreśvara temple at Sindhiyā Ghāṭ.[38] It is interesting to note that except for this ancient Kātyāyanī image none of the other Durgā icons is (as far as I could determine) a Mahiṣāsuramardinī. Kātyāyanī receives daily *pūjā* and does otherwise not draw special attention. In 1995 the nude Kātyāyanī was decorated in the morning of the sixth day only with some flower garlands, but after 4 pm adorned (*śṛṅgār*) and "beautified" with a mask. The mask is neither particular nor obligatory for Navarātri, but dependent on the priest's choice.[39] It may be applied also during other festivals, or on ordinary days by request of devotees who give a donation. Unmarried women praying for good husbands specially come to worship Kātyāyanī. Even on the sixth day (1995) mainly women are present. They wave incense, oil-lamps, and camphor light towards the goddess while standing beside the *liṅga,* the main *mūrti* of the temple. Traditionally this Śiva temple is known to grant children, a favour not attributed to the Devī, but rather to the "*svayaṃbhu*" *śivaliṅga.*[40] The temple stands out for its daily *saptarṣi-āratī* in the evening performed by the *pūjārī* and devoted lay Brahmans who are also the managers and main patrons of the temple (Zeiler 1998: 41–45; 49–61).

Kālarātri, the Seventh

The seventh Durgā, worshipped in the Kālikāmandir (Kālikāgali), is represented by an awesome black *mūrti* with a snout and a huge red tongue. The icon visualizes well the most terrifying and bloodthirsty aspect of the goddess. Of course, no blood sacrifice is taking place, ritual slaughter being a rather despised practice in Varanasi nowadays. The *pūjārīs,* two learned Brahman brothers, put a special emphasis on their purely vegetarian "*sāttvika pūjā*". The temple is not very crowded even during the early morning hours of the seventh day, except for the small platform on which men sit closely to meditate, to pray, to repeat holy names and to read the *Durgāsaptaśatī.*

Mahāgaurī, the Eighth

On *aṣṭamī* we find an almost equally large crowd like on the first day. Annapūrṇā who is nowadays the governing queen of Kāśī, is worshipped as Mahāgaurī. In contrast to Śailaputrī, Annapūrṇā has a strong brahmanical flavour. Her magnificent temple in the heart of the city, two minutes away from her spouse Śiva-Viśvanātha, is spacious with its two stories, the pillared side porch and the many subsidiary shrines. During Navarātri, the side porch is crowded in the early morning with *sādhus* and Brahmans in white, who come to pray and do *japa,* to recite the *Durgāsaptaśatī* and Vedic texts. Annapūrṇā's sanctum (*garbhagṛha*) in the centre of the ground flour can be circumambulated. On the eighth day thousands of people come for this *pradakṣiṇā,* moving around the sanctum eleven or even one hundred and eight times. The crowd is so large that a bridge is constructed to pass above the moving body of devotees in order to reach the porch in front of the *garbhagṛha.* Looking down from the bridge the sight is rather impressive: it looks as though *one* collective body and the whole temple is rotating. In the back of the porch stand ladies, facing the circumambulating crowd. They hold their Sarīs spread out, so that the people moving by can throw rice, barley, sugar globulets and coins inside. Also a few men and children stand begging. This rite is a mimetic enactment of Śiva as a beggar and Annapūrṇā as goddess of nourishment who is implored in the *Annapūrṇāstotra,* attributed to the famous philosopher ascetic Śaṅkara, to grant sustenance and alms.

During Navarātri, Annapūrṇā gets redressed several times every day. On *aṣṭamī* 1995 she wore a silver mask and a gorgeous brocade Sārī in the morning, a yellow metal mask and a rich red Sārī at noon, and again another yellow metal mask in the evening, but no cloth, since the black stone idol does not show her nude. In the evening a white metal alm-begging Śiva icon was placed near Annapūrṇā, which again recalls the myth.

During the day several "Mahālakṣmīpūjās" were offered onto a yellow metal *śrīcakra* (*meru* form) in close vicinity to the *garbhagṛha.* On ordinary days they are called "Lakṣmīpūjā". These are privately sponsored *arcanas,* performed by a South Indian *sādhu* who chants the 108 or 1000 names of Annapūrṇā and throws flower petals and vermillion on the sacred diagram which is otherwise kept in the sanctum.[41] Pilgrims, particularly from southern India, have them performed for family welfare.

The temple has a strong South Indian influence.[42] South Indian pilgrims never miss to visit Annapūrṇā who is also very popular in Tamil Nadu, where she is often worshipped as supreme Mother of the World. Some identify her with Lalitā-Tripurasundarī, the queen of the *śrīcakra* and female Absolute of the tantric Śrīvidyā tradition. An oral legend holds that the ascetic philosopher Śaṅkara installed the Kāśī Annapūrṇā on a *śrīyantra.* Besides the movable *śrīcakra* on which Lakṣmīpūjās are performed, there is a huge three-dimensional metal *śrīcakra* (around 60 cm in diameter) in a glass container in front of the porch of the sanctum. It was donated by a South Indian lady in 1994. An important subsidiary shrine, close to the entrance, contains a black stone *śivaliṅga* on which again the *śrīcakra* is engraved. Inspite of the presence of so many *śrīyantras,* no *śrīcakrapūjā* proper (*āvaraṇapūjā*) is performed and none of the officiants seems to be a tantric initiate. There are further southern connections. In 1977 the old Annapūrṇā idol, whose hands were broken, was replaced by a new idol donated by the Śaṅkarācārya of Śṛṅgerī who also performed the consecration rites.[43]

The Annapūrṇāmandir and -maṭh, since 1969 under the leadership of the Mahant Goswami Tribhuvanpuri,[44] is rich. It owns many masks and images of Annapūrṇā which can be seen by a donation of 2500 Rupies. The golden image in a shrine on the first floor can be seen only through the three days of Annapūrṇā's own festival (Annakūṭa).

Also Annapūrṇā is not likely the original eighth Durgā, but rather the small Bhavānī(śaṅkar)-image in the Rāma temple adjunct to Annapūrṇāmandir.[45] The location and *mūrti* of the original eighth Durgā is a matter of dispute: an alternative mentioned is Saṅkaṭādevī (Eck 1982: 360; Kedarnath Vyas 1987: 153). Nowadays hardly anyone visits Bhavānī's shrine. Saṅkaṭā on the other hand belongs to Kāśī's most prominent Devīs, but the number of devotees on the eighth day does not exceed the usual number throughout the year.[46] If any of these should have been the original Mahāgaurī, as scholars suspect, no devotee is aware of it. Mahāgaurī Annapūrṇā has stolen the show completely.[47]

Siddhidātrī, the Ninth

Local tradition calls Saṅkaṭādevī also "Siddhidātrī". Some say, Siddhidātrī's temple does not exist anymore. The seat of the ninth Durgā seems to be disputed, too. Today two minor goddess shrines in the Chowk area are claiming to be her residence, namely Siddhimātā (Siddhimātā Galī, CK 60/29) and Siddheśvarī (CK 7/124). The *Āj* identifies Siddhidātrī CK 7/159) as Siddhimātā and a banner across the lane of Siddhimātā's shrine affirms her identity as ninth Durgā.

In 1995 a naked lady ascetic performed severe austerities at the site by having buried herself under a heap of earth filled with barley grains. Were it not for this spectacular vow, the temple would have been almost empty even on the ninth day. As far as I could see, the least number of people pay *darśan* to the ninth Durgā.[48] Compared with the theological importance given to the ninth Durgā, she demonstrates best how much living and scholarly tradition may differ.

Generally one can detect a decline of visitors on the last four days, except for the eighth. I suppose this declining interest is due to the activities in the temporary *paṇḍāls* during the last three nights of the festival. To go there has become more attractive and popular than the *darśan* in the traditional temples.

In the light of the findings presented, I should specify what I have called "the local bias of the Navadurgā". The bias is in fact not only local, but incorporates a diverse set of cultural spaces of more than one ethnic community, including diverse social stratas of male and female ritual specialists. It is hard to see any connecting link regarding this plurality of cultural spaces. Even the prolonged opening hours differ. The spacious Śailaputrī site is open 24 hours on the first day[49], whereas the small Citraghaṇṭā shrine closes on the third day at 5 p.m. instead of the usual 1 p.m. Even though each Durgā of the day—except for the huge faced awesome Kālarātri—wore a mask for beautification in 1994 and 95, and my informants pretended that some masks are destined particularly for this day, also the masks are not really a common element, nor are they a rule. Some goddesses wear a mask everyday, some only for the festival's sake. Some temples own only one mask, others own many masks made of costly materials. Part of the wealth and prestige of a temple can be determined by the number, quality and material of the masks donated. These masks are there for beautification (*śṛṅgār*), they have no other function, I was told. The black stone icons are no longer considered attractive.

The empirical diversity of shrines leads to the question as to why they have been selected. I will suggest that the mental map underlying the selective processes is the spatial orientation of the nine days' pilgrimage route which forms a triangular pattern. Against the background of the *Kāśīkhaṇḍa* 72 and a cartographic representation of the 19th century, I tend to add that one ideal scheme replaced another ideal one.

Textual Tradition and Historical Shifts

There are three (at first sight very confusing) Navarātri passages in the *Kāśīkhaṇḍa*. If we expect to find a rationale explaining the selection of the Navadurgā sites, we will be disappointed. We may only infer that the present Navadurgā cycle seems to be the end of a longer, rather complex process.

The first passage, *Kāśīkhaṇḍa* 45.34–54, indicates that the Navarātri celebration might not have included any Navadurgā group at all in former times, but was celebrated by worshipping the 64 Yoginīs (45.47–54). Although the highly brahmanic rituals prescribed (name invocation with the Praṇava Oṃ added and 108 *havanas* with *ghee* and *guggulu*) do not sound Yoginī like, it is probable that the Yoginī worship predates the Navadurgā.[50] None of the Yoginīs' names (most of them theriomorphic) coincide with the Navadurgā group known today, but they have been unified at a later date in the goddess Caūsaṭṭī (literally "sixty-four") whose temple in the Bengalitola area may be visited in addition to or instead of those of the Navadurgā.[51]

The second Navarātri passage, *Kāśīkhaṇḍa* 70.23–42, comes closer to the Navadurgā cycle, since we find two of the Durgās mentioned: Śaileśvarī (70.37), identical with today's first Durgā Śailaputrī, and Citraghaṇṭā (70.38–40, 42), today's third Durgā,[52] but the festival is primarily associated with a single goddess called Viśvabhujā (-Gaurī) (70.22–25) who is forgotten today:

> O sage, Viśvabhujā-Gaurī, stationed in front of Viśālākṣī,[53] is always destroying the great obstacles of those devoutly attached to the holy territory (*kṣetrabhaktijuṣāṃ*). For the augmentation of all desired things, the autumnal Navarātri pilgrimage (*yātrā*) of Goddess Viśvabhujā should be assiduously celebrated (...). Those by whom Goddess Viśvabhujā is eulogized and worshipped in Vārāṇasī are never harassed by the masses of obstacles. (KKh 70.22–23, 25)

It is noteworthy that in-between the verses on Viśvabhujā and Śaileśvarī/Citraghaṇṭā nine other goddesses—the group of Mātṛkās mentioned above—appear (70.26–36). However, Vārāhī who starts the list, is only the first of a large number of goddesses including Śaileśvarī and Citraghaṇṭā and ending with Durgā-Devī who "protects the southern part" of the holy city (70.97). All goddesses of chapter 70 are precisely localized. Some of them are mentioned individually, others are grouped together. Many are introduced by a relational mental mapping which alludes to the directions of the compass. Chapter 70 thus poses questions concerning the celebration of Navarātri instead of answering them: Was Viśvabhujā alone worshipped during the nine days? Or was she rather worshipped along with the nine (Mātṛkā) goddesses, or even along with a larger number of female deities including Śaileśvarī, Citraghaṇṭā and the Durgā "in the south"? The latter is given much prominence in the following two chapters (KKh 71–72) and seems to relate to yet another spatial scheme.

The third Navarātri passage, *Kāśīkhaṇḍa* 72.81–91, poses similar problems, but adds new and better structured material. It stands out by its clear association with the Durgā myth and a pregnant spatial concept. While *Kāśīkhaṇḍa* 70 ends with the "Durgā in the south", *Kāśīkhaṇḍa* 71–72.1–80 proceeds with the narrative of Durgā's victorious battle along with her 90 million ("nine cores") of Śaktis represented by a list of names starting with Trailokyavijayā and ending with Jvālāmukhī (72.3–13). *Kāśīkhaṇḍa* 72.81ff. finally outlines Durgā's worship in Kāśī. It prescribes a pilgrimage or procession (*yātrā*) during *āśvina*-Navarātri and characterizes Durgā and her Śaktis as city protectors. When properly worshipped by *mahāpūjā* and *mahābali* (72.84) the goddesses are said to destroy obstacles and evils (KKh 72.82–84), to bestow intelligence, *siddhis* and family wellbeing, to delete the sins of future births, and also said to protect Kāśī as *digdevatās* (divine guardians of the directions) from all sides (72.88, 101f.). Bathing in the *durgākuṇḍa* and worshipping Durgā along with her Śaktis, the foremost (*mukha*) of whom being Kālarātri (72.87–88), seems to be the main element. However, the verses immediately following (72.89–91) introduce a list of further nine Śaktis who are also said to be the city protectors, and most importantly, who are associated with the eight cardinal directions and the centre:

> Similarly, there are nine other Śaktis who protect this territory [Kāśī] from thousands of oppressive torments. They are the following deities (protectors) of the quarters in due order: Śatanetrā, Sahasrāsyā, Ayutabhujā, the great Aśvārūḍhā, Gajāsyā, Tvaritā, Śavavāhīnī, Viśvā, and Saubhāgyagaurī. They have been posted in the eight directions beginning with east and (the last one) in the middle. These deities protecting Kāśī should be assiduously worshipped.[54]

From this it would seem that eight of these nine directional Śaktis form a circle around the city and constitute along with the ninth goddess Saubhāgyagaurī in the centre a mandalic scheme. The concluding portion of chapter 72 refers back to its beginning by stating that Durgā's numerous Śaktis, starting with Trailokyavijayā, are "protecting Kāśī all around in the four quarters" along with the eight Bhairavas, the 64 Vetālas, and the hundred crores of Bhūtas (72.3–13, 101–112).

Chapter 72 discloses with new goddesses and a clearer spatial scheme a Great Goddess theology which also may underlie *Kāśīkhaṇḍa* 70: there is one Durgā, there are nine Durgās, and still numerous other manifestations. The various systems of space orientation and large catalogues of Devīs as well as the divergent central goddesses and sets of nine lead of course also to further questions: Did several forms of Navarātri celebrations exist, each being a self-referential system, and each including different goddesses? Were possibly all these systems internally organized like the nine Śaktis of *Kāśīkhaṇḍa* 72? Were they conceived as externally related, comparable to the concentric arrangement of the *pañcakrośīyātrā* (circumambulating the city) and other *kāśīyātrās* (pilgrimage routes inside the city)?[55] Undoubtedly, the present Navadurgā cycle is an immense reduction of complexity compared with the host of Devī's mentioned in the *Kāśīkhaṇḍa*.

One thing is certain: the *Kāśīkhaṇḍa* addresses the inhabitants of Kāśī (*kāśīvāsins*). In all three Navarātri passages it is said that a lot of problems would befall the citizens if they did not perform the worship. I suppose, the main rationale in later times was to unify a larger set of goddesses and to select a Navadurgā list known to more people than only *kāśīvāsins*. Its deliberate choice may have come about by the demand of pilgrims, clever Banarsi Paṇḍā priests who ventured to make Varanasi more "cosmopolitan", or wealthy migrants who wanted to find their familiar Navadurgās in the holy city. Some of the Kāśī Durgās have different names than those of the pan-Hindu list. This holds good for the past and the present. Today Annapūrṇā is worshipped as eighth Durgā Mahāgaurī and in former times it was Bhavānī who attained this role. Such local deviations strongly suggest an adoption theory. A goddess called Śaileśvarī was already known. Being the first to follow the Mātṛkā list in *Kāśīkhaṇḍa* 70, it was not a big step to identify her with the first Durgā Śailaputrī, whereas Citraghaṇṭā could be comprehended as (onomatopoeic) equivalent to the third Durgā Candraghaṇṭā.

Particularly chapter 72 is interesting if compared with the present situation. The "southern" Durgā of the *durgākuṇḍa*, representative of the battle queen Durgā and leading city protectress, has become the fourth Durgā Kūṣmāṇḍā and one of the most powerful and popular deities of Varanasi. Kālarātri, Durgā's major mythical co-fighter, and today's seventh Durgā, resides in a smaller temple and attracts much less crowds. We should avoid, however, quick conclusions. If the Durgā(kuṇḍ)mandir equals in importance the site in the *Kāśīkhaṇḍa* it was "re-invented" that way: in its present form it was established in the 18th century by the Bengali queen Rāṇī Bhavānī who re-animated a dilapidated shrine. The pious widowed queen spent much of her time in Banaras generously donating also to the re-establishment of the *pañcakrośīyātrā* (Rodrigues 1993: 188). The 18th century was a time of a strong Hindu revivalism, in which many of the city's forgotten temples were restored (ibid. 190). This is not the only example where non-Banarsis contributed a good share to the "Kāśī Navadurgā cycle", which may explain its "pan-Hindu" feature. Some of its Durgās may not even have existed before, some may have been known by different names, and others like Kūṣmāṇḍā may have changed their representation. From the *Kāśīkhaṇḍa* one might gather that originally only the pond was representing (Kūṣmāṇḍā-)Durgā, from a later legendary account (Rodrigues 1993: 189) it could be concluded that her seat had become a Pīpal tree. Today she is represented by a beautiful "self-manifested" (*svayaṃbhu*) silver mask.

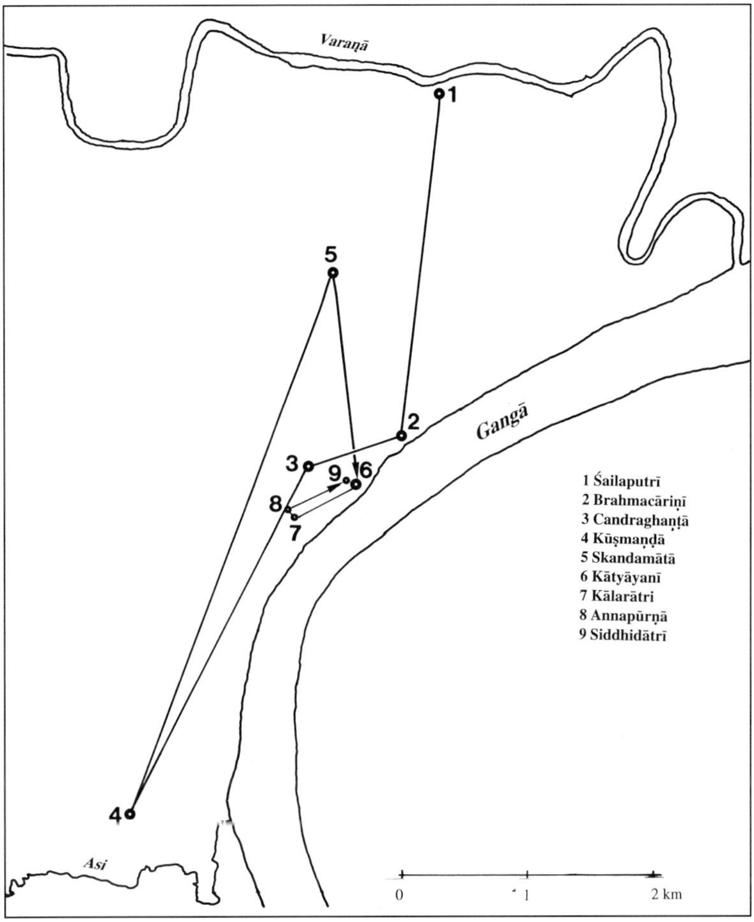

Fig. 3: Processional route of the present Banarsi Navadurgā Cycle (drawing by N. Gutschow and A. Wilke)

More interesting regarding visualized space, however, are the nine directional goddesses who are forgotten today. They disclose a strong sense of space orientation and suggest that periphery and centre played a decisive role in former Navarātri celebrations. Today we no more find such a clear spatial scheme and certainly not a circular structure. The Navadurgā shrines are not spread in the eight directions of the compass. Obviously, an older spatial model (imagined or real) got dismissed.

Selective Processes: From Circular to Triangular Processional Route

Kāśīkhaṇḍa 72.89–91 speaks of nine protective goddesses or *digdevatās* suggestive of a *maṇḍala*. The *Kāśīkhaṇḍa* does not tell us whether at all or how exactly this narrative relates to the “90 million” protective goddesses with whom the chapter starts and ends.

Much less do we learn how chapter 72 connects with chapter 70. While *Kāśīkhaṇḍa* 70 speaks of definite Kāśī goddesses, each of them having her own character and each being localized in a particular place, *Kāśīkhaṇḍa* 72 presents only a list of names except for Durgā at the *durgākuṇḍa*. The precisely localized goddesses of *Kāśīkhaṇḍa* 70 are benefactresses and protectresses of their individual worshippers, whereas the groups of goddesses of *Kāśīkhaṇḍa* 72 are specifically city protectresses. There are two narratives of protective goddesses: the first links the Durgā myth with a definite location (*durgākuṇḍa*) and mentions (Kūṣmāṇḍā-)Durgā, Kālarātri as well as "nine crores" of Śaktis (headed by Trailokyavijayā) who are the guardians of the holy territory "all around in the four quarters". The interposed second narrative condenses and formalizes the same spatial scheme. In the place of the "nine crores" it introduces nine additional goddesses who personify the eight directions of the compass and the centre. The instruction to start in the east with Śatanetrā and to end in the centre with Saubhāgyagaurī conveys the idea of a topographic *maṇḍala*.

Whereas the text leaves us with missing links, Kailāsanātha Sukula's pilgrimage map of 1876 provides a solution.[56] The map distributes both groups of Śaktis mentioned in *Kāśīkhaṇḍa* 72 outside the "ring road" encircling the city, grouping Trailokyavijayā and the rest in sets of eight times twelve Śaktis, whereas Saubhāgyagaurī is localized right in the centre. It is noteworthy that the map in fact traces two central Saubhāgyagaurīs, one of whom occupying the same space as Viśvagaurī.[57] The latter is hardly another one than the Viśvabhujā-Gaurī of *Kāśīkhaṇḍa* 70. The map seems to suggest that the different systems, relating to a number of goddesses inside and outside the cityspace, converge in the centre.

Sukula's cartographic representation does not seem to know the present Navadurgā: Brahmacāriṇī and Kātyāyanī are absent altogether, whereas the rest bears slightly different and more ancient names.[58] However, a conclusion about the existence or non-existence of the "Kāśī Navadurgā Cycle" in the 19th century is as little possible as a definite knowledge about the "reality" of the sites Sukula represents, since he is not interested in the living tradition. He rather converts the *Kāśīkhaṇḍa*'s verbal glorification of Kāśī into a visual medium. Presenting an ideal landscape, his "written map" is in fact "improving" the text: his persuasive interpretation of an outer circuit of 96 Śaktis (KKh 72.2–13) reinforces the textual mandalic pattern of nine directional Śaktis (KKh 72.89–91). By his visual construction of a protective ring around the city, he may or may not preserve a cultural memory, but he definitely creates a strong concept.

We do not know whether the nine Śaktis ever alluded to concrete locations or to a ritual diagram[59] or to both. Such correlations of diagrams and landscapes do exist, for instance in Bhaktapur, Nepal (Levy 1992: 154f.). There, we still have a living tradition, whereas none of the nine Kāśī Śaktis plays an important role in the religious life of today anymore. It is noteworthy, however, that for the contemporary Banarsi Pandit Kedarnath Vyas (1987: 153) it is a proven fact that they actually did exist. While stating that most of the nine Śaktis are completely extinct nowadays, he traces some still present in Varanasi: Aśvārūḍhā in the Jaitpurā Bāgeśvarī-*mandir* (site of Skandamātā), Viśvā (under the name of Viśvabhujā!) at Dharmkūp, and Saubhāgyagaurī in vicinity to Kāśī's most famous Śiva temple (Ādiviśvanāth).

The various pieces of information give some hints about the selective processes which possibly determined the present Navadurgā cycle. Even considering it as a later adoption, the cycle retains some older features found in the different *Kāśīkhaṇḍa* Navarātri accounts. Supplemented by further observations pertaining to goddesses or places and concepts, we get the full list of the present cycle except for Brahmacāriṇī.

> *Goddesses:* 1. Sailaputrī (=Śaileśvarī), 3. Citraghaṇṭā, 4. (Kūṣmāṇḍā-)Durgā at Durgākuṇḍ, 7. Kālarātri;
>
> *Places:* Accepting Jaitpura as seat of Aśvārūḍhā we get 5. Skandamātā prefigured. Considering that the Mātṛkā list of *Kāśīkhaṇḍa* 83 relates to Śiva-Vīreśvara we have traced the seat of 6. Kātyāyanī whose ancient Mahiṣāsuramardinī image made it easy to identify her as one of the Durgās. Saubhāgyagaurī's central position in close proximity to Śiva-Viśvanātha was taken over by 8. Mahāgaurī Annapūrṇā. The disputed seats of 8. Mahāgaurī and 9. Siddhidātrī (centre and Chowk area) possibly reflect Saubhāgyagaurī's double identity and Viśvabhujā's former importance.[60]
>
> *Concepts*: The importance given by *Kāśīkhaṇḍa* 70.63ff. to "the mother of the universe" Mahālakṣmī and her worship on *aṣṭamī* may explain the "Mahālakṣmī-pūjās" offered to 8. Mahāgaurī. The various accounts of nine goddesses were compressed into one group of nine. The protection of boundaries was retained by the "Durgā in the South" and extended to the north with Śailaputrī, whereas the (real or imagined) circular shape was given up.

Although today's Navarātri hardly looks like the one of the 12th century Gāhadavāla inscriptions, this reconstruction makes it likely that some ancient places have been retained and given new names. It does not explain, of course, the puzzle why (at least today) insignificant goddesses are included in the Navadurgā cycle, and important ones like Saṅkatādevī not.[61] Why their selection and sequence? Why Brahmacāriṇī's addition? I propose that the rationale is the mental map of a new goddess theology mirrored in a new ideal spatial model which replaced the circular scheme. What remained was the conversion of the temporal sequence of nine days into a processional route (*yātrā*) alluded to by *Kāśīkhaṇḍa* 72.82–91. There is still a demarcation of boundaries which retains the aspect of city protection: it is a spatial fact that contemporary worshippers of the Navadurgā cover the whole area of Vārāṇasī, from the river Varaṇā which forms the northern natural border (site of Śailaputrī) to the river Asi in the south (site of Kūṣmāṇḍā). In contrast to the older model, there is, however, a strong concentration on the heart of the city, particularly during the last four days. If we draw a line on the map of Varanasi following the course of the Navadurgā on the nine days, we find that the first five days present a figure whose furthest corners extend from Varana to Asi, enclosing the whole city space, whereas the last four days form a second figure which encompasses the centre of the city. With the bird's-eye view we may discern an even more interesting structure, if we distinguish the three times three days as single units. Marking them separately, we get three triangles: two elongated larger ones—the first extending from north to east and centre (1st–3rd Durgā), the second from south to north west (4th–6th Durgā)—and finally a third very compressed one in the centre of the city converging with the centre of the second triangle and heading its eastern angle (7th–9th Durgā).

Both circle and triangle are existing cultural codes depicting different theologies. While the circle fits well to the theology of city protectors, the triangle is apt to match the theology narrated by the names starting with Śailaputrī. The triangle is well known as main symbol of the Goddess, representing her *yoni* or womb, and her creative, protective and destructive powers. Her most important diagram, the *śrīcakra,* consists of nine intersecting triangles, five facing down, four facing up. The coincidence of the numbers four and five with the structure of the processional days may be random,[62] but I suppose the triadic/triangular pattern is more than a mere accident. It is a common device to associate the Great Goddess with all sorts of triads and to visualise her by triangular structures. A triadic theology is underlying (particularly the South Indian) Navarātri celebration during which Durgā is worshipped in her triple form as Mahākālī (first three days), Mahālakṣmī (second set of three days) and Mahāsarasvatī (last three days). The code "three times three" as well as the triangle as major pattern are thus symbolically meaningful in many ways. We do know of other incidents of sacred geography charged with the imagery of the number three and the triangular form. In Varanasi itself, the three goddess sites Durgākuṇḍ, Lakṣmīkuṇḍ and Vāgīśvarī are known as the "small" triangle (*trikoṇa*) of the triadic Great Goddess Mahākālī, Mahālakṣmī, Mahāsarasvatī. Vindhyācāl, the famous pilgrimage place in the vicinity of Varanasi, is supposed to be the "big" triangle, consisting of the three Devī shrines Aṣṭabhujā, Vindhyāvāsinī, and Kālīkhoh.[63] In other parts of India, too, we can find the same phenomenon of sacred topography: the three and a half most famous goddess seats of Maharashtra form a triangle which corresponds with the borders of Maharashtra (Jansen 1995: 80f., pl.xiii).

None of these examples is very ancient. In fact, additional *pīṭhas* were added to the ancient ones to get the triangular shape.[64] The *vindhyātrikoṇayātrā* was probably established in the early 18th century on the base of a sophisticated Great Goddess theology; the earliest historical reference is as late as 1830 (Humes 1996: 57, 60f., 64). That means: all places of such kind have been intentionally charged with meaning at a later stage of history by adding a half, one or two new goddess sites in order to get associated with the highly symbolic number three and triangular form. I guess, this holds also true for Varanasi. It may be a mere coincidence again, that we get the first of the three triangles only because of the Maharashtra dominated Brahmacāriṇī shrine and the second because of Kātyāyanī who has no temple on her own. The small Brahmacāriṇī temple (missing in the tentative reconstruction above) is the only one with the picture and name of the Durgā above the entrance—maybe a sign of its eagerness to assure to be one of the Navadurgā shrines. The Śrīvidyā practised therein has certainly a very pronounced triadic theology.

It would be worthwhile to investigate the history and patrons of the temples. The Brahmacāriṇīmandir was established by a Maharashtrian Brahman, if we trust the *pūjārīs* claims. Non-local sponsorship is certainly noteworthy regarding the most prominent temples: the Annapūrṇāmandir was established in the 16th century by the

Rāja of Poona, the Durgāmandir in the 18th century by a Rāṇī of Bengal. For the promotion of certain shrines patronage has been decisive no doubt. New ideas may have been promoted thereby, too. I recall that none of the Navadurgā temples of today is very old, even though the sacred site itself and even the *mūrti* might date back to the time of the *Kāśīkhaṇḍa* or even earlier (like Kātyayanī).

My hypothetical answer to the puzzling question why such a diverse set of goddesses, including some insignificant ones, have been selected to form the Kāśī Navadurgā Cycle, is a new space organisation. Other goddesses did not fit as much into the triangular scheme. Of course, such explanations are very speculative. There are no pilgrims' books like the ones of Vindhyācāl which would proof a deliberate sacred geography. There may be much more concrete reasons, unknown to me, which led to the present cycle, or no reasons at all. Possibly, the sites have gained their role just by accident. However, regarding the well-documented love of Banarsis for sacred geography, the existing instances of triadic theology and triangular pilgrimage routes, and finally the pregnant triangles constituting at least the first six days, the ideal map suggested has some plausibility.

Concluding Remarks

The Navadurgās starting with Śailaputrī is pan-Hindu and this rationale might have been exactly the reason why this group has become the Kāśī cycle of goddesses. It is not the only example to depict Varanasi as little India. The famous list lends a further "cosmopolitan" touch to it. Taking part in the living tradition, we get a polymorphic picture. Each Durgā has a unique life and identity of her own. The Navadurgā constitute a multitude of cultural and even ethnic spaces: folk and Brahman-oriented, local and pan-Hindu, Banarsi, South Indian, Gujarati, Maharashtrian. Not only Durgās' names and iconographies differ, but also the importance of the temples and the mode of rituals. Some of the temples belong to the most famous Devī shrines of the city, others are hardly known and not visited during other times of the year. Some belong to a prestigious Mahant, be it a lay person (Durgāmandir) or a renouncer (Annapūrṇā), others to non-Banarsi ethnic groups (Brahmacāriṇī) or to a private owner (Citraghaṇṭā). The officiants are Brahman priests (all but for Śailaputrī), including a lady (Skanda-mātā), but we do also find a tantric Brahman priest (Brahmacāriṇī) and tantric non-Brahman priests (Śailaputrī). Local and pan-Hindu goddesses are dynamically unified by the Navadurgā concept and its spatial coordination. It is the particulars, the local, ethnic, and personal colourings and the special rituals of each of the nine goddesses, which deserve most attention.

These particulars obviously have been an issue of constant negotiation. The diversity of rituals and importance of each Navadurgā shrine is something fluid. Citraghaṇṭā, once an important goddess, has faded out. Bhavānī has been supplemented by Annapūrṇā, Bāgeśvarī by Skandamātā and only Navarātri discloses today Bāgeśvarī's existence. Some of the temples have strong non-Banarsi influences, such as Brahmacāriṇī with her Maharashtrian and Gujarati worshippers, or Annapūrṇā with her South Indian bent. Some of the goddesses include features of a folk festival *(mela)*, such as Śailaputrī and to a lesser extent also Skandamātā.[65] Śailaputrī with her Nīm tree has still strong traits of a folk goddess, whereas those of Kūṣmāṇḍā-Durgā who supposedly once resided in a Pīpal tree are forgotten. Kūṣmāṇḍā still accepts blood-offerings, but Kātyayanī, Kālarātrī and the rest do not. Brahmacāriṇī receives the highly sophisticated Śrīvidyā (*śrīcakra*) worship and Annapūrṇā has a popularized, exoteric *śrīcakra* ritual. Quite a number of Devīs are associated with particular women's *vratas*. But only two Navadurgā temples belong to the most important Devī seats of the city: (Kūṣmāṇḍā-)Durgā and Annapūrṇā, presumably because they had royal sponsors whom Citraghaṇṭhā on the other hand has lost. While writing this article the ritual landscape may look different than in 1994–5.

The paper has dealt with both, the plurality and empirical space on the one hand, and the uniform overstructure and mental space on the other hand. The possible unifying traits are worthy to be considered, too, since they do lend an extra meaningfulness, a sense of an unfragmented whole. I have suggested that the unificatory principles may be seen in the names functioning as narrative theology and in the processional route forming a triangular shape or *yoni* projected on the city space. In trying to answer the riddle: Why have these highly diverse goddess shrines been selected and not other ones? I have proposed that the sites have been chosen because of their distribution in space, to form a sacred geography. A triangular pilgrimage route was invented to replace an older circular route.

Notes

1 I use the English transliteration for proper names of places and persons concerning modern India. My special thanks go to Jörg Gengnagel and Niels Gutschow for their valuable suggestions to improve the thesis of this paper.

2 In fact, it did skip my attention while staying in Varanasi for a scholarship at the Banaras Hindu University in 1991–92, but in 1994 and 1995 I returned for fieldwork studies on this form of celebration. I am greatly indebted to Prof. Rana P.B. Singh (BHU) and Pandit Kedarnath Vyas for their guidance in 1994. My thanks extend to Vagish Shastri (Acharya B.P. Tripathi), Mrs. Tripathi and Varsha Tripathi for their information and their help in interviewing the *pūjārīs* in 1995. My presentation is rather a "work in progress", because to check the findings again could not be done so far. I owe, however, some additional information on Brahmacāriṇī and Kātyāyanī to Xenia Zeiler who participated in the festival in 2002.

3 Citraghaṇṭā (KKh 70.38ff.) became a synonym for the third Durgā Candraghaṇṭā (*Durgāsaptaśatī kavaca; Āj)* also known as Caṇḍaghaṇṭā (Mishra 1973: 23) and Caṇḍakāṇḍā (*Tattvanidhi*-list, cf. Nagaswamy 1982: 133f.). In the present shrine of the third Durgā both names, Citraghaṇṭā and Candraghaṇṭā, appear referring to two separate *mūrtis* (see below).

4 Alternatively we find the names Siddhidāyanī, Siddhimātā, Siddhayogeśvarī or Siddheśvarī.

5 See Ravi and Rana Singh in this volume.

6 The meditation verses were added in 1995 from the third Durgā onwards. They are in corrupt Sanskrit and do not give the textual source except for the fifth Durgā whose *dhyāna* is said to be taken from the *Vāmanapurāṇa*. According to Vagish Shastri all the *dhyānas* stem from the *Devīpurāṇa*.

7 My references are based on *Kāśīkhaṇḍa*, ed. K. Tripāṭhī, and *Skandamahāpurāṇa*, Vol. 4, Delhi 1986 (reprint of the Bombay edition 1967). All texts referred to coincide with Tagare's English translation.

8 See Eck (1982: 360; 174; 388n.26); Kedarnath Vyas (1987: 152f.); Ravi and Rana Singh in this volume. It should be noted that the scholars present the Navadurgā list as a given fact and refer primarily to KKh 72.81; 72.82–87 (Durgā worship in Kāśī) wherein none of the list's names appear. Likewise, additional references given by Eck (KKh 72.89ff.: nine directional Śaktis; KKh 45.48ff.: Yoginī Navarātri worship), and Singh (KKh 71: Durgā myth) do not contain the list.

9 This dating has been suggested by Dr. N.P. Joshi (Zeiler 1998: 12,fn.13), and ascertained by Dr. Joachim Bautze (personal e-mail communication by Xenia Zeiler, 25.6.02).

10 Caukhambā edition 1984: 26; Gītāpress edition with Hindi commentary, *saṃvat* 2051: p.19f.; Sriramakrishnamath edition (*Devīmāhātyam*) with Tamil translation 2001: 7f. The 1983 edition by Kāśīnātha Śāstrī contains nice drawings of the Navadurgā (p.25–33).

11 As second group known by that time Mishra mentions a very fierce set of nine goddesses: Ugracaṇḍā, Pracaṇḍā, Caṇḍogrā, Caṇḍanāyikā, Caṇḍā, Caṇḍavatī, Caṇḍarūpā, Aticaṇḍikā, and Rudracaṇḍā.

12 For a number of diverse lists see Nagaswamy 1982: 133f.; Liebert 1986: 194; Mishra 1973: 23; Levy 1992: 154ff., and *Skandapurāṇa* 1.2.47,24–78.

13 Vārāhī near Kratuvarāha, Śivadūtī, Aindrī (south of Indreśa), Kaumārī (in the vicinity of Skandeśvara), Maheśvarī (south of Maheśvara), Nārasiṃhī (near Nirvāṇa-Narasiṃha), Brahmī (west of Brahmeśa), Nārāyaṇī (west of Gopī-Govinda), and finally Gaurī in the form of Virūpākṣī (north of Devayānī). The "Seven Mothers" (Saptamātṛkā) Vārāhī, Aindrī, Kaumārī, Maheśvarī, Nārasiṃhī, Brahmāṇī, and Vaiṣṇavī (synonymous to Nārāyaṇī) appear in the third episode of the *Devīmāhātmya* (8.12–62), and Śivadūtī is Durgā's messenger. Virūpākṣī is not part of the myth, but semantically appropriate: the name connotes a wild and dark form of Gaurī, correlative with the mythical Pārvatī (also called Gaurī) who assumed a wild and dark form (Kālikā) to fight.

14 Brahmāṇī, Vaiṣṇavī, Raudrī, Vārāhī, Narasiṃhikā, Kaumārī, Māhendrī, Cāmuṇḍā, Caṇḍikā. Raudrī (replacing Maheśvarī) as well as Cāmuṇḍā and Caṇḍikā connote Durgā in her most frightening, war faring aspect.

15 In contrast to the nine Durgās we find a complete list of the nine Gaurīs (KKh 100.68–72), worshipped during *caitra*-Navarātri until today. But this fact does not necessarily mean that only the Gaurīs were known, in chapter 70 also only two of the Gaurīs are mentioned (70.4–22, 62ff.).

16 According to the Śaiva Āgamas and Tantras sound creation prefigures the material universe.

17 These explanations of the third and fourth Durgā were suggested by Vagish Shastri who also offered the alternative interpretation of the list mentioned below.

18 On the ferocious virgin goddess and the "tamed" golden spouse goddess see E. Masilamani-Meyer and C. Vogelsanger 1993: 117. According to the third episode of the *Devīmahātmya* (5.40ff.) only Pārvatī's split image, the black Kālikā/Kauśikī, destroys the demons. The eighth name of Durgā also recalls the myth of Pārvatī toiling hard to get her dark skin changed to become Gaurī, "the Golden" or "Fair" one (*Śivapurāṇa*, *Vāyavīyasaṃhitā* 24.23–25.48; 27.1ff.).

19 It is said that Śiva himself worshipped her and received eighteen *siddhis* whereby half of his body became female. Only Śiva-Ardhanarīśvara is supposed to incorporate all eighteen powers, the nine of the goddess and the nine of the god. I am not aware of a textual basis, but got the information from Pandit Vagish Shastri. See also Ravi and Rana Singh in this volume.

20 It is noteworthy that only Kālarātri and Kātyāyanī are directly associated with the Durgā myth, while the other names are rather epithets of Śiva's spouse. Also the *Kāśīkhaṇḍa* associates most Devīs with Śiva forms (diverse *liṅgas*) and the Durgās with the Bhairavas (72.90ff.).

21 The best known example is found in the *Jābālopaniṣad,* wherein the external location Varanasi transmutes into an inner experience.

22 I have visited the temples during 1994 und 1995 Navarātri twice a day, morning and evening, and observed that in the early morning around sunrise they were frequented the most. The first Durgā in particular attracts a large crowd (see also Rodrigues 1993: 200, fn. 21). In no the other Durgā temple, I noticed an equally heavy body of police to keep the crowd in line, nor political parties like Śiva Sena or Samāj Saṃgaṭhan offering food to the poor and health care for the sick.

23 The sanctum equals the second of the five main types of Banarsi temples given by Verma 1971–84: 194–205.

24 Two of the priests and Dr. Nagarika Suraksha who was ordered by the District Magistrate to care for security and health on the Śailaputrī grounds.

25 Raja Ram Sagarkar (interviewed in September 1995 by myself and in July 2003 by X. Zeiler).

26 There is a metal *śrīcakra* in front of the *mūrti* and a painted one on the wall. Maharashtra has been an important Śrīvidyā place and also Banaras has been a centre of Śrīvidyā at least since the Maharashtrian Brahman and most eminent Śrīvidyā philosopher Bhāskararāya settled in the city (17th cent.). Although Karpatri Swami popularised Śrīvidyā in the 20th century Varanasi, the school has never left its esoteric circles of initiates. The *śrīcakra* is also present in the Durgākuṇḍ- and Annapūrṇāmandir, but no *śrīcakrapūjā* proper *(āvaraṇapūjā)* is done. In the former, the first (exoteric) *pūjā* in the morning is offered to the *yantra* which is placed in front of the goddess of the sanctum. For the latter see below.

27 *Garbha* dance was done in 1995 at least. The association of Durgā worship and Kṛṣṇa tradition, including the dance theme, is not uncommon in Śākta circles (see *Devībhāgavatam* 9.45.47.50). Better known is the myth of the demon-slaying goddess Vindhyāvāsinī who saves the child Kṛṣṇa from his murderous uncle.

28 This ritual, as also the next, was noticed by Xenia Zeiler in 2002.

29 The participants had been invited by the Vidyarthi Sahadstasava Samiti a month before and ritually prepared themselves. In earlier years the ritual was performed in a more distant place (e-mail communication by Xenia Zeiler 19.10.02).

30 The temple at Lakhīcautarā Lane belongs to Pandit Saha who also gets the *prasāda*. It is an area where many Gujaratis as well as Banarsi Brahmans live.

31 See H. Bakker in this volume.

32 The present Durgāmandir has a tall spire (*śikhara*) established by Rāṇī Bhavānī in 1760, later patrons added porch and quadrangle. The temple compound covers a large area with subsidiary shrines surrounded by rest houses for pilgrims.

33 According to the priest consulted, animal sacrifice on private request may take place at any time of the year. It is performed in the courtyard of the temple where a sacrificial pole is erected or inside the temple premises in front of Kālī's image, but never in front of Durgā.

34 Eck 1982: 167; Humes 1993: 186. Several of my informants confirmed that no image exists, only the empty space or the nail on which the silver mask is fixed. A *pūjārī* claimed that the image is somewhere else and offered changing versions of its iconography: whereas the first corresponds with the Kūṣmāṇḍā type of the *Tattvanidhi* (two hands: the right one bearing a vessel of ambrosia or liquor (*amṛtakalaśa*), the left one a vessel of blood), the second pretends it to be a Mahiṣāsuramardinī type with ten hands.

35 Durgādevī has many masks, but two special ones: one of them she is wearing during Annakūṭa and the other one at special (sponsored) *pūjās*.

36 Its features were hard to determine due to size, colour (black stone with *sindhur*) and age. I was allowed to take a photograph only after its "beautification" with flowergarlands and a large mask.

According to the head *pūjāriṇī* the image has two hands with fists pressed against the belly. Another lady attendant claimed that the goddess is sitting with her son on a lion (= Skandamātā's iconography), whereupon the *pūjāriṇī* changed her description saying, that Bāgeśvarī has no son, but is doing *pūjā* to a Śiva-*liṅga* in front of her. Neither description matches the representations (9th–17th cent.) of Vāgīśvarī (sometimes identified with Bagalāmukhī) published by Bautze 2000: a female figure sitting on a lion, and tearing the tongue from (one or) two male figures prostrated at her pedestal.

37 She was the only goddess where I was not allowed to take a photograph. Even the *Āj* (1995) portrays her only in a drawing while giving photos of the other eight Durgās.

38 It is an area were mostly merchants and craftsmen of the lower middle classes live. For a detailed study of this temple see Zeiler 1998.

39 I owe this information to Xenia Zeiler who could not ascertain a mask in Navarātri 2002, but *Durgāsaptaśatī* recitation and *havana* on *navamī*.

40 The *Śivapurāṇa* (*Śatarudrasaṃhitā* 10.27ff.) already mentions this feature and until today the temple attracts pilgrims from all over India for this reason (Zeiler 1998: 5f.;15, fn.25; 46).

41 The Sādhu who lives in the Maṭha in the upper floor of the temple is given *dakṣiṇā* and in addition ten rupees for the temple authorities.

42 One version of its *kaṭhā* relates it to Kanchi, another to Assam. For the latter version see Eck 1982: 162. The *Annapūrṇā Vrata Kathā* (Sanskrit with Hindi translation, or Hindi only) is sold in small shops in the Viśvanāth galley.

43 The souvenir volume shows the new statue without mask and cloth garments (Caturvedī 1977: 1). The black stone had been sent by the king of Nepal.

44 There is an agreement that for ten days per month the worship in the ground floor goes to a local Paṇḍā. However, these Paṇḍās have no right for *bhog* and *ārtī*. During the 20 days of the Mahantji's period a signboard announces his absolute right.

45 Personal communication by Kedarnath Vyas. Also in the *Kāśīkhaṇḍa* Annapūrṇā is known as Bhavānī (Eck 1982: 164, 160).

46 On the contrary, Vindhyāvāsinī's small shrine adjunct to the large Śaṅkaṭhā temple seems to be more attractive during Navarātri.

47 The Annapūrṇā temple was established (or reconstructed?) in the 16th century by the Rājā of Poona (cf. Gutschow & Michaels 1993: 81) and reconstructed (again?) in 1830 by the Mahārāja Baji Raw Peshava. Diana Eck (1982: 164) gives with the early 18th century an even later date of the temple's installation. On the other hand, Kāśī's Annapūrṇā is already known in the *Devībhāgavatam* (11.18.60–71) which seems to prove that the site is more ancient.

48 In 1995 *aṣṭamī* and *navamī tithi* coincided and this was not in favour of the ninth Durgā either.

49 Its normal opening hours are 4.30 to 12 a.m.; 3.30 to 9 p.m.

50 According to Diana Eck (1983: 51;157ff.) the Yoginīs belong to oldest deities of Kāśī (see also Gutschow & Michaels 1993: 95). By the 10th century the Yoginī cult was widely spread in North India (Mishra 1973: 31).

51 Caūsaṭṭīdevī's identification with the Navadurgā is visually symbolized: on each of the nine days Caūsaṭṭī wears another mask.

52 It is not clear from the text whether these two goddesses are associated with the autumnal or rather spring Navarātri, the latter being inserted in the verses on Citraghaṇṭā (70.41). Also other passages allude to a closer connection of the two festivals than today's performance.

53 This important Kāśī goddess is nowadays still worshipped as fifth Gaurī during spring Navarātri. My English rendering as well as the one below follows largely Tagare's translation (1997) of the *Skandapurāṇa*.

54 *rakṣanti kṣetram etad vai tathā 'nyā navaśaktayaḥ / upasargasahasrebhyas tā vai digdevatāḥ kramāt // śatanetrā sahasrāsyā tathā 'yutabhujā parā / aśvārūḍhā gajāsyā ca tvaritā śavavāhinī // viśvā saubhāgyagaurī ca sṛṣṭāḥ prācyādimadhyataḥ / etā yatnena sampūjyāḥ kṣetrarakṣaṇa-devatāḥ //.* The list of Śaktis is followed by the eight Bhairavas to whom worship should be offered in the same way.

55 These *yātrās* have been dealt with by Singh 1993 and Gutschow & Michaels 1993: 102–144.

56 The map has been published by Gengnagel & Michaels 2001. Noteworthy is Jörg Gengnagel's introduction "a. The outer circle" discussing Sukula's interpretation of KKh 72.2–13 to relate to 96 (8 x 12) Śaktis encircling the city space (see above). Rana P. B. Singh 1993: 39–43 and also Ravi and Rana Singh in this volume postulate an "outer circle" (*caurāsīkrośīyātrā*) of 96 goddesses and an "inner circle" of 9 goddesses and visualize it by a diagram (see Fig. 1).

57 See index numbers 28 and 36 (Gengnagel & Michaels 2001) as well as KKh 72.91 and 100.69.

58 We find Śaileśvaradevī but not Śailaputrī, Durgādevī at Durgākuṇḍ but not Kūṣmāṇḍā, Kālikādevī but not Kālarātri, Annapūrṇā and Bhavanī but not Mahāgaurī.

59 Groups of eight goddesses spread in the cardinal directions and a ninth in the centre are often visualized in tantric diagrams. For an example including the Mātṛkā see *Devībhāgavatam* 9.50; an example including the Durgās of the Śailaputrī list is found in the *Kālikāpurāṇa* 54.43–45; regarding the Yoginīs 54.39. I thank Mark Dyczkowski for pointing out this connecting link which is missing in the passage from the *Kāśīkhaṇḍa* to the present Navadurgā.

60 For a goddess called Siddheśvarī in proximity to Saubhāgyagaurī and Viśvagaurī see number 36 (Gengnagel & Michaels 2001).

61 This goddess is known by the *Kāśīkhaṇḍa* as Vikaṭā (83.26–34; 97.40) and appears on Sukula's map with both names (Gengnagel & Michaels 2001: index number 37). The triangular pattern suggested below, may only explain her exclusion as eighth (but not as ninth) Durgā. The exclusion of other prominent Devīs is easier to explain: Viśālākṣī belongs to the Gaurī cycle (spring Navarātri) from the times of the *Kāśīkhaṇḍa* (KKh 100.67–73) till today. Kāmākhyā/Kāmākṣā in the South West, whom some identify with the ancient *yoginī* Kāmākṣī, is presumably too remote from the Navadurgā pilgrimage route.

62 It should be noted that also other groups of Kāśī goddesses (see Ravi and Rana Singh in this volume) possibly reflect Śrīvidyā influence: the nine Gaurīs incorporate Lalitā, the enlarged group of sixteen ends with Trilokasundarī (both well known names of the Śrīvidyā goddess; sixteen (15+1)= number of Nityās and of syllables of the Śrīvidyā *mantra*); the groups of Mātṛkā (worshipped in various Śrīvidyā rituals, too); the Mahāvidyā goddesses Ṣoḍaśī, Tripurasundarī, Rājārājeśvarī and Lalitā (these names are synonymously used in Śrīvidyā).

63 Pilgrimage guides like Miśra, *Vindhyā-Trikoṇa-Māhātmya,* promote this idea. See also Humes 1996: 49–76, and Singh 1997. Singh postulates that the three Devī shrines are associated with three Śiva shrines to form a hexagram.

64 Regarding Vindhyācal, Vindhyāvāsinī is the original *pīṭha,* whereas Aṣṭabhūjā and Kālīkhoh were built much later. The pilgrims' handbooks, however, present the three places as a triangular unity which forms a pilgrimage route. Regarding the Bhavānī of Tuljapūr, half a *śaktipīṭha* was added to the original three old *pīṭhas* to get a triangular shape.

65 The street leading to her shrine was also full of booths with devotional paraphernalia, toys and jewelry.

References

Texts

[Śrī] Annapūrṇā Vrata Kathā, [Hindi] ṭīkā Śrīśivadatta Miśra Śāstrī, Annapūrṇā-stotrasahitā. (without year). Vārāṇāsī: Prakāśak Ṭhākurprasād.

Kālikā Purāṇa, Pt II. Text, Introduction and Translation in English by Biswanarayan Shastri. 1991. Delhi: Nag Publishers.

Kāśīkhaṇḍaḥ Maharṣivyāsapraṇitaḥ Rāmānandapraṇitayā "Rāmānandī" vyākhyayā; Nārāyaṇapatitripāṭhīpraṇitayā "Nārāyaṇī". 1991; 1992; 1996; 1998. 4 vols. Ed. Karuṇāpati Tripāthī. Varanasi: Sampurnand Sanskrit University.

[The Śrīmad] Devī Bhāgawatam, transl. Swami Vijnanananda. [3]1986. New Delhi: Oriental Books Reprint (first publ. by Panini Office, Allahabad).

Devīmāhātmyam. 2001. Madras: Sriramakrishnamath.

Durgāsapataśatī. 1984. Ed. Divakar Mishra. Vārāṇasī: Caukhambā.

Durgāsapataśatī. 1983. Ed. Kāśīnāth Śāstrī. Vārāṇasī: Pharma Bābū Baijanāth Prasād Bukselar.

Durgāsapataśatī. saṃvat 2051. Hindi anuvāda by Śrīrāmanārāyaṇadattajī Śāstrī. Gorakhpur: Gītāpres.

[Śrī] Brahmacāriṇī Mantrapuṣpāñjalim (stuti-pustaka). (without year). Varanasi: Brahmacāriṇī Mandir.

Skandapurāṇa. Pt. 10–11 [Kāśīkhaṇḍa]. 1996; 1997. Transl. and Annotated by G.V. Tagare. Delhi: Motilal Banarsidass.

[The] Skandamahāpurāṇam. 1967. Bombay: Kṣemarāja (reprinted 1986. Delhi: Nag Publishers).

Secondary Literature

Awasthi, A. B. L. 1978. *Studies in the Skanda Purāṇa.* Part II. Lucknow: Kailash Prakashan.

Bautze, Joachim K. 2000. Vāgīśvarī. In: Maurizio Taddei & Giuseppe de Marco (eds.), *South Asian Archaeology 1997.* Rome: Istituto Italiono per l'Africa e l'Oriente, pp. 1229–1248.

Caturvedī, Sītārām et al. (eds.). 1977. *Śrī Kāśī Annapūrṇā navamūrti pratiṣṭāpana.* Vārāṇasī: Śrī Kāśī Annpurṇā Devī Maṭh.

Coburn, Thomas B. 1988 (1984). *Devī-Māhātmya. The Crystallization of the Goddess Tradition.* Delhi: Motilal Banarsidass (first publ. Cambridge 1982).

Eck, Diana. 1983 (1982). *Banaras: City of Light.* Princeton: Princeton University Press.

Gengnagel, Jörg & Axel Michaels. 2001. *Kailāsanātha Sukula's Mirror of Kāśī (Kāśīdarpaṇa, 1876). A historical map of Vārāṇasī based on the Kāśīkhaṇḍa and related texts.* Electronic Publication of the Varanasi Research Project I. Heidelberg: South Asia Institute. (*www.benares.uni-hd.de*).

Gutschow, Niels & Axel Michaels. 1993. *Benares: Tempel und religiöses Leben in der heiligen Stadt der Hindus.* Köln: DuMont.

Humes, Cynthia Ann. 1993. The Goddess of the Vindhyas in Banaras. In: B. R. Hertels & C. A. Humes (eds.), *Living Banaras*. Albany: SUNY, pp. 181–204.

— 1996. Vindhyavāsinī: Local Goddess yet Great Goddess. In: John S. Hawley & Donna M. Wulff (eds.), *Devī. Goddesses of India*. Berkeley: University of California Press, pp. 49–76.

Jansen, Roland. 1995. *Die Bhavani von Tuljapur*. Stuttgart: Steiner.

Levy, Robert I. 1992 (1990). *Mesocosm. Hinduism and the Organisation of a Traditional Newar City in Nepal*. Delhi: Motilal.

Liebert, Gösta. 1986 (1976). *Iconographic Dictionary of the Indian Religions*. Delhi: Satguru Publications.

Masilamani-Meyer, Eveline & Cornelia Vogelsanger. 1993. Pārvatī, Kālī, Māri—verschiedene Formen der Devī. In: Cornelia Mallebrein (ed.), *Die anderen Götter. Volks- und Stammesbronzen aus Indien*. Köln: Ed. Braus, pp. 112–125.

Mishra, Vibhuti Bhushan. 1973. *Religious Beliefs and Practices of North India during the Early Medieval Period*. Leiden: Brill.

Miśra, Rājanāth. (without year). *Vindhya-Trikoṇa Māhātmya*. Mīrzāpur: Mudraka-Śāṅkaro Pres.

Nagaswamy, R. 1982. *Tantric Cult of South India*. Delhi: Agam Kala Prakashan.

Rodrigues, Hillary P. 1993. Some Puranic Myths of the Durgākuṇḍa Mandir in Vārāṇasī. *Purāṇa* 35, 2: 185–201.

Singh, Rana P.B. 1993. Varanasi: The Pilgrimage Mandala, Geomantic Map and Cosmic Numbers. In: Rana P.B. Singh (ed.), *Banaras (Varanasi): Cosmic Order, Sacred City, Hindu Traditions*. Varanasi: Tara Book Agency, pp. 37–64.

— 1997. Sacredscape, Cosmic Territory, and Faithscape: Goddess Territory of Vindhyachal, India. *National Geographic Journal of India* 43, 3: 237–63.

Verma, T. P. 1971–84. The Temples of Banaras, *Bhāratī* 15: 194–205.

Vyās, Kedarnāth. 1987. *Pañcakrośātmaka Jyotirliṅga Kāśīmāhātmya*. Vārāṇasī: Kedarnāth Vyās, Jñānavāpī.

Wilke, Annette. 1993. Durgā-Pūjā in Benares, *Unipress* (Bern) 78: 19–23.

Zeiler, Xenia. 1998. *Der Vīreśvara-Tempel in Vārāṇasī*. Magisterarbeit Humboldt-Universität Berlin.

Sunthar Visuvalingam and Elizabeth Chalier-Visuvalingam

Bhairava in Banaras: Negotiating Sacred Space and Religious Identity

For Gerda, Kurt and A.S.K.
Dedicated to John Irwin

Banaras, like Jerusalem, is the manifestation in urban space of the religious identity of those who claim it as their sacred centre. The holy city indwells believers (and non-believers), even (and often especially) those who do not dwell within its territorial confines. Though perceived as transcending space and time, preceding and modeling creation itself, the terrestrial city reflects the vicissitudes of political power and the historical evolution of the tradition itself. Whether the contending tradition splits off from within (Buddhism/Christianity) or imposes itself from without (Islam), the uneasy accommodations in the sharing of sacred space—punctuated in the latter case by regular outbursts of increasingly politicized violence—offer deep insights into how religious identities are perpetually negotiated and re-constructed.

The Gangetic plain was once occupied by Tibeto-Burman and Austro-Asiatic (Muṇḍa) populations, whose archaic cosmogony may still be excavated beneath the archeology of later historical cults, such as the Fish-Womb conjunction (*matsyodarī-yoga*) and the marriage of Lāṭ Bhairõ. Before Banaras became the Hindu sacred city par excellence, elements of this cosmogony had been assimilated into Buddhism as the funerary *stūpa* and the cosmic pillar, just as they have been in the Kathmandu Valley. The Muslim conquest, with its iconoclastic imposition of the mosque, *īdgāh* and *mihrab*, "merely" attempted to "reorient" this religious axis (westwards) towards Mecca, whose role is itself largely modeled on Jerusalem. These overlapping spaces were also witness to syncretizing practices that reveal shifting religious identities. Far from obliterating this (inner) landscape, secularization has resulted in the city's material future becoming a hostage to its unresolved religious past. Even the ongoing cycle of indiscriminate violence inscribes utopian political agendas into a common underlying sacrificial paradigm that would constitute the core of the sacred city. An attentive reading of how religious identities have been negotiated in Banaras may help us (re-) write the contested text of history so as to visualize and facilitate the prospect of a shared future.

As the secret transgressive identity of the royal Śiva in the form of Kāśī Viśvanāth, Bhairava is of central significance to the holy city of Banaras (Chalier-Visuvalingam 2003). The divine magistrate (*kotvāl*), who inflicts his liberating metaphysical punishment (*bhairavī yātanā*) at the sacrificial pillar, is also the brahmanicide who has violated the most sacred laws of the Hindu tradition. The marriage of this Lāṭ Bhairō to the adjacent "maternal" well used to be celebrated with "lower-caste" Muslim participation. Likewise, the Hindus used to participate in the marriage of the "decapitated" martyr Ghāzī Miyā̃, whose posthumous worship has assimilated the "Hindu" symbolic universe of that very cult of human sacrifice that this proselytizing Muslim warrior had sought to extirpate. Despite a century of more or less peaceful coexistence, the contested ground around Lāṭ Bhairō erupted, shortly after the imposition of British rule on Banaras, into the Hindu-Muslim riots of 1809, which has provided the conceptual model for the understanding of subsequent communalism in the Indian subcontinent. The ritual space that united the central Viśvanāth temple with the peripheral pillar became the processional route for mutual mob violence that resulted in the "apocalyptic" felling of the world-pillar. Despite this phase of religious polarization, Banarasi society was able, within the space of a couple of years, to have the British-imposed house tax repealed through successful passive resistance predicated on caste-based autonomy and Hindu-Muslim unity. In today's context of growing Wahhabi "fundamentalism" among Muslims and the aggressive claims of "Hindu nationalism" over disputed sacred spaces, this paper invites all concerned South-Asians and well-meaning foreign scholars to join us in a "pilgrimage" across time and space to an excavation of Lāṭ Bhairō. Permanent reconciliation between Hinduism and Islam will be achieved only when, by reducing the inner distance between Mecca and Banaras, the questions posed by (the mutilated stump of) the world-pillar—which still straddles the boundary between the two religions—are finally resolved.

Introduction: Hindu-Muslim Representations of Communal Space

The ancient site and cult of Lāṭ-Bhairava (popularly known in Hindi as "Lāṭ Bhairō") in Banaras offers crucial insights not only into the roots of the communalism that now threatens to tear apart the social fabric of Indian polity but also into the formative processes of religio-cultural identity. The so-called "War of the Lāṭ" in 1809 between Hindus and Muslims, which resulted in the felling of this sacred stone pillar, and the subsequent Hindu-Muslim peaceful joint agitation against the excesses of colonial rule, have together furnished the interpretative lens for two significant attempts to understand the role of religion, politics and colonial history in the communalization of Indian society (Pandey 1990; Freitag 1989c). Previous colonial writings had understood the conflict as a primarily religious one and then proceeded to construct irreconcilable "essences" out of Hinduism and Islam forever at war with each other. That lower-castes of Hindus and Muslims jointly celebrated the marriage of the Lāṭ with the adjoining well, as also the Bharat Melāp episode of the Rāmāyaṇa, has led Freitag, on the contrary, to deny the religious character of the 1809 riots and instead seek their cause primarily in the intrusion of colonial administration, with its alien mentality, onto the pre-existing communitarian space. Given the subsequent Hindu-Muslim unity against the "house-tax" imposed by the British, she even goes so far as to affirm that "Hindu" and "Muslim" did not then exist as distinct identities. Gyan Pandey likewise tends to minimize the role of religion and instead focus on the agency of the subaltern (i.e.

"lowest", including in particular the untouchable) castes in organizing anti-British resistance on behalf of all Banarasis. Sorely lacking is a more complex, non-reductive, analysis of the articulation between religious identity and political action that is capable of clarifying the sometimes bewildering shifts in social allegiance.

Not only was the initial violence sparked off by a dispute over and within shared sacred space, the ensuing riots erupted along a symbolic trajectory that united this spatially peripheral site of the Lāṭ (pillar) with the religious centre of Banaras, and were moreover marked by evident ritual and even processional notations. It was the felling of this pillar that drew the rest of the population, the Hindu and Muslim higher castes, into the inexorable violence of what had become a plainly religious conflict. Even the mutual participation of (especially the lower-caste) Hindus and Muslims in each other's festivals is more readily explicable in terms of syncretism and shared religious values, than in terms of some purely secular civic consciousness. Such politico-economic attempts to understand the etiology of this communal outburst and the psychology of the subsequent reconciliation betray a hopelessly inadequate appreciation of the role and power of symbolic structures in securing individual commitment and orientating group action. Traditional perceptions of political and even economic activity are rather anchored in, to some extent even determined by, religious values both explicit and implicit. Ethno-geography[1] is better equipped to explore, decipher and restore the qualitative aspects of sacred space in expressing religious beliefs and in determining human representations of self and the other.

Having already demonstrated elsewhere in greater detail the common sacrificial paradigm underlying both Hinduism and Islam through the syncretism evident in the overlapping cults of Lāṭ Bhairõ and Ghāzī Miyā̃ (Visuvalingam 1992), and shown how a similar religion-based acculturation process was already in its terminal phases between Newar Buddhism and Hinduism (Visuvalingam 2003), this paper presents and analyzes the sacred marriage in its original "pagan" context before exploring how this central mythico-ritual theme was appropriated within an otherwise iconoclastic framework. These shared symbolic themes were woven around intricate scenarios of death, sexuality and rebirth—developed in divergent but not necessarily exclusive directions by each religion—that would seem to be rooted in and addressed to the innermost mechanisms of the human psyche. We have focused our efforts here primarily on the "economy of violence" within this paradigm, the logic of its containment and redirection by the Hindu and Muslim religious frameworks as expressed especially through the ritual organization of space, its recorded visible effects on the social fabric of the sacred city of Banaras in the wake of colonial rule, and culminating in some informed speculation on its possible future course. In the larger context of the recent revival of Hindu-Muslim claims and clashes over other holy sites in India, Lāṭ Bhairõ provides some invaluable insights into the manner in which human violence is channeled into and regulated by symbolic archetypes.

The Pillar and the Well: The Marriage of Lāṭ Bhairõ

The origin myth of Bhairava (the terrifying aspect of Śiva) found in the Purāṇas, attests to the intimate and indissoluble link between Banaras and Bhairava (Chalier-Visuvalingam 1989, 1994a). After having emerged from the pillar of fiery light (*jyotir-liṅga*) to violently cut off the head of Brahmā, the "skull-bearing" (Kāpālika) Bhairava wandered about for twelve years to expiate his brahmanicide. Finally he reached Banaras where the skull of Brahmā, and with it the sin of brahmanicide, fell into a tank appropriately named the "liberation of the skull" (*kapāla-mocana*). Yet even after his absolution, the "Black" (Kāla) Bhairava remained at Kapālamocana as the "sin-eater" (*pāpa-bhakṣaṇa*) to devour the impurities of pilgrims to the city of final liberation. Paradoxically, Bhairava, the (ex-)criminal, also reigns as the policing magistrate (*kotvāl*) in Banaras, entrusted with the duty of preserving its sanctity not only by barring its access to sinners but also by punishing those who indulge in sins within the confines of the holy city. The "punishment of Bhairava" (*bhairavī-yātanā*) burns up the accumulated sins of seekers of liberation and is inflicted on everyone at the moment of death in this "great cremation-ground" (*mahāśmaśāna*). This punishment was administered at a pillar (*lāṭ*) whose stump, now called "Lāṭ Bhairõ", still stands beside the present-day Kapālamocana tank where it is worshipped as the representation (*liṅga*) of Bhairava (Chalier-Visuvalingam 1986, 1994a).[2]

Topography and Identity of Lāṭ Bhairõ: Executioner, Victim and Pillar of the World

The Lāṭ Bhairõ temple is one the eight temples of Bhairava (see Fig. 1, for their spatial disposition), collectively known as *aṣṭa-bhairava*, i.e., "the eight-fold Bhairava" (Chalier-Visuvalingam 1994a). The Lāṭ Bhairõ is a pillar that was almost completely leveled during the Banaras riots of 1809. Today this Hindu icon is a mere stump, 3 feet thick and 7 to 8 feet high, that stands on a slightly elevated stone platform (Fig. 2) in the midst of a Muslim "prayer ground" (*īdgāh*), where the devout of both faiths continue to pray and offer their respective sacrifices (Fig. 3). Entirely encased under protective copper sheeting installed after the riots by the British District Magistrate, the pillar is separated from the *īdgāh* only by a small enclosing brick wall. Just outside the wall and to the north is the adjoining "well of Bharata" (*bharatakūpa*), the youngest brother of Lord Rāma. To the south of this terrace, 5 or 6 meters below, is a large tank named Kapālamocana, a strong well-built structure with stairs and foundation of solid stone (see Fig. 4). Hindus bathe here for the tank is reputed to cure women of sterility and bathing daily for 40 days can even remove leprosy. There are also some sacred trees: the branch of a *tulsi* tree that was uprooted during the riots, and particularly a *pippal* and a *nīm* tree, whose "marriage" all over India is a Hindu prolongation of the Vedic sacrificial symbolism of (the union of) the *aśvattha* and the *śamī* trees.

The shrine is located in the north-eastern part of the present city at the junction of the Grand Trunk Road with the road leading to Sarnath, and is almost a mile west of the confluence of two sacred rivers, the Varaṇā and, of course, the Gaṅgā. The open praying area of the *īdgāh* is bounded on the west by a wall with the niche (*mihrāb*) indicating the orientation (*qiblah*) of Mecca. To the stump of the original Lāṭ, once famous among the Hindu population both for its antiquity and sanctity, is normally affixed a small mask of Bhairava. The divinized magistrate of Banaras appears to supervise the work of the cudgel-bearing policemen who are permanently posted in the vicinity to prevent the outbreak of fresh communal tension.

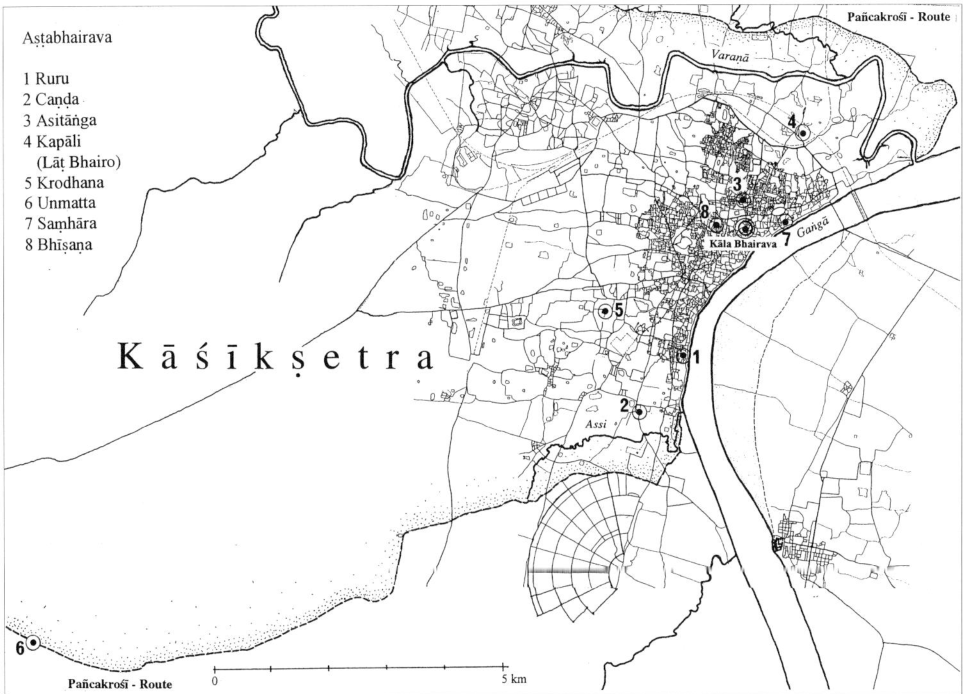

Fig. 1: Current spatial disposition of the eight Bhairavas in Banaras (drawing by Niels Gutschow).

The evidence points to an "Ashokan" pillar, no doubt the one that Hiuen Tsang in 636 A.D. saw standing before a Buddhist *stūpa*.[3] The Hindu Māhātmyas (traditional texts that extol the virtues of a pilgrimage site) refer to a "pillar of the great cremation-ground" (*mahā-śmaśāna-stambha*) standing at the present location of the Lāṭ (Sukul 1977: 120–21, 260–01; Eck 1983: 196–97). The *Kāśīkhaṇḍa* (97.64–6), which reflects the post-Islamic adaptations of the mid-14th century, speaks of the "great Rudra" (Mahārudra, the fierce Rig-Vedic prototype of Śiva) residing with his consort Umā (not in an adjoining temple but) in the pillar itself, near the "Lord of the Skull" (Kapāleśa), and refers to the adjacent Kapālamocana. The Kāpālikas, who were adepts of the Soma doctrine (now understood as overt sexual rites modeled on Śiva's union with Umā), generally haunted the cremation-grounds and were adepts of human sacrifice. Stories still circulating among the Muslims of the surrounding area tell of the Muslim warrior,

Ghāzī Miyā̃, having eradicated the regular human sacrifice at a temple of Somnāth that is said to have existed near the confluence of the Varaṇā and the Gaṅgā. This may refer to the Mahārudra temple ("confused" or, rather, identified symbolically with the fabulous temple of the same name in Gujarat that was actually destroyed by his uncle, Mahmud of Ghazni), probably a Kāpālika cult centre, which Sukul claims once existed around the pillar. It must have been devastated along with the rest of the city by Quṭb-ud-din Aibak, chief general of Muhammad Ghori, at the end of the 12th century.

Fig. 2: Lāṭ Bhairõ pillar with mask of Bhairava attached.

The French lapidary, Tavernier, saw the pillar in 1665, during the reign of Aurangzeb, within the walled garden—with many sculptures and beautiful architecture—of a mosque. Its shaft, which was 32 to 35 feet high and all of one piece, terminated in a pyramid with a large sphere. Since Bhairava functioned as Sin-Eater at both the Mahāśmaśāna-Stambha where as Kotwal he executed the ultimate punishment, and also at Kapālamocana where as Kapālin he was freed of the ultimate crime of Brahmanicide, it is perfectly logical that, in the wake of the Muslim occupation of Oṃkāreśvar, the heart of Hindu Kāśī, Kapālamocana had come to be (re-)identified with Lāṭ Bhairõ, where the Kapālin remains as executioner, victim and pillar of the world (Sukul 1977: 71–72, 250, 346–48).

Fig. 3: *Īdgāh* with well and Lāṭ enclosure in foreground: the well in the open praying area of the *Īdgāh* bounded on the west by a wall with the niche (*mihrab*) indicating the orientation (*qiblah*) of the Mecca.

The Devotees of Lāṭ Bhairõ: From Brahmanicide Skull-Bearer to Vegetarian Brahman

The (Mahārudra temple around the) Mahāśmaśāna Stambha must have been the haunt of Kāpālikas and Pāśupatas in the pre-Muslim period. Through a general evolution well attested elsewhere in North India, even in the major Bhairava temples of Banaras and Ujjain, the post-Muslim Lāṭ seems to have been in the religious custody first of the Nāths (Jogis), then the Gosains and finally the Brahmans (Chalier-Visuvalingam 1989: 159, 205–10; 1994a). According to the available genealogy, the rituals were first of all—we have concrete evidence only from the 17th century—performed by householders of the Gosain (Gosavi) caste. The migrations of Gosavis, who functioned as merchants and bankers in the 18th and 19th century, were connected with the sale of cattle and are intimately related to the spread of Kāla Bhairava. These mendicant soldier-traders, who were the largest owners of urban property in Banaras during the late 18th century and

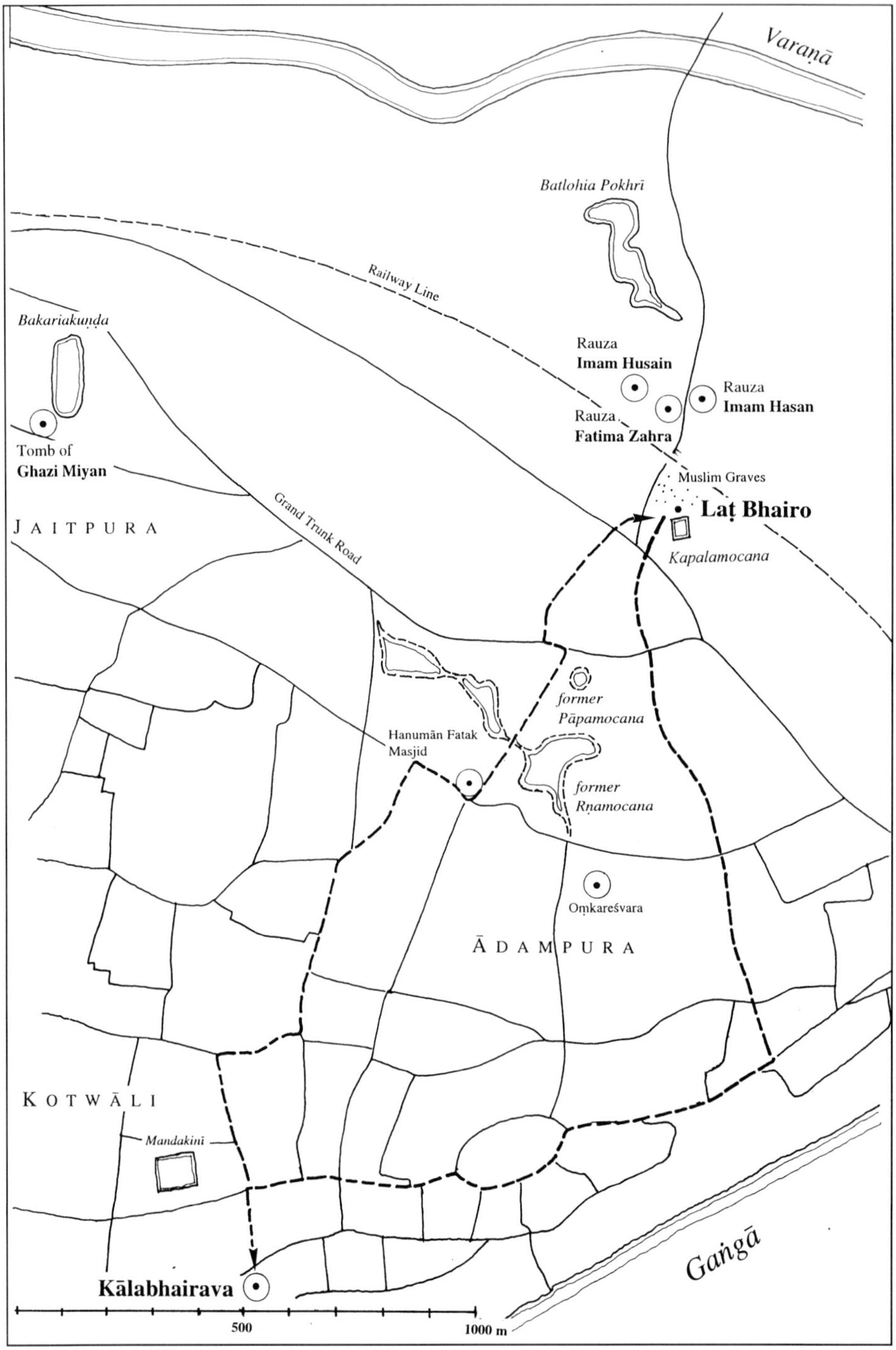

Fig. 4: Processional route of Lāṭ Bhairõ's marriage (drawing by Niels Gutschow).

Fig. 5: The *jīvatputrikā* vow (for "long living sons"), which must be performed near a pool, in this instance the Kapālamocana tank. In fact, the ceremonies are conducted on the *īdgāh* itself.

constituted (along with other mendicants?) one fourth of the city's inhabitants in the early 19th century, were a bulwark to the politico-religious authority of the Bhumihar Maharajas of Banaras. True to the ethos of the Bhairava cult, however, this "military caste" was organized around the guru-disciple relationship and recruited its members according to ability and without any regard to caste-distinctions. Brokers between diverse social groups, they maintained the pilgrimage routes and rhythms on the one hand and conserved their ties with the orders of regular *sādhus* like the Daśanāmīs on the other (Freitag 1989c: 6, 24–25). The Gosains connected with Lāṭ Bhairõ are clearly linked with the Kānphaṭā Jogis, its preceding custodians, for their names generally ended with Nāth. None of the Gosains once connected with this temple are now alive; they must have surrendered the care of this temple in the fifties due to the lack of heirs. The last Gosain was a widow named Bhagavatī Devī who was apparently "too friendly with the neighbouring Muslims" and also not managing the temple very well. Because of this, a Committee was formed in 1958 and its direction has been always in the hands of a Brahman. The earlier temple committee has now been replaced by a trusteeship. An official board of 8 trustees named the "Lāṭ Bhairõ Prabandhak Samiti" has been duly registered under the Registration of Societies Act n. XXI of 1860. This social development corresponds to the progressive purification of the Hindu mode of worship from human sacrifice to bloodless vegetarian offerings, leaving it to the Muslims to perform the intermediate goat-sacrifices during their (Baqr) Īd celebrations which the "brahmanized" Hindus now find rather distasteful (Kumar 1989: 157–58; Robinson 1877: 114). Just as the classical reform of brahmanical ritual had all but eliminated the violence from the earlier Vedic sacrifice even while retaining its symbolic framework intact, the underlying Tantric framework of the Hindu temple cult has been likewise purified even while leaving its topology and significance largely intact.

Organization and Ritual at Lāṭ Bhairõ Today: Vestiges of a Shattered Universe

The temple is very well managed and the daily rituals and the festivals are regularly performed. The devotees, though belonging to every strata of society, are mainly from the neighbouring area of the holy city. However, there is a small hard-core group of male devotees who particularly venerate Lāṭ Bhairõ and enjoy being with "Baba" as he is fondly called. According to the trustees, it is only in 1965 that a wall was built with the agreement of both Hindus and Muslims in order to avoid communal tensions. Nevertheless, the atmosphere remains tense, there are a great number of court cases from both sides, and a police presence day and night around the area. The well had remained closed after its destruction by the Muslims during the riots of 1809 and had almost disappeared until 1895, when the Court ordered the Muslims to pay 100 rupees as penalty and to rebuild the well (Upādhyāya *samvat* 2028: 5). The water of this well, which is supposedly different from that of the Gaṅgā and the Varaṇā flowing some distance away, is used for washing not only the utensils needed for the ritual but also (the temple of) Lāṭ Bhairõ itself (Upādhyāya *samvat* 2028: 52). According to the local myth, Dakṣa came to this "well of Sītā" (the chaste wife of Rāma) to unwillingly give away his virgin daughter Satī to Śiva in marriage (*kanyā-dān*). The well, usually called Bharat Kūpa Jananī "the mother, the well of Bharata", reflects the role it has taken on during the Rām Līlā festival. If we omit the name Bharat as being a later accretion, we have the name "mother (in the form of a) well" (*kūpajananī*) and for many devotees the well is indeed the symbol of the Mother, recalling the maternal waters of the Ṛgveda.

The daily rituals performed at Lāṭ Bhairõ are quite similar to that of other Śiva temples, involving decoration and adoration, as well as offering to the deity. The Lāṭ and other images in the temple are washed with water drawn from the adjacent well, and the adoration with lighted oil-lamps (*ārati*) is performed before the Lāṭ, then the well and finally the Kapālamocana tank. Apart from curd and jaggery (brown sugar candy) the Lāṭ is also offered horse-bean fried in oil, which is a substitute for meat in the mortuary rites, and some wine; sometimes fried fish is offered as well. Praises of Lāṭ Bhairõ, Kāla Bhairava and the "Boy" (Baṭuk) Bhairava are sung and finally the one hundred and eight names of Bhairava are recited. As at other Śiva temples, the principal days are Tuesday and Sunday, and also the 8th and the 14th of every lunar month. The main festivals celebrated at the Lāṭ today are:

a) the marriage of Lāṭ Bhairõ with the well celebrated every year on the 15th (full-moon) day of the waxing fortnight of Bhādrapada (August–September).

b) *jiotiya* (*jīvatputrikā-vrata*). A vow observed by women for the longevity and welfare of their children on the 8th day of the waning fortnight of Āśvina (September–

October) particularly at the Lakṣmī Kuṇḍ just west of Godaulia on the Luxa Road (photo Fig. 5).
c) the episodes of the 3 days just preceding and following the new moon of Āśvina, during the 350 years old "original" (*ādi*) Rām Līlā, that still unfolds in this part of the city for about 21 days from the 9th of the waning fortnight of Āśvina till the 6th of the waxing fortnight of Kārttika.
d) Fair with enactment (*līlā*) of a fight between the opposing tribes of Kols and Bhils on the 12th or 13th of the waning fortnight of (Āśvina) September–October. In recent printed programs of the Rām Līlā, this event is mentioned as the first of the above episodes staged around the Lāṭ. The fair was already obsolete and attracting no more than a few dozen people in the mid-19th century when Sherring visited Banaras (Sherring 1868: 195).
e) Śivarātri, Navarātra, Dīvālī, Bhairavāṣṭamī, Holi, etc. (the most popular Hindu festivals).

Hindu festivals are not isolated celebrations but form a meaningful complex, spanning across time (the ritual calendar) and space (sacred geography), that is best deciphered as a transposition of the (semantic connections and semiotic processes underlying the) brahmanical sacrifice (Chalier-Visuvalingam 1991; Visuvalingam 2003). The product of a bygone economy, the supposedly "obsolete" Vedic religion has spawned a symbolic universe of which the Hindu will always remain a prisoner until he has understood its innermost mechanisms and resources. The task of the anthropologist is partly a therapeutic one: free the reader's mind by inviting participation in reconstructing the whole.

The Marriage of Lāṭ Bhairõ: Death and Rebirth of the Royal Sacrificer

While all these festivals are noteworthy, the focus here will be on the marriage of Lāṭ Bhairõ with the well, which in a significant manner "incorporates" the central themes of the remaining occasions that also celebrate Bhairava, directly or indirectly, as the paragon of Hindu sacrifice. The metal "crown" (*mukuṭa*) of Bhairava is kept in the house of Bhageluram Tokedar, one of the Trustees, who is no doubt the major patron of the festival. This "crown" is actually a huge mustached head with two faces like that of a warrior (Fig. 6). It is apparently of quite recent origin but is said to be made of a consecrated mixture of eight minerals (*aṣṭa-dhātu*) as per the prescriptions of the *Kāśīkhaṇḍa*.

Late in the afternoon of the full-moon day, the crown is carried by some devotees on a red palanquin from the house of this trustee at Viśveśvarganj in procession to Kāla Bhairava, who is considered the younger brother of Lāṭ Bhairõ. The modern temple is located in the Kotwalpuri district in the maze of lanes between Chaukhamba Lane, the "Main Street" of pre-modern Banaras, and Maidagin Park. The procession does not go inside the Kāla Bhairava temple but merely stops in front to allow one of the Kāla Bhairava temple priests to worship with the typical waving of lighted oil-lamp (*ārati*) and with vegetarian offerings before the crown. As the only Gosain among the otherwise Brahman priests, he is the solitary vestige of an earlier ritual organization when the Kāla Bhairava temple was in the hands of the Nāth ascetics, themselves the successors of the skull-bearing Kāpālikas. It is in his house, just adjacent to the temple, that the capital of the Mahāśmaśāna Stambha is still preserved and worshipped as the "Discus-handed" (Chakrapāṇi) Bhairava. From there the procession proceeds through first Jatanabar, back to Viśveśvarganj, then through Kajimandi, Baluabīr, Hanumān Phaṭak, Teliyana (Lāṭ Bhairõ Bazaar) and Jalalipura and finally skirts the Kapālamocana tank to reach the Lāṭ around midnight (see the map of the procession Fig. 4). It halts at least 50 times to perform very simple rituals for devotees, much (mostly locally produced) wine is drunk during these stops. The "marriage party from the bridegroom's side" (*barāt*) is large and accompanied by elephants, musicians, and youth performing acrobatic dances with swords. In former times, the procession was joined especially by singing prostitutes.

Fig. 6: The Lāṭ with a yellow silk garment and crowned with Bhairava's mustached head.

The make-shift temple complex—the Lāṭ, the well and the tank—is heavily decorated, music is played and, in short, everything is readied to receive the bridegroom who arrives in the form of the large head of Bhairava. The well is draped with a red sari in the manner of a Hindu bride, the Lāṭ with a yellow-silk garment. Though the whole area is under the close vigilance of the police, there are a large number of people rejoicing with all the fanfare of a modern Indian marriage. Finally the "crown" is taken out of the red palanquin by four men to be carried in three circumambulations around the well, then around the *tulsi* tree. It is then mounted with great efforts on the top of the pillar, and worshipped in a mode that is not substantially different from the daily ritual.

Though during some other festivals such as Bhairavāṣṭamī or Śivarātri, one may offer meat, wine and other intoxicants, the offerings during the marriage must be purely vegetarian. The priest tolls the four bells, then the devotees play cymbals and two of them play the *ḍamaru*, a small hourglass-shaped double-faced drum associated with Śiva. He performs *ārati* in front of the Lāṭ, the eight Bhairavas (in the order enumerated in the accompanying map of the temple) and at the corner consecrated to the goddess. Then the priest, followed by the devotees, leaves the enclosure of Lāṭ Bhairõ in order to perform the *ārati* in front of the well and also in front of the Kapālamocana tank. Finally the priest returns, stopping near the *tulsi* tree, to the Lāṭ, where the devotees join him in singing different hymns in praise of Bhairava. The only difference between the daily ritual and that of the marriage is the approximately hour-long fire-offering (*havan*) which is essential for all Hindu marriages. The main offerings during this ritual are rice, grain, semolina, ghee (clarified butter) and wine. The priest puts a religious mark of ashes (*tilak*) on the forehead of the male devotees and a *tilak* of vermillion for the female devotees. However, because it is very late in the night, the participants are mainly males. When the *havan* is over, the priest ties the Lāṭ and the well together with a garland in the Hindu manner of a bridegroom and a bride. The following morning there is a simple feast for all the devotees, consisting principally of rice and pulse boiled together into a porridge which is cooked inside the precincts of Lāṭ Bhairõ. The crown is finally dismounted and brought back to the trustee's house under heavy police escort.

The fecundating powers of Lāṭ Bhairõ's "phallus" (*liṅga*) are revealed in Banaras eight days after the celebration of the Lāṭ's marriage—hence on the 8th of the still inauspicious waning fortnight of Āśvina—as part of the vow for "long living sons" (*jīvatputrikā*), which must be performed near a pool, in this instance the Kapālamocana tank. In fact, the ceremonies are conducted on the *īdgāh* itself and couples—those recently married and also those already blessed with children—make a circum-ambulation around the Lāṭ. The day immediately following the Jīvatputrikā Vrata is Mātṛ Navamī, which is reserved for the propitiation of deceased women particularly mothers. The well of the chaste Sītā is, after all, the site where Dakṣa gave his virgin daughter Umā in marriage to the terrible Rudra, only that she might eventually throw herself as Satī into his sacrificial fire. Dakṣa-Prajāpati's consequent decapitation in the midst of his sacrifice by Vīrabhadra is itself a variant of Brahmā's beheading by Bhairava. Since Bhairavāṣṭamī, the birthday of Bhairava, is the only other day when the "crown" is again mounted on the pillar, though without any fanfare, Bhairava's wedding would appear to be symbolically identical with his very birth. But the full moon of Bhādra also marks the beginning of the fortnight of the waning moon of Āśvina, "the fortnight of the manes" (*pitṛ pakṣa*) when the most potent funerary rituals are performed for the departed ancestors. An execution stake on the cremation ground, particularly at this most inauspicious moment, would be the last place any orthodox Hindu would wish to celebrate, much less consummate, his marriage. For the devotees, however, "Baba is beyond this kind of consideration".

Though it does not speak of a marriage as such, the *Kāśīkhaṇḍa* (100.99) already singles out the full moon of Bhādra for particular worship of the Kula Stambha. It is possible that the marriage has been revived in its present form only in the early 20th century (1906?), and that the procession itself has perhaps been influenced by similar Muslim practices during Muharram (Kumar 1988: 212, 224), but the comparative evidence suggests that it derives from pre-Muslim times. Indeed, it could be easily demonstrated that the ritual framework of the traditional Hindu marriage, despite its "auspicious" public character, is riddled with such notations of violent death, just as the death ceremonies are likewise impregnated by (the symbolism of) sexual union. Unconstrained by spatiotemporal and pragmatic limitations, the events of the human life-cycle are condensed as it were within the marriage of the royal divinity.

Lāṭ Bhairõ and Kāla Bhairava: The Hidden Identity of Periphery and Centre

Just as the Indra pole is erected in front of the royal palace at the centre of Kathmandu and then carried in procession after its felling to the cremation-ground on the southern fringe of the city to be devoured by the flames of Pacali Bhairava (Chalier-Visuvalingam 1996: 271–72), the marriage of the Lāṭ in Banaras, celebrated within the same time-span as this "Vedic" festival for the renewal of Nepali kingship, likewise sets into motion the hidden ("transgressive") identity of the otherwise immobile centre and periphery of the sacred city. The present Kāla Bhairava temple would have been, according to Sukul (1977: 203–04), the site of the pre-Muslim pillar (*liṅga)* to Lord Bhairava (*bhaireveśvara*) with an adjacent "well of Bhairava" (*bhairava-kūpa*), which is precisely the configuration now duplicated at the Lāṭ. It is physically impossible to transport the Lāṭ to be immersed in the motherly well. There is, however, the practice in Rajasthan of keeping images of Bhairava in stepped wells, and the folklore of Banaras abounds in *liṅgas* being hidden in sacred wells. Between the modern Viśvanāth temple and Aurangzeb's mosque, which is itself a Muslim transformation of the earlier 16th century Viśveśvar temple, stands the Gyānvāpī well. "In the beginning" long before the descent of the Ganges when there was no other water on earth, it was dug out by Śiva himself with his trident in order to cool the *liṅga* of Viśveśvar with its water. According to legend, the *liṅga* was preserved from Aurangzeb's desecration only by being thrown into the deep waters of the well. The royal inscription of the Queen of Indore, who sponsored the construction of the present Viśvanāth temple, makes no mention of her having established a different *liṅga*.

In their post-riot "memorial"[4] of November 25, 1809, the Muslims charge that some Hindus had corrupted the Superintendent (Mutwali) of the Viśveśvar mosque. The Hindus began to worship the well and share their offerings with the Mutwali pretending that Viśveśvar had concealed himself in the well. Hindus likewise "worship with the utmost faith a stone fountain" in a mosque compound in the Daranagar quarter, and "so also was the [Lāṭ] of [Feroz] Shah converted by them into the [Lāṭ] of [Bhairõ] and the lower order of Hindus worshipped it" (Robinson 1877: 113–4). The position of the motherly Bharat Kūpa beside the Lāṭ on the *īdgāh* indeed replicates the situation of this "well of liquid wisdom" between the Viśvanāth temple and the Gyānvāpī mosque. The Maṇikarṇikeśvara *liṅga*, which stands underground at the bottom of a deep shaft, could at one time be reached by a tunnel originating on the cremation (Maṇikarṇikā) *ghāṭ* (stone landing with broad steps). In the Purāṇas, this "Lord of Maṇikarṇikā" is mysteriously said to be in the middle of the Maṇikarṇikā tank (*kuṇḍ*) itself (Eck 1983: 246). It is this "well of the Jeweled Ear-ring" beside the cremation *ghāṭ* that has the greatest cosmogonic significance in Kāśī for, at the beginning of time, the city itself floated upon its primordial waters (Eck 1983: 238–251). After all, the true form of Kāśī is the Śiva-liṅga itself, no different from the pillar of light from which Bhairava himself was born. The motif of the "*liṅga* in the well" encapsulates and fuses the image of sexual union with that of the embryo in the womb, the very "union" (*yoga*) that was also celebrated as the "fish-wombed" (*matsyodarī*) conjunction (Chalier-Visuvalingam 1989: 179, 221 n.77), when the entire city was transformed into a primordial mound engulfed by the backwards flowing flood-waters of the "menstruating" Gaṅgā: it was precisely at such a moment that the Kāpālika Bhairava is said to have plunged into her "amniotic" waters to be freed forever from the sin of brahmanicide. Not only is the marriage of the "peripheral" Lāṭ with the well replicated, in this way, at the centre of (and elsewhere within) the urban ritual space, the entire sacred city is as it were enveloped by and becomes an expression of this mystical union.

It is the concern for general fertility that dominates the popular perception of these "marriage" festivals in Nepal and even Banaras, but the themes of human sacrifice and of ultimate liberation still remain in the background. The "punishment of Bhairava" undergone by all sinners who are privileged to die in Kāśī is hence modeled on the decapitation of criminals, with the Lāṭ itself assuming the symbolic role of the Vedic sacrificial pole (*yūpa*).

Marriage of Lāṭ Bhairõ and Ancient Indian Cosmogony: Pre-Islamic Universal Kingship

Through a systematic analysis of the Newar New Year festivals centreed around the raising and felling of a wooden pole called the *liṅga* (Chalier-Visuvalingam 1989, 1994) and a parallel decipherment of the folk-cult of the trickster-god Kāttavarāyan's "marriage" to a sacrificial stake in Tamil Nāḍu (Visuvalingam 1989: 438–45), we have already shown that the marriage of Lāṭ Bhairõ is the vestige of a royal cosmogony re-enacting the symbolic death and rebirth of the king as sacrificer par excellence. The Indra festival of Kathmandu is still celebrated on the same day, and the Bisket festival in its sister-city of Bhaktapur is centred around the raising and felling of a pole identified with the decapitated Kāla Bhairava, who is said to have come from Banaras. The New Year at Bhaktapur indeed celebrates the sexual union of this cosmogonic pole with the tutelary goddess of the city in her fierce form, a marriage that is also a sacrifice and a rebirth accompanied by ritualized violence. Identified with both the royal Indra and the transgressive Bhairava, the sacrificial pole exhibits all the ambivalence that may be expected of an *axis mundi* that is simultaneously the execution stake and the central pillar of the world. The mythico-ritual processes they set in motion, even visually through processional routes, consist largely in uniting spaces—pure and impure, profane and sacred—that are otherwise rigorously held apart: the nuclear temple, the royal palace and the cremation-ground. Both these Newar festivals reveal a complex inner structure that not only juxtaposes public rituals of death, marriage (or sexual union) and rebirth, around the theme of kingship, but also a web of interconnections that encompass other festivals at other times and places within the Newar society. Though the Newars, in keeping with local Hindu myths attested elsewhere in India, affirm repeatedly that (Lord Viśvanāth in the form of) Bhairava came to Nepal from Kāśī, the royal dimension of his cult has been largely submerged in his native city. Islamic iconoclasm destroyed that sacrificial nexus between Hindu kingship, the divine pantheon, temple organization and cosmogonic festivals that still constitutes the symbolic core around which the complex social hierarchy and the highly differentiated community life of the Newars has been articulated (Chalier-Visuvalingam 1994a).

The Bisket marriage is obligatorily accompanied by violence: not only the collision between the chariots of Kālī and Bhairava, with reminiscences of the latter's decapitation to appease the blood-thirsty goddess, but also the public enactment of ritualized conflict between the two halves of the city of Bhaktapur during the fierce tug-of-war to drag the royal chariot into their own quarter. Kathmandu is likewise divided into an upper and lower half around the royal palace with such conflict erupting during specific festivals that are said to have been instituted by the reputed "founder" of this present-day capital city, king Guṇakāmadeva. Though his explicit purpose is said to have been to prevent his subjects from revolting, this "culture-hero" would have unified the city by instituting these rites through a cathartic mechanism for the release of the diffused inchoate violence that might otherwise tear apart the social fabric: a strategy, rather, of "divide-rule-and-unify". This fundamental duality of the Newar urban agglomeration is best understood as a retention of the cosmo-ritual opposition between the two moieties (*mithuna*) of a tribal society, vestiges of which may be found even in the Ṛgvedic vocabulary (root *mith-*). This would also explain the mysterious tug-of-war around the Lāṭ between Kols and Bhils that was supposedly connected with the Rām Līlā festival. Not only is it likely that the original pillar at this spot might have been pre-Aśokan, as claimed by John Irwin (1983), but the sacred character of Banaras and of the marriage of Lāṭ Bhairava itself may have its roots in the pre-historic tribal religion of the Gangetic plain (Visuvalingam 2003). "Ritualized", but very real, conflict typically erupts when Hindu and Muslim processions collide head-on during festivals, just as it does between rival groups of Hindu ascetics as during the Kumbha Melā.

Why do Hindus cling tenaciously to the derelict remains of a stone pillar situated precariously within a Muslim enclave at a site so peripheral to their sacred city otherwise so rich in wholly functional and more centrally located temples and outdoor shrines? Of course, here is where they undergo, after death, the metaphysical "punishment of Bhairava" without which no salvation is possible in Banaras. The mythico-ritual complex that binds Lāṭ-Bhairõ to Kāla Bhairava and ultimately to Kāśī Viśvanāth is a projection of Hindu sacrificial ideas centred around (a universalizing divine) kingship. Though the Mahārāja of Banaras has long lost even his local political power and most of his royal privileges, Kāśī Viśvanāth still conserves, at the symbolic level, the sense of social unity and coherence. What is more, Bhairava represents the hidden transgressive aspect of the royal deity that accounts for the peripheral location of this "sacrificial stake" in space. Similarly, the impure goddess who conferred (tantric) power on the Newar king were spatially dispersed and relegated to the shrines of the "eight-mothers" (*aṣṭa-mātṛkā*) beyond the territorial limits of the urban agglomeration. Whereas the Otherness of the transcendent God of the Judeo-Islamic tradition is expressed through His being beyond representation through any human image, the Other is represented for and by Hindus instead through the image of the criminal slayer of the pure Brahman, relegated by his very impurity to a subdued, subordinated role at the religious periphery of the conventional consciousness (Visuvalingam 2003). The question now is: what has become of the innate disaggregating violence that the Vedic sacrificial logic sought to contain, so tenaciously and often hopelessly, within the context and from the perspective of Islam? Why did Abrahamic iconoclasm demolish the surrounding temple complex only to leave the Hindu Bhairava's imposing "stake" uncontested in the middle of a prayer-ground where Muslims continue to perform their own sacrifices?

Ghāzī Miyã and the Sacrificial Pole (*quṭb*) of Islam: Otherness in the Heart of Tradition

Though the devotees of Lāṭ Bhairõ are no longer able to explain Bhairava's marriage to the adjacent well, they invariably refer to a similar practice of their predominantly Muslim neighbours in the same Alavipura (= Adampura + Jaitpura) ward. Not only is the pillar beside a Muslim cemetery with a central sepulcher (*rauza*) made up of "Buddhist" architectural remains (Sherring 1868: 306–07), but the custodians of the Muslim tomb-shrines elsewhere in the city generally "digress" into the story of Ghāzī Miyã when likewise questioned about the particular saint of their locality. On the first Sunday of the solar month of Jyeṣṭha (falling between 14th and 21st May), they celebrate—like Muslims elsewhere in North India and western Nepal—the annual wedding procession (*barāt*) of Ghāzī Miyã's tragic marriage with Johara Bibi. It moves from the Jaitpura crossing to the domed mausoleum in the Sālārpūr *muhalla*, within the same (Jaitpura) quarter, which houses (the replica of) his "tomb" at whose head is likewise a high pillar.

Salar Masud's Martyrdom and the Islamic Conquest: The Marriage of Ghāzī Miyã

Ghāzī Miyã was born into "history" as Salar Masud, the nephew of Mahmud of Ghazni, at Ajmer in 1014. As his desire for martyrdom was as intense as his proselytizing zeal, he headed the Muslim warriors in their numerous incursions into the Gangetic plain, until he was felled in battle in 1033 at the tender age of 19 by the Hindus at Bahraich. When Muslim domination over North India was permanently established towards the end of the 12th century, his tomb was rediscovered. It became such an important pilgrimage site that, already by the 13th century, the poet Amir Khusru could speak of the whole of Hindustan being embalmed by the fragrance from the perfumed tomb. The ballads (*sohila*), which are sung by low-caste Muslim musicians (*dafali*) belonging to a fraternity devoted to his cult, make Bahraich itself his birthplace. Most significant of all is the repeated identification by the *sohila* of the city of Ghāzī Miyã's tomb with Mecca and Medina (Gaborieau 1975: 300, 306). Such were the material difficulties that in Akbar's reign, the doctors of religion (*ulema*) even declared that the *hajj* to Mecca was no longer obligatory for Indian Muslims. Ghāzī Miyã was cursed even before his birth to be martyred on his wedding-day. On the fateful day, he has to exchange his marriage garments for armor and the wedding music becomes martial as he rides out to battle. He annihilates the aggressors and it is only while returning that he is killed by the arrow of a survivor.

While heading for Bahraich in 1034–35, Salar Masud had dispatched a portion of his army and its retinue under Malik Afzal Alavi towards Varanasi (Sukul 1974: 152–5, 1977: 24–26). The invading contingent was thoroughly defeated on the northern outskirts beyond the boundary wall of the city at the site where the Masjid Ganj-i-Shahidan ("treasure of martyrs" mosque) now stands near the Kāśī Railway station (the mosque was unearthed only during this construction activity). The Muslim civilians, with their women and children, were permitted to settle down in that area as townsmen and over the following century peacefully served the Hindu kings even as soldiers. After Quṭb-ud-din Aibak had devastated the city in 1194, destroying nearly one thousand temples, the Muslim locality was renamed "Salarpur" or "Alavipur/ Alaipura" (which today includes the two wards of Adampura and Jaitpura). During the reign of Feroz Shah Tughluq, the famous Arahi-Kangra mosque, the Chaukhamba and Gola ghat mosques, many others in Alavipur, and almost the entire building scheme around the Bakaria Kuṇḍ were constructed, generally on the site of and with the materials obtained from demolished Hindu temples. The Tughluq dynasty patronized the, by now already famous, cult of Ghāzī Miyã and Feroz Shah Tughluq made the pilgrimage to Bahraich where he had his hair cut.

The Muslims' own post-1809-riot "memorial" (see Robinson 1877: 119) even claimed that the Lāṭ Bhairõ pillar was in fact "the structure of [Feroz] Shah, like the pillar [Lāṭ] at Allahabad, Delhi and other places, and which the [Hindus] state to have been erected by their own forefathers. But, be that as it may, it was not an object of their worship entitled to any great veneration like the temples of [Viśveśvar] and [Bhairõ Nāth]; for no account of this pillar is to be found in any of their orthodox books. The style of worship of the Hindus is this, wherever they find set up (a pillar) they call it, at the incitement of their priests, a place of their worship, and after sometime has elapsed they consider it as a place of worship of the highest sanctity" (ibid.). The same source notes, however, that "for some years the lower classes of [Hindus] and [Muslims] have annually celebrated the marriage of the [Lāṭ], and have divided the offerings between them" (Robinson 1877: 113–4). The latter fact was still reluctantly admitted by the legal custodians of the *īdgāh* when we interviewed them with John Irwin (in the early 1980s). Whereas their caste-fellows living in Madanpura resort to the Gyānvāpī mosque, the illiterate weaver (Julaha) community of Alaipur generally congregates at this Lāṭ *īdgāh*. Rather than immigrant Muslim weavers subsequently seduced by Hindu idolatry, they would be "Hindu"(-ized Buddhist) castes that continued to worship the pillar even after their conversion *en masse* to Islam by the hard core of Ghāzī Miyã's original followers (Kumar 1988: 50). By leaving the aniconic Ashokan pillar standing intact before the *īdgāh* when it tore down the surrounding pantheon of Hindu idols, Islamic iconoclasm had re-inscribed the continuing Hindu worship within a Mecca-centred framework. Its spectrum of signification could have thus ranged from that of a divinity proper for the devotees of Bhairava to that of a mere victory monument for the uncompromising legalists. The Hindu-Muslim cult of Ghāzī Miyã merely confirms that Islamic proselytism has succeeded through a judicious blend of violent imposition of symbolic (architectural) structures and syncretizing accommodation that operates on the common ground occupied by both religions.

Some of the most significant finds from the "archaeology" of Lāṭ Bhairõ are hence remnants of what had been incorporated by the Muslims into their baroque "frieze" about this martyred warrior (Visuvalingam 1992). In a common legend, Ghāzī Miyã removed his own head to avoid seeing and being seduced by the hundreds of naked women sent by the king's astrologer in order to destroy the power of his purity and thereby render him an easy sacrificial victim. In Bahraich, (a representation of) the long-haired head of the saint is carried in procession on the tip of a lance.

As Gaborieau (1975: 314) points out: "In India, poles whose summit is ornamented with an effigy of the head of the martyred hero are taken out in procession; in Nepal, it is the pole itself which receives the blood of kids offered to obtain rain; no doubt that in these rites, it is the saint himself who is represented by the poles through a symbolism which is widespread in the Muslim world". Gaborieau (1975: 313) adds that the pole itself, "in the rites meant to obtain rain, appears as a sort of phallic symbol uniting heaven and earth". Ghāzī Miyã, the martyred youth, is not just the lord of rain and the harvests, his tomb dispenses all boons, particularly sons to the childless. It is thus not so much the martyr's union with Allah that is the popular focus of the Muslim cult but rather the regenerative forces unleashed by his tragic marriage that begins to be celebrated in India even 2 to 3 days before the Sunday festival. Nevertheless, the participation of the Hindus is an individual affair and, at the strictly ritual level, the marriage of Ghāzī Miyã remains as specifically Muslim as the marriage of Lāṭ Bhairõ is Hindu.

The martyrdom of the Muslim warrior and the punishment of the Hindu god serve as the two poles of a common ideology of self-sacrifice based on the identity of the killer and the victim. Despite the still unresolved tensions between Islam and Hinduism on the social and doctrinal levels, such "hybrid" legends reveal an uncanny understanding of the cults of both Ghāzī Miyā̃ and of Lāṭ Bhairõ. Indeed, among the foremost devotees of Ghāzī Miyā̃—who, like Ḍulha ("bridegroom") Dev, is also believed to have been killed by fire on the eve of his wedding—are the untouchable Ḍoms of Banaras who specialize in the "human sacrifice" of the crematory ritual for the entire Hindu community from the furthest reaches of the Indian subcontinent. The colloquial wisdom still refers (in Hindi) to the cremation (ground) as the "place of the bride" (*ḍulhan kā sthān*) and as the "last marriage" (*ākhirī śādī*). Even now, the Pakistan-based Lashkar-e-Taiba terrorist group that has been carrying out suicide attacks in Kashmir (and, more recently, even in Delhi) prepares its martyrs by assimilating their impending death, through ceremonies and ritual notations derived from this symbolic universe, to the consummation of a mystical marriage. In fact, this Muslim group recruits its members primarily from the Mujahir community, that is, Punjabi refugees of the 1947 partition from "Hindu" India.[5]

The Tenth of Muharram and Ghāzī Miyā̃: The Shia-Sunni Marriage in India

A more searching and even-handed analysis, however, reveals that the marriage of Ghāzī Miyā̃ is not only the Islamic co-opting of a pre-existing Bhairava-cult shared by Hindus, Buddhists and tribals, but just as much a transposition of Iranian popular cult and Shia esotericism, especially as crystallized in the celebration of (the tenth of) Muharram (Visuvalingam 1992: 42–46, for details and references). The contrast and identification between marriage and death is underlined in all the songs of lamentation and constitutes the very essence of Muharram; there is even a syncretic version of the Ghāzī Miyā̃ ballad, focused instead on the martyrdom of Hasan and Husain, that serves as an Indianized founding legend for this ten day festival of mourning! The celebration of the death anniversary of any Muslim saint is technically designated by the Arabic term *urs* that originally signifies "marriage" so much so that the union of the soul with God has become, in both popular religion and esoteric literature, a sexual union that is consummated only in death. In the vicinity of Lāṭ-Bhairava are (now separate) Shia and Sunni complexes comprising the "tombs" (*rauza*) of the Imams Hasan and Husain along with that of their mother Fatima. It is most significant that Johara (Zohra or even Zahra), the name of their common wife, is just a variant of the epithet "the Radiant or the Resplendent" that permanently characterizes their mother Fatima (al-Zahra).

Outside of India, Muharram is not celebrated by the Sunnis, who are rather likely to be the target of Shia wrath and vengeance. The popularity of Muharram among both Shias and Sunnis has indeed been expanding throughout the last century in Banaras, but since the 1931 Hindu-Muslim riots the Hindus of the sacred city have stopped participating in it (Kumar 1988: 215–16). Though the festival continued to be the occasion of Shia-Sunni conflict in India, the transposition of the sacrificial marriage to the Ghāzī Miyā̃ cycle served, in part, to facilitate and legitimize a common front against the infidel Hindu majority. The all-night narration of his legend even during a normal marriage ceremony—before a backdrop of painted representations of his (battles and) martyrdom—confirms that Ghāzī Miyā̃ provides a role-model for the Indian Muslim.

The Marriage of Lāṭ Bhairõ and Ghāzī Miyã: Sacrifice and Conflict in-between Hinduism and Islam

The joint celebration by lower Hindu-Muslim castes of the marriage of Lāṭ Bhairõ and Ghāzī Miyã has been the outward expression of the common sacrificial paradigm underlying both Hinduism and Islam. Ghāzī Miyã continued to play the role of the sacrificial victim not only for the Hindu-Buddhist (nominal) converts but also in some measure for the Muslims. By blurring the sharp distinction between the Muslim polity (*umma*) and its Other, the gradual mass conversion of entire castes posed a dilemma for Islam: should the "regressive" Hindu ways and superstitions of these freshly harvested souls be tolerated in the expectation of their eventual assimilation or should they be simply excluded and penalized as covert Hindus? Whatever their allegiances of the moment, these "Mohammedan Hindus" would have probably been unable, for the most part, to decide the question even in their own self-perceptions. Indeed, the marriage of Bhairava, once reduced to a largely non-anthropomorphic form, offered an as it were ready-made mould for the overt widespread resurgence of a similar identification of death to a mystical marriage within Islam, and in a manner that tended to dissolve the Sunni-Shia divide in the new Indian context. The cult of Ghāzī Miyã could thus win the devotion of not just Hindu lower-castes, like the untouchable Ḍoms, and converted Brahmans, like Baba Khalil Das Caturvedi, but also that of Muslim rulers, like Feroz Shah Tughluq, and cultural elite, like Amir Khusru. The "Hindu" appropriation of Ghāzī Miyã had paradoxically facilitated the resurgence and celebration within Indian Islam of a fundamental "mystical" dimension that was not just the preserve of Sufi saints and Shia Imams, but firmly anchored in popular folklore and in the Meccan ritual traditions (Visuvalingam 1992: 46–49).

The Islamic strategy of exteriorizing violence (*jihad*) onto those outside the larger universal community (*umma*) was no more effective in its containment within than was the Hindu approach of circumscribing it internally through the (transposed) mechanisms of the sacrifice. In either case, the dualistic "deep-structure" could resurface under any "rational" pretext, especially where ritual mechanisms were no longer available to keep the violence within bounds.[6] Under normal conditions in Banaras, the celebration of the festivals of Muharram and Barawafat, which have been consistently growing over the recent decades, is characterized rather by intense but sportive competition organized by the various clubs (*anjuman*) between the neighbouring Muslim wards (*muhallas*) themselves (Kumar 1989: 158–163). The stubborn persistence of the dualistic mechanism into our own times—whether contained within integrative (royal) festivals, diverted to further partisan political agendas, distorted by systematic economic exploitation, exacerbated by modern racism, generalized in our age of enlightenment into intense individual competition for climbing the ladder of "success" or perverted into fissiparous outbursts of crime pure and simple—only proves that the ideology of pacifism urgently needs to be supplemented by adequate techniques for confronting, neutralizing and transmuting the innate violence that masquerades under the most varied panoply of "rational" motivations.

The joint participation of Hindus and Muslims in each other's cults and festivals should not obscure the intense ideological struggle—even where peaceful and mutually accommodating—between the rival religions on the symbolic level for the heart, mind and soul of India. The Hindus could not remain oblivious of the living visual testimonies to the systematic razing of their religious architecture (c. 1660s) by Aurangzeb, who had sought unsuccessfully to impose an Islamic city called "Muhammadabad" upon their socio-religious centre (Freitag 1989a: 3). Having now lost their political supremacy in India, the Muslims, on the other hand, were not willing to submit to Hindu acculturation to the extent of surrendering the divergent world-view encoded into their own ritual practices. Having served to inscribe its continuing Hindu worship in the direction of Mecca, the Lāṭ could now just as well serve to gradually reintegrate the lower classes of Muslims within the symbolic universe of Hinduism. This intrusion of the Hindu sacrificial paradigm into the heart of Islamic observance and this encirclement of still idolatrous indigenous worship by the Meccan-oriented Abrahamic faith, was reflected in ritual space by Lāṭ Bhairõ standing, with visible persistency, as an aniconic column within the Muslim prayer-ground. Since the peripheral pillar had always been ritually linked to the sacred centre occupied by Kāśī Viśvanāth, alias Kāla Bhairava, and since its new Meccan orientation was likewise an article of faith for the larger Muslim community, any local religious dispute, even if initially occasioned by purely incidental or personal disputes, was apt to set the whole symbolic apparatus into motion sucking, like a semiotic "black hole", the entire sacred city into its gravitational field. The really significant question, for our purpose here, is whether the fault lines of such an eruption would amount to no more than a stereotyped replay of the religious antagonism that had originally pitted Salar Masud against the Hindu Bhairava, or would the centuries of relatively peaceful coexistence, cultural commerce and religious symbiosis have imperceptibly succeeded in effecting gradual but durable shifts and readjustments in the inner topography of sacred space? Would the Islamic approach of preserving community by exteriorizing violence onto the Other have gradually resulted in a reciprocal hardening of Hindu attitudes, even across internal caste-barriers, vis-à-vis the Muslims, or would the brahmanical approach of interiorizing and circumscribing violence within a universal commitment to self-restraint have softened and reshaped Muslim sensibilities vis-à-vis the larger social fabric of the city? Taken together, the Lāṭ Bhairõ riots of 1809 and the "Gandhian" civil disobedience against British rule of 1911 not only underscore the continuing power of religious values in shaping the course of (trans-)national history but also reveal the impossibility of reducing the social identity and agency of the actors to some inevitable "fatality" of the "religious" violence of that initial large-scale encounter between Hinduism and Islam.

Lāṭ Bhairõ, the Scapegoat of the Lord of the Universe: From Hindu-Muslim Riots of 1809 to "Gandhian" Civil Disobedience of 1811

Within the space of hardly 16 months from October 1809 to January 1811, Banaras witnessed a city-wide conflagration between Hindus and Muslims, the first serious "communal" riot since more than a 100 years prior to colonial rule, that was transformed, even before it had wholly subsided, into "civil" strife between the police and the military, before it culminated in a remarkably peaceful and sustained opposition to British rule that united all castes of Hindus and Muslims. The challenge for the anthropologist, historian and student of comparative religion is to avoid reducing socio-religious processes to politico-economic causes (or vice versa) and transforming collective instincts into rational aims (or vice versa), but instead provide an explanatory framework that takes all three acts of this human drama into account (Visuvalingam, Hindu-Muslim Relations, for details).

The "War of the Lāṭ": Paradigm for Hindu-Muslim Conflict and Partition?

The "Lāṭ Bhairõ riots" of 1809 (Pandey 1990: 28; Freitag 1989b: 203) have played a crucial role in colonial historiography not only because of their gravity and magnitude—comparable, we are told, only to the Kanpur outbreak of 1931—but also because they are among the first to be recorded in the colonial period (Pandey 1990: 29). Robinson (1877) has given a very detailed report using first-hand British accounts and the "memorials" written by the Hindus and Muslims themselves shortly after the Lāṭ Bhairõ riots, so called because the pillar was the destructive focus of the three-day carnage between Hindus and Muslims.[7] In 1809, the conflagration was sparked off by a trivial incident: in fulfillment of a vow upon his recovery from illness, a Hindu of the Nagar, a lower caste, was attempting to replace the makeshift mud dwelling of Hanuman on the contested ground between the *īdgāh* and the pillar with a permanent stone enclosure.

The Muslim "memorial" (Robinson 1877: 114) observes that earlier legal proceedings against a similar alteration on the shared sacred space were never determined and hence taken as a victory by the Hindus. It claims that the Hindus had even placed an image of Rāma and Lakṣmaṇa into the pulpit of the *īdgāh* for worship during the *Dasehra* celebrations of the previous year, and that they were again guilty of the same disrespect this year.[8] At first the Muslim weavers were content to appeal to the law-officer (*qadi*) and agreed not to disrupt the Bharat Milāp celebrations but to refer the dispute to the court immediately after the *Dasehra* holidays. When the festival was over on 20th October, they instead held a mammoth protest meeting during their Friday prayers, in the excitement of which they polluted the Lāṭ and its surroundings by overturning Hanuman's pedestal, uprooting the adjacent *tulsi* tree, and beating the pillar itself with shoes. Freitag (1989b: 211; 1989c: 39) indicates that the Muslim weavers had continued to participate in this (?) Bharat Milāp as in previous ones, which would only confirm the initially localized character of the dispute. Their objection to the repeated installation of Rāma on their pulpit nevertheless suggests not only the primarily "civic" character of their participation in the Rām Līlā but also the primarily religious character of the now evolving confrontation. The conjuncture of events on the *īdgāh* around the Lāṭ so faithfully reflects the overlapping disposition of Hindu-Muslim sacred space between the Viśvanāth temple and the Gyānvāpī mosque, that Robinson's account mistakenly displaces the Muslim demonstration to this mosque within the old Viśveśvar enclosure, and shifts imperceptibly to the defilement of the pillar which is in fact quite a long way from the religious centre of the modern city (see Fig. 1). Though worshipped primarily by the lower castes of Hindus, the "pillar of the great cremation ground" was indeed quite central to the symbolic significance of the sacred city, and this sacrilege had occurred precisely when the Hindus, particularly the Rajputs, "were seeking justice for the slaughter of a cow" (Hindu "memorial", Robinson 1877: 108).

By daybreak the whole Hindu community had heard of the sacrilege, and a crowd of mostly Rajputs began to assemble at the (physically still intact) Lāṭ, so much so that the Muslim Kotwal (title to the local superintendent) had to arrive on the scene with a large contingent of police in order to disperse them. The Muslim "memorial" (Robinson 1877: 115) claims that during the night some Hindus had already broken open the door of the pulpit. The Hindu "memorial" claims that the Kotwal, seating himself at the foot of Lāṭ Bhairõ, invited Hindu representatives to appear before him to air their grievances. On complying, Rattan Singh and three associates were simply taken to the police-station where they were beaten without any enquiry. The remaining 200 or so Hindus and the even greater number of Muslims present began to pelt each other, such that the Imambarrah and some idols were damaged in the process. Apprehending the worst as the whole city was becoming restless, the acting British magistrate (Mr. Bird) had already deployed 2 companies of sepoys (indigenous Indian soldiers in the pay of the colonial administration) to protect the Muslim places of worship. Anticipating retaliation by the Hindus, the outnumbered Muslim weavers, who were at the forefront of all these manifestations, then decided to take preemptive action by sacking the temple of the king of the gods, Viśvanāth himself. Had the attempt succeeded, it would have certainly resulted in the utter annihilation of the Muslim community in Banaras. For the local dispute between Hindu and Muslim low-castes, who had been as much united as separated by the pillar, was already becoming a full-scale confrontation between Hinduism and Islam. The Hindus led by the Rajputs, whose attempts to assemble at the Lāṭ had been thwarted by the district magistrate and the army, fell back and regrouped to bar the route of the Muslims who were advancing in a Muharram-like procession with raised standards and crying "Hasan, Husain!" Outnumbered and beaten back by the better armed Hindus at Gāi Ghāṭ, the 7 or 8 thousand weavers retreated leaving behind their dead. To revenge their defeat, the Muslims slaughtered a cow on one of the holiest *ghāṭs* and mingled its blood with the sacred waters of the Gaṅgā and, according to Heber (1828: 429, 431), the sacred well itself was subjected to the same sacrilege.

Having triggered off this irreversible *quid pro quo* and now exulting over their short-lived triumph, the weavers simply went into hiding in their quarters. Such was the horror of the sacrilege that the Hindus, even on receiving the news, would not visit the defiled spot and kept milling around the area of the Viśveśvar temple. The Rajputs, who had already been incensed by the issue of cow-slaughter, counterattacked only on the following day (Robinson 1877: 98–99). They had even begun to demolish the tombs around the Dargah of Fatima, the mother of the Imam Husain, and would have proceeded to do the same to the tomb of Prince Jewan Bukht, held in the highest veneration by the Muslims, had not Mr. Bird checked them in the nick of time. On re-entering the city, Bird saw that "multitudes of armed Hindus were assembled in every quarter directing their rage chiefly against the lives and properties of the weavers and butchers. The Gosains were busy delapidating the [Gyānvāpī Masjid] and had set fire to it. Several bazaars were in flames, and the whole quarter of the Julahars [weavers] was a scene of plunder and violence" (Robinson 1877: 100). The available details on the evolution of the riots suggest a cathartic eruption of self-consuming violence that exploited every possible fissure in the social fabric before falling back to more normal modes of self-regulation. The Hindu-Muslim conflict over the Lāṭ would seem at first sight to reflect, at least in part, the social tension (Robinson 1877: 93) between the low-castes of Muslim weavers and butchers, who initiated the agitation, and the higher Hindu castes grouped around the "aristocratic" Rajputs.

The reconstruction of various Hindu temples, including Viśveśvar, begun by the Rajputs under Akbar's rule was prohibited under Shah Jahan, in whose reign 76 temples under construction were demolished. Then, under Aurangzeb, the Rajputs were reduced to helpless witnesses of the transformation of Bindu Mādhava, Viśveśvar and the Lāṭ Bhairõ into imposing mosques. Regaining political control of the Hindu holy places of Mathurā, Prayāg, Kāśī and Gaya from the Muslims was central to the Maratha strategic alliances with the Nawab of Awadh in the 18th century. Even after Banaras passed under British suzerainty, the Peshwa proposal to rebuild Viśveśvar on its old site after paying compensation to the Muslims did not succeed, not only because the local Brahmans considered the mosque site to be already polluted but also because the English refused. The present Viśvanāth temple was rebuilt by Rani Ahilyabai of Indore in 1777 and modern Banaras is largely a creation of the Marathas, who had no doubt a special pride in the Hindu religious architecture of the sacred city, to whose rebuilding they had contributed so much after 1738 A.D. The period preceding the riots may have seen the reformistic (Wahhābī) wave of Islamic self-consciousness among the weavers coincide with a heightened self-confidence and assertiveness conferred by recent economic prosperity (Pandey 1990: 96–107). Their very readiness to resist the encroachment of Hindu idolatry upon their *īdgāh*—by defiling the Lāṭ in whose cult they had otherwise been participating—suggests that in the first place their own pillar never—or no longer—had for them the same sanctity that it had for the Hindus. It is precisely within a syncretic context—rather than among the orthodox of either faith whose practice is generally segregated from each other—that the growth of such a reformistic consciousness may be expected to extract its first casualties over shared spaces and symbols. The Hindu "memorial" (Robinson 1877: 108) seems to perceive this initial desecration of the Lāṭ not so much as a spontaneous act by the weavers but as instigated by their leaders. Fuming over the suppression of their otherwise customary cow-slaughter and subjected to repeated encroachment, the weavers and butchers had replayed the 1752 march headed by the *kazi* on the Viśvanāth temple, which was the supreme consecration of a re-assertive Hindu polity. In retrospect, whatever the contingent immediate causes of the conflict and the changing politico-economic conditions, a more fundamental logic of religious identity—both denied and exacerbated by the advent of (not just colonial) modernity—seems to have been shaping the long-term relations of the two communities.

The Hindu "memorial" makes counter-claims on the sacred sites of the city (Robinson 1877: 106). The *īdgāh* here had no particular sanctity but was esteemed by the Muslims only because it marked the former ascendancy of Islam over the religion of the Hindus, whereas the Kapālamocana tank and (what was left of) the Lāṭ was of the highest sanctity to the Hindus. The district magistrate (now Mr. Watson) hence proposed to hand over the whole site to the Hindus as part of an overall policy of separating the two communities—to prevent future clashes—even at the price of totally excluding one or the other at disputed sites like Viśveśvar. However, Mr. Bird opined that both Hindus and Muslims had suffered so severely that neither would molest the other. "Government adopted his counsels and no alteration whatever was made to the original position of the parties. Permission was given to both alike to repair damages, and according to their respective religious customs each purified their violated altars. The Hindus held high ceremonies, and with prayers and Ganges water the fragments of the Lāṭ were restored to their original sanctity and reverently buried" (Robinson 1877: 106). But it was not until June 1810, "when the Hindus reconsecrated their outraged shrines and the veneration paid to the original pillar was transferred to the mutilated relic, that the first riot can be said to have actually concluded" (Robinson 1877: 102).

But even as "the original disturbances marked only by shocking religious outrages had subsided in June 1810 [...] a singular feud had sprung up between the military and the police. The result was long succession of petty affrays, but also a fortunate diversion of the popular attention from religious matters. The sepoys carried on a guerrilla warfare in the streets of the city against the police, and in either body Hindus and [Muslims] were indiscriminately mingled" (Robinson 1875: 92). The sepoys had not only persistently defied a magisterial order against the carrying of arms in Banaras, but also ridiculed the police for their earlier role, thus leading to a long succession of affrays in August and September 1810 (Pandey 1990: 40).

Human conflict, it would seem, has a logic of its own whereby innate "irrational" impulses readily find "rational" causes to justify their expression before the violence consumes itself, is diverted into more manageable channels, and/or surrenders its destructive energy to other constructive social imperatives. On the other hand, a religious outlook predicated on non-violence could just as well inhibit the recourse to deadly force by those whose authority, like that of the Brahmans, is otherwise most likely to be undermined by such external provocation. Taken together, the flow of events in Banaras from late 1809 to early 1811, pose a problematic of human violence that can be reduced neither to blind religious frenzy nor to rational concerted action. By dissolving the religious divide between Hindu and Muslim, the police-military skirmish prolonged the violence even while diverting it into a fresh context that refocused public attention on social issues that could be more readily addressed through political means and administrative action. The containment of violence through its ritual enactment, resorted to internally by both Hindu and Muslim societies, seems to have re-emerged here, spontaneously, before "normalcy" could be restored to the social fabric.

House Tax Revolt of 1811: Paradigm for Hindu-Muslim Reconciliation and Concerted Action?

Even before the city had been quietened down through a partial reorganization of the city police in October 1810, "the House Tax Regulation (XV of 1810) had been extended to Banaras, and from the ashes of the sepoy-police agitation, the phoenix of riot rose in all its original strength", so much so that "the 10th of January 1811 found Banaras, as the 1st of January 1810 had found it, seething with clamorous mobs and troops holding the city" (Robinson 1877: 92). In the wake of measures to ensure greater control over collective gatherings, this ill-advised house tax was introduced in December 1810 to provide greater revenues for funding the expanded state apparatus. The Banarasis protested that this additional burden would make it impossible for them to continue with their own charitable practices of maintaining numerous traditional institutions for the widows, Brahmans, mendicants and the poor who filled the sacred city. They were even willing abandon their city for good rather than submit to this violation of the basic principles on which it had been established (Freitag 1989c: 43–50). Most unpopular both for its amount and its novelty (Heber 1828: 432–6), the proposed tax provoked the whole population of Banaras and its neighbourhood to sit in the passive resistance of a unanimous self-mortifying "strike" (*dharna*). The magistrate, Bird, noted that "men of all classes and description, from the highest to the lowest, whether [Muslims] or [Hindus], [Julahars], [Rajputs] and [Gosains] included, were all of one mind, and engaged by oath to promote the common cause". The "religious orders" including "men of rank and respectability" exerted their full influence in favor of the agitation. Heber even wrote of religious tracts (*dharma-patris*) "issued by the leading Brahmans as being central to the process of mobilizing the people" (Pandey 1990: 41ff.). One British official observed that "instead of appearing like a tumultuous and disorderly mob, the vast multitudes came forth in a state of perfect organization: each caste, trade and profession occupied a distinct spot of ground, and was regulated in all its acts by the orders of its own [pañcāyat—traditional 'deliberative assembly']" (Pandey 1990: 47, 197).

The British did not attempt to disperse the strikers nor the subsequent (abortive) march on the Governor at Calcutta, nor did they immediately yield to the civil pressure. Once the rebellion had exhausted its momentum, however, they quietly repealed "the obnoxious tax, and thus ended a disturbance which, if it had been harshly or improperly managed, might have put all India in a flame" (Heber 1828: 436). This civil disobedience from the grassroots, indeed, became "an important object lesson for Banaras and other dissident towns" for "similar events occurred during 1815–17 in other urban areas when the British attempted to introduce a more general regulation for [the 'watchman'] tax" (Bayly 1983: 320, 21). Even where such protest eventually took on a violent turn, as in Bareilly in 1816, the "resistance of the mob at Banaras [was] in everyone's mouth" (Bayly 1983: 324). The crucial development, which seems to have escaped the otherwise exemplary impartiality of the British reporters, is that the Hindus and Muslims of Banaras were now united in opposing their economic exploitation and political domination by foreigners.

The Felling of the World Pillar: An Islamic Fulfillment of Vedic Cosmogony?

The "pillar of light" (*quṭb minār*) had such a hold over the politico-religious imagination of a succession of Indian Muslim rulers—ranging from Quṭb-ud-din Aibak, Alauddin Khilji and especially Firoz Shah Tughluq to Akbar—that they sought to concretize it visibly before their Indian subjects in various architectural forms often annexed from or imposed upon pre-existing Hindu-Buddhist monuments (Irwin 1986, 1987, 1989; Visuvalingam 1992: 50–51). His Hindu mother and his awareness of the cosmogonic significance of the Indian pillar-cult notwithstanding, the iconoclastic Feroz Shah would have had sound Islamic justification for (re-)erecting a(n even Ashokan) pillar at the *īdgāh* at Banaras, so long as the monument was not treated as a divinity in itself. In some of the Shiite traditions, moreover, the link between God and the Imams is visualized as being a pillar of light descending from heaven upon the (typically martyred) Imam, which only serves to identify his station even further with the "*axis mundi*" or "pole" (Quṭb), the Perfect Man (al-Insān al-Kāmil) of (Sunnite) Sufism. The inevitability of the socio-religious confrontation hence did not preclude—from the very beginning—a certain complicity between Hinduism and Islam in the symbolic interpretation of the violence to which the Lāṭ was subjected.

After all, the Muslim lower-castes had connived at the Hindu worship of the world-pillar, participated in celebrating its marriage and even claimed it as their own, so much so that the Banaras myths of Ghāzī Miyā̃ reflect as profound an understanding of its function as the Hindu theologem of the "punishment of Bhairava" (*bhairavī-yātanā*). For the Hindu mythico-history, on the other hand, the leveling of the Lāṭ was as inevitable as the Kali Yuga, which would be redeemed only by a "barbarian" (*mleccha*) messiah (Kalki), a role which was readily fulfilled for certain Indian (especially Bengali) Muslim innovators by the Prophet Muhammad. More than just the tendency of Sunnis and Shias to close ranks within a single community (*umma*), the Indian cult of Ghāzī Miyā̃ represents the symbolic implantation of this egalitarian Islamic ideal within the heart of (not only popular) Hinduism. Though interrupting this process of syncretic assimilation at the folk level, even the spread of the iconoclastic Wahhābī ethos—which cannot be judged in terms of the mere numbers of its adherents (Kumar 1989: 162–3) nor be reduced to its Arabian trappings—thus tends in its own way to transform the Indianized symbol into a universal social reality.

The chaotic birth-pangs of a new order based on the abolition of the caste-system were already being jointly rehearsed by both Hindus and Muslims during the festivals of Ghāzī Miyā̃ and Muharram all over India, and by the Hindus themselves in their own festivals both before and after the arrival of the Muslims. Such festivals could easily be (re-)interpreted as an exteriorization of the (temporary but) necessary abolition of caste-distinctions within closed Tantric circles, as in the esoteric Kaula cults of Bhairava, whose leading theoreticians were all Brahmans like Abhinavagupta. From the Hindu perspective, the Muslims were merely guilty of "hastening or forcing the end". Faced with the *fait accompli* however, the Hindu "memorial" simply translated the event into a re-enactment of a sacrificial embryogony: "it has been ascertained that the Lāṭ notwithstanding all these attempts, did not fall till they sprinkled it with the blood of a cow and her young, which they got from a [garden] and dragged, tied by the neck to the spot. On this outrage the [capital] on the [revered Bhairõ Lāṭ] spun round and tumbled and the Lāṭ burst and fell to the ground. They cast the cow which they had slaughtered into the tank of [Kapālamocana] which is near the Lāṭ and completely defiled it" (Robinson 1877: 109). And like the fallen pole of the Indra festival, the Lāṭ itself is said to have been thrown into the Gaṅgā about half a mile away, whereas the physical probability is that the sandstone largely crumbled under the heat of the fire (Sherring 1868: 191, 306). Since the issue of cow-slaughter which was propagandized by Ārya Samāj inspired Cow-Protection Societies was subsequently destined to play a central role in "mobilizing the Hindu community" (Pandey 1990: 158–200) against the Muslim minority, it is pertinent to recall how ambivalently this central theme was handled not only by the Banaras orthodoxy but also by Hindu folklore. The leaflets urging all Hindus to take concerted action repeatedly assimilated cow-slaughter to the horrible act of matricide, incest and "marrying one's mother to a Muslim" (Pandey 1990: 185) represented here in purely negative terms. However, the same "sacrilege" is instigated with wholly positive results by Mallanna, alias Malkhān, who is identified with the solar (Mārtaṇḍa-)Bhairava himself, and worshipped by both Hindus and Muslims in the Deccan (Visuvalingam 1989: 452).

Considering that Lāṭ Bhairõ had always been the "scapegoat" of Lord Viśvanāth by taking on the latter's impurities (the same that the caste-Hindu inevitably incurred through processes such as birth and death), it is more than just a curious coincidence that the Muslim mob that set out on a Muharram-like procession to desecrate the nuclear temple of the royal divinity ended up felling the pillar instead on their way back after the foiled attempt. Not only did these weavers participate in the marriage of the Lāṭ, their fateful politico-religious action was (pre-)inscribed within the ritual space connecting the dual foci of the Hindu cosmogony, channeling their aggressivity willy-nilly into the unwittingly prescribed role of "sacred executioner" that had always been the transgressive prerogative of Bhairava, the reason for his very birth. This was followed by an eruption of the very violence that the sacrificial mechanism was supposed to contain into a conflagration that well-nigh consumed the entire city and seems to have provided the occasion for settling scores even between co-religionists. This violence, whose religious dimension had earlier compromised the policemen who were responsible for maintaining civil order, broke loose before long from its religious confines, narrowly defined, betraying its quasi-ritualized dualistic pattern in the conflict between the police and military, with Hindus and Muslims standing shoulder-to-shoulder on either side, even as its fury was consuming itself out. Yet, the self-consciously non-violent city-wide demonstration against colonial rule was just as much inspired by religious ideals—the defense of Brahmans and fakirs—and had already revealed itself not only in the non-participation of certain castes (beginning with the Brahmans and the Hindu Maharaja himself) in the earlier riots but also in individual acts of principled courage like the Brahman's bodily intervention to save the infant of the Muslim cleric of the Aurangzeb mosque. For their part, the highly caste-conscious Muslim weavers chose to follow the example of the Brahmans and the (earlier violent)

Gosains who had occupied the *ghāṭs* of the Ganges whose holiness, and that of the sacred city, they had themselves defiled in quite recent memory by polluting it with the blood of a cow slaughtered for the very purpose. What we seem to be witnessing is the ability of internalized religious values to operate, depending on the changing social circumstances, in opposing ways: by reinforcing predefined boundaries between Self and the Other or by transgressing these boundaries, especially in contexts where core values are not being compromised, to make common cause with those same others. The recent history of genocide and ethnic violence bears eloquent testimony that a secular (or "profane") world, supposedly impervious by right to all religious consideration, is no less vulnerable to exclusionary violence that often flows into stereotyped "ritual" moulds. The post-modern challenge is not so much the suspension of religious identity (or at least its quarantining from socio-political processes), but rather the unleashing of the (re-) constructive potential of each tradition and its harnessing within a shared non-denominational civic space. The challenge and responsibility of serious scholarship, it seems to us, is to provide the sort of ethno-historical analyses that would facilitate such a therapeutic process by appealing to the devout of all hues and also to the well-intentioned skeptic entrenched in a "secular" faith.

Conclusion: Visualizing Space and (Re-)Constructing Identity

The encounter between Islam and Hinduism has also served as a pretext, in the present paper, for exploring the articulation between two principles in the theory of religion: how the reigning idea-value-intention encoded into the religious history of a community could eventually outgrow, challenge and survive the restrictive injunctions and prohibitions that had earlier been its indispensable supports, and how the conceptual recovery of this transgressive dimension within the various traditions could help extricate their respective approaches to "*communitas*" from the increasingly reified symbolic structures within which they have been till now contained (Visuvalingam 1992). Together, they could help us isolate and formulate a common and not so obvious principle of self-determination that seems to emerge within the field of the complex social interaction of Hindu hierarchy and Islamic egalitarianism with each other and with the encroaching (colonial regime as the carrier of) modernity. Shifting the focus to this principle of spiritual-cum-political autonomy could serve as a bridge to reconcile the otherwise conflicting traditions and eventually even facilitate their self-transformation.

Unlike the secular ideologies of today the socio-religious paradigms simultaneously encode a commitment to certain transcendental and even esoteric aims which address themselves to universal human aspirations. The identification of death and sexual union, especially as revealed in the syncretic cults of Lāṭ Bhairõ and Ghāzī Miyā̃, is very much a part of this common symbolic core shared by Hinduism and Islam. Ultimately, this equation does not make sense except as the mythico-ritual projection of a lived experience of "initiatic death" that has come to grips with and inwardly transformed those primary, largely unconscious, undifferentiated energies that are channeled into otherwise structured expressions of human sexuality and violence.

These scenarios of the sacred thus also reveal the shared concern for containing and productively re-directing an innate human violence. But the differing requirements of a universalizing polity unified around an egalitarian ideal, and of a hierarchic society, based on the opposition of the pure and the impure and strained by regional, inter-caste and sectarian tensions, resulted in a divergence of emphases in the shared sacrificial paradigm. Primarily directed outward through holy war (*jihad*) in propagating the monotheistic creed which encoded the universalizing egalitarianism of the Islamic polity, violence in Hindu society was instead repressed through the ascetic-brahmanical ideal (*ahiṃsā*) only to be ultimately turned back upon itself through (the multiple transpositions of) a Vedic paradigm centred on the identification of killer and victim in an (at least symbolic) act of human sacrifice. The difficulty of subordinating the inherent logic of human violence to even religious constructions of such power and magnitude is revealed in the constant resurgence of a "primitive" dualistic pattern which tends to feed upon and ritualize other sources of (economic, political, clan-based, etc.) tension within both these societies. Even the global conflict between Hinduism and Islam on the doctrinal level has been undermined, distorted and compounded by this dualistic deep-structure, by the fact that neither tradition has been wholly successful in domesticating violence for its own purposes. It could however be argued that, without such mechanisms of relative control offered by the domain of the sacred, the increasingly global society of today runs an even greater risk of succumbing to a generalization of violence.

Well before the arrival of Islam, there was already in place an extensive mapping of Hindu sacred geography, with its intricate network of pilgrimage routes, that had sought to construct the (more than merely) symbolic unity of India (Bayly 1983: 125–44). Like the daily *namāz* in the direction of Mecca, the centrality of Banaras to this all-powerful construction of the Hindu religious imagination has been encoded even into the initiation rites, the ceremonies of marriage and the funerary practices of Hindus from Nepal to South India. Conversely, the urban political economy, demographic patterns and architectural landscape of Banaras was intimately related to its undisputed status as the major pilgrimage centre of India and integral to the subcontinent's economic unity. The holy city is the pulsating heart of a distinctively brahmanical "consciousness" which circulates out through the arteries of the commercial routes into the remotest villages and increasingly informs the outlook of an otherwise heterogeneous religious culture. As the home of a religious elite with various means of directly shaping popular values, the sacred centre is the dynamic catalyst in a socializing process that can continue to function even independently of state patronage: a socializing process which (aspiring) Caliphs and Rajas could harness and promote for their own political ends, or defy only at the risk of de-legitimizing their own rule. Pilgrimage, as a "performative act" that (perpetually re-) created, sustained and locally reduplicated the symbolic centrality of Banaras and Mecca, was also fraught with political implications. "The British were extremely sensitive to the fact that Varanasi was considered a sacred city. With the constant coming and going of numerous pilgrims, the news of any excessively violent repressive actions on the part of the British was sure to be widely circulated and resented" (Heitler 1972: 251). Repealed in late 1811, the House Tax "which might have put all India in a flame" was re-imposed in 1813 in a modified form on Dacca, Patna and Murshidabad: "not until 1860 did the British dare to institute a house tax in Banaras" (Freitag 1989c: 50 n.87; Heitler 1972: 244). It was precisely this recognition of its centrality to Hindu civilization that impelled Aurangzeb's earlier politico-religious move to impose "Muhammadabad" on Banaras, and the Meccan-orientation of the Jama Masjid on the temple of Viśveśvar. The Lāṭ Bhairõ riots remain significant because their symbolic focus—their occurrence in the heart of Hindu tradition, the centrality of the pillar to the sacred significance of Banaras, the interpenetration in its cult of Hinduism and Islam and, especially, the re-inscription of its historical felling into a politico-religious mythology—obliges and still allows us to address fundamental issues. An adequate deciphering of the semiotics of space, identity and collective aspiration embodied by this once fiercely contested "pillar of the world" might, hopefully, better equip us to address today other more pressing questions, fraught with international geo-political implications, such as posed by that other sacred city par excellence: Jerusalem.

Notes

1 By this term we mean the systematic attempt to understand how the inhabitants of a given territory traditionally visualized their "living space", as opposed to the (already floundering) attempts of contemporary "geo-political" discourse on (the boundaries of) the nation-state to strip, or at least ignore, the symbolic properties of this "Lebensraum".

2 The field-work was carried out in 1984–89 with the help of Om Prakash Sharma to whom we express our gratitude. We likewise thank Dr. Niels Gutschow for providing all the maps, plotted with characteristic cartographic precision, for this paper. For more details on and analyses of Ghāzī Miyã and Muharram (1992), dualistic conflict as a universal paradigm ("Divide, Rule and Unify"), and the Lāṭ Bhairõ riots ("Hindu-Muslim Relations"), see references under Visuvalingam.

3 The cosmogonic significance of the cult of pillars and poles in South Asia and elsewhere first came to our attention with the visit of John Irwin (1983, etc.) to Banaras to complete his research on the Lāṭ. A clear résumé [partly by the editor] of the contents of his various papers may be found in Irwin (1990). Limitations of space have prevented the detailed treatment of not only the successive relocations of Hindu sites, after the Muslim occupation, in the sacred geography of Banaras, but also the properly Buddhist aspects of the pillar and its cult.

4 This is how the official records refer to the separate petitions, each narrating its own version of the underlying causes and salient events of the riots, submitted by the two aggrieved parties to the British authorities.

5 See Miriam Abou-Zahab "La conception du martyr chez les *mujāhidīn* pakistanais au Cachemire", talk given on 10th March 2003 to Marc Gaborieau's seminar series on Indian Islam at the Centre for Indian Studies in Paris.

6 Ritualized "sectarian" conflict (Hanafites versus Shafi'ites, Ni'mati versus Haydari, etc.) would re-surface in Iranian cities especially during the tenth of Muharram (Visuvalingam, "Divide, Rule and Unify").

7 Neither Freitag nor Pandey seem to have been aware of Robinson's report based on first-hand accounts that John Irwin (1983) had sent to us in Banaras from the India Office Library.

8 Scholars who insist that "communalism" (as they define it) is only a recent "construction" may note that exactly the same sequence of symbolic encroachments and litigation is at the root of the current polarization of the Hindu and Muslim communities over the Babri Masjid versus Ram-Mandir dispute.

References

Bayly, Christopher Alan. 1983. *Rulers, Townsmen, and Bazaars: North Indian Society in the Age of British Expansion, 1770–1870*. New York: Cambridge University. Press.

— 1985. The Prehistory of "Communalism"? Religious Conflict in India, 1700–1860, *Modern Asian Studies* 19, 2: 177–203.

Chalier-Visuvalingam, Elizabeth. 1986. Bhairava: Kotwāl of Vārāṇasī. In: T. P. Verma et al. (ed.), *Vārāṇasī through the Ages*. Varanasi: Bharatiya Itihas Sankalan Samiti, pp. 241–60.

— 1989. Bhairava's Royal Brahmanicide: The Problem of the Mahābrāhmaṇa. In: Alf Hiltebeitel (ed.), *Criminal Gods and Demon Devotees*. Albany: SUNY Press, pp. 157–229.

— 1994a. *Terreur et Protection: Bhairava à Banaras et à Katmandou. Étude des mythes, des rites et des fêtes*. Thèse de Doctorat d'État, Paris: Université de Paris X.

— 1994b. Union and Unity in Hindu Tantrism. In: Hananya Goodman (ed.), *Between Jerusalem and Banaras*, Albany: SUNY.

— 1991. Le Roi et le Jardinier: Pacali Bhairab de Katmandou. Manuscript, incorporated in Visuvalingam (2003). (*www.svabhinava.org/pacali/default.htm*).

— 1996. (with S. Visuvalingam) Bhairava and the Goddess: Tradition, Gender and Transgression. In: Axel Michaels et al. (eds.), *Wild Goddesses in India and Nepal*. Bern: Peter Lang, pp. 253–301.

— 2003. Bhairava: Terreur et Protection. Mythes, rites et fêtes à Bénarès et à Katmandou. Series: Dieux, Hommes et Religions. Bruxelles: Peter Lang. D.Litt. thesis (1994a) reworked and reduced for publication.

Eck, Diana. 1983. *Banaras: City of Light*. London: Routledge and Kegan Paul.

Freitag, Sandria B. 1989a. Introduction: The History and Political Economy of Banaras. In: Sandria B. Freitag (ed.), *Culture and Power in Banaras: Community, Performance and Environment 1800–1980*. Delhi: Oxford University Press, pp. 1–22.

— 1989b. State and Community: Symbolic Popular Protest in Banaras's Public Arenas. In: Sandria B. Freitag (ed.), *Culture and Power in Banaras: Community, Performance and Environment 1800–1980*. Delhi: Oxford University Press, pp. 203–228.

— 1989c. *Collective Action and Community: Public Arenas and the Emergence of Communalism in North India*. Berkeley: University of California Press.

Gaborieau, Marc. 1975. Légende et Culte du Saint Muslim Ghāzī Miyã au Népal Occidental et en Inde du Nord, *Objets et Mondes* 15, 3: 289–318.

Heber, Bishop Reginald. 1828. *Narrative of a Journey through the Upper Provinces of India from Calcutta to Bombay (1824–1825)*. 3 vols. London: John Murray.

Heitler, Richard. 1972. The Varanasi House Tax Hartal of 1810–11, *Indian Economic and Social History Review* 9, 3: 239–257.

Irwin, John. 1983. The Lāṭ Bhairõ at Banaras (Varanasi): Another Pre-Ashokan Monument? *Zeitschrift der Deutschen Morgenländischen Gesellschaft* 133: 321–352.

— 1986. Akbar and the Cosmic Pillar. In R. W. Skelton et. al. (eds.), *Facets of Indian Art*. London, pp. 47–50.
— 1987. Islam and the Cosmic Pillar. In: *Dedicated to Herbert Härtel: Investigating Indian Art*. Berlin: Museum für indische Kunst, pp. 133–45.
— 1989. Islam and the Cosmic Pillar [another version of the same paper]. In: K. Frifelt & P. Sorensen (eds.), *Scandinavian Institute of Asian Studies, Occasional Papers 4, Papers from the Eighth International Conference of South Asian Archaeologists in Western Europe, held at the Moesgaard Museum, Denmark, 1–5 July 1985*. London: Curzon Press and USA: Riverdale Co., pp. 397–406.
— 1990. Origins of Form and Structure in Monumental Art. In: Karel Werner (ed.), *Symbols in Art and Religion: The Indian and the Comparative Perspectives*. London: Curzon Press, pp. 46–67.
Kāśī Khaṇḍa (Skanda Purāṇa). 1961 ed. (in Sanskrit) Gurumandala Granthamalaya N. XX. Vol. IV. Calcutta.
Kumar, Nita. 1988. *The Artisans of Banaras: Popular Culture and Identity, 1880–1986*. Princeton: University Press.
— 1989. Work and Leisure in the Formation of Identity: Muslim Weavers in a Hindu City. In: Sandria B. Freitag (ed.), *Culture and Power in Banaras: Community, Performance and Environment 1800–1980*. Delhi: Oxford University Press, pp. 147–170.
Pandey, Gyanendra. 1990. *The Construction of Communalism in Colonial North India*. Delhi: Oxford University Press.
Robinson, Phil. 1877. The Banaras riots of 1809–1811, *Calcutta Review* 65: 92–119.
Sherring, M. C. 1868. *The Sacred City of the Hindus: An Account of Banaras in Ancient and Modern Times*. London: Trubner and Co.
Sukul, Kubernath. 1974. *Varanasi Down the Ages*. Patna: Kameshwar Nath Sukul.
— 1977. *Varanasi Vaibhava* (in Hindi). Patna: Bihar Rastrabhasa Parisad.
Upādhyāya, Baikuṇṭanātha. samvat 2028. *Śrī Kāśī-Khaṇḍ (adhyāya 30 va 31): Śrī Kāla Bhairava, Śrī Kapāla Bhairava aur Kapālamocana Tīrth* (in Hindi). Varanasi: Shri Bhrgu Prakashana.
Visuvalingam, Sunthar. 1989. The Transgressive Sacrality of the Dīksita: Sacrifice, Bhakti and Criminality in the Hindu Tradition. In: Alf Hiltebeitel (ed.), *Criminal Gods and Demon Devotees*. Albany: SUNY Press, pp. 427–62.
— 1992. Between Mecca and Banaras: Towards an Acculturation Model of Hindu-Muslim Relations, *Islam and the Modern Age* 24, 1: 20–62.
— 2003. (with Chalier-Visuvalingam). A Paradigm of Hindu-Buddhist Relations: Pacali Bhairava of Kathmandu. *Evam – Forum on Indian Representations* 3: 106-170. New Delhi.
— Hindu-Muslim Relations in Colonial Banaras. (Manuscript). (*www.svabhinava.org/1809riot/default.htm*).
— The Marriage of Lāṭ Bhairõ: The Untold Tale of Sex and Violence in Hinduism. (Manuscript).
— Divide, Rule and Unify: Religious Dualism and the Dialectics of Human Violence. (Manuscript).

Maps

AXEL MICHAELS

Mapping the Religious and Religious Maps: Aspects of Transcendence and Translocality in Two Maps of Varanasi

Introduction

Mapping is "visualizing, conceptualizing, recording, representing and creating space graphically" (Cosgrove 1999: 1). It leads to a number of complexities and uncertainties which seem to be overcome by criteria of "representational accuracy, formal clarity and ease of use".[1] However, map makers or cartographers who want to express religious spaces face a special problem: How can religious notions such as merit, grace, transcendency or liberation be represented on maps? How is the religious to be depicted? How can the infinite be limited and located? How is Vaikuṇṭha or Kailāsa to be mapped (cp. Scafi 1999)?

It is generally accepted that maps are extracts of a totality of space based on conventions (and even fashions) as well as on cartoliteracy: symbols, signs, legends, must be understood by the users of maps as reductive representations of another, spatially broader and wider reality. Moreover, maps are themselves a culturally determined medium: From the early studies of Jomard and von Humboldt, who sought through facsimile reproduction and critical philology to bring to scholarly attention the role the maps had played in Europe's progressive "discovery" of the globe and organization of its varied spaces, to the critical collections of historical maps by scholars and archivists such as A. E. Nordenskiöld and R. Almagia, a history of cartography was constructed over the course of the nineteenth and early twentieth centuries. Governed by Enlightenment beliefs in disinterested observation, scientific calculation and objective representation as markers of universal rationality and progress. Mapping was figured as a form of literacy, a sign of civilization.; it was thus restricted to a fairly closely defined form of knowledge collation and archiving (Cosgrove 1999: 8–9).

The early mapping of the religious world in Varanasi[2] was certainly far removed from such "Enlightment beliefs in disinterested observation, scientific calculation and objective representation as markers of universal rationality and progress". It struggled with different questions: How can omni-, trans- or polylocality that holy places often demonstrate be represented on maps? As Diana Eck (1983: 39) aptly remarked, Hinduism does not claim singularity. On the contrary, notions of omnipresence and transcendency or translocality are perhaps much stronger in Hinduism than in other religions. This raises a major problem for map makers: Which visual form is appropriate for the representation of omnipresence? In a world, where the gods can manifest themselves almost anywhere, maps could be an inappropriate reduction of the holy.

As it is well-known, Kāśī (as well as other holy places in India and abroad) claims that the totality of space, i.e. the universe or all worlds, *is* represented at any spot or territory of its encircled holy fields. The so-called Pañcakrośī temple which has been studied by Niels Gutschow and Rana P. B. Singh is an apt example for such a claim: All deities and all *tīrthas* are represented in the niches of its walls. Through such representations the view is expressed that a *tīrtha* is both in this world as well as in other world(s), in short: a crossing.

However, such claims imply a number of formal problems with regard to the basic features of maps (cp. Cosgrove 1999: 9ff.), i.e. (a) scale, (b) framing, (c) selection and (d) coding, to which I would add (e) centrality (Michaels 2000). For, the question arises: should a map have a centre? If so, where should it be? Which deity should occupy it?

All these questions lead to certain patterns of condensation and extension and to methods of selections, additions and omissions that have to be analyzed in order to understand the specific spatial representations better. I will demonstrate this by comparing two maps of Varanasi from second half of the 19th century: a map called *Kāśīdarpaṇa* of Kailāsanātha Sukula (1876) and a map *Kāśīdarpaṇapūrti* of Kṛṣṇacandra Śarmā (1877).

The Maps of Kailāsanātha Sukula and Kṛṣṇacandra Śarmā

Kailāsanātha Sukula's *Kāśīdarpaṇa*

In the year 1876 A.D. Kailāsanātha Sukula, father of the historiographer Kuberanātha Sukula (Sukula 1977), published the map *Kāśīdarpaṇa*, "Mirror of Kāśī" (Fig. 1). Kailāsanātha was the author of the map, but he is not its cartographer who remains unmentioned. The map was printed with three or four block prints probably on cotton in the Vidyodaya Press (Mudrālaya) of Ḍhuṃḍhirāja Śāstrī. It was published on the 3rd day of the bright half of Bhādrapada 1933 V.S. and registered according to a publishing act of 1867. Given the above mentioned criteria, Kailāsanātha Sukula's map which has been studied and published by several scholars[3] shows the following features:

1. *Scale*: It is difficult to determine the scale of the map. In a way it presents a bird's eye view of Benares, the Ganges being disproportionally large.

2. *Framing*: The map has the form of a circle or *maṇḍala* and encircles an imagined holy field of Varanasi, the *kāśīkṣetra*. The frame of the circle separates the inside from the outside, depicting a mandalic model of Kāśī. The outer ring corresponds to the *pañcakrośī* which is mentioned in the North and the East. Beyond the circle, 69 Śaktis are mentioned which are related to the eight directions, as well as 8 Bhairavas and 8 Durgās of which, however, only one is mentioned, and (8 x 8 =) 64 Vetālas, of which three are mentioned and five are not explicitly listed. Inside of the circle is a mixture of

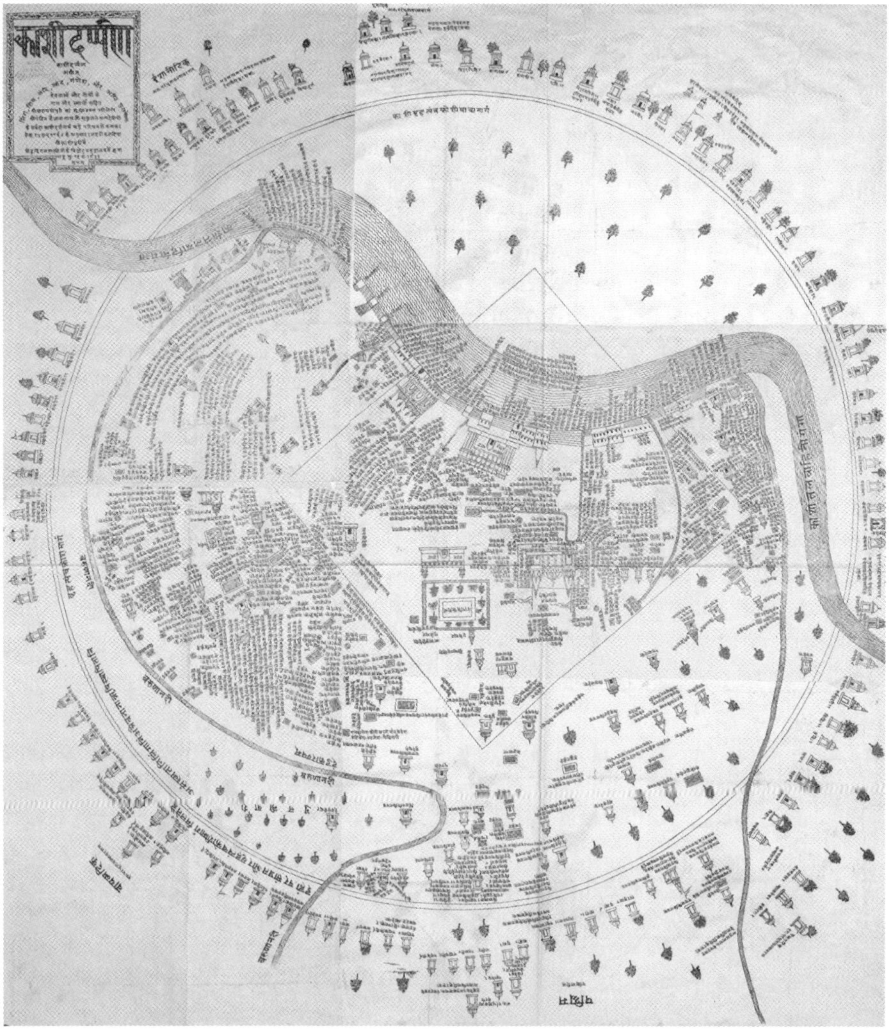

Fig. 1: *Kāśīdarpaṇa* by Kailāsanātha Sukula (1876).

water places called *tīrthas, kuṇḍas, kūpas, vāpīs, sāgara, hrada, liṅgas*, sun and moon (*āditya, candra*), Gaṇeśas, demonic beings (*gaṇa* etc.).

3. *Centrality*: However, it is difficult do say what the centre of the map could be. The centre of the procession and the map is certainly not the Viśvanātha, nor is it the Madhyameśvara, approx. 1.5 km north of the Viśvanātha temple. We still do not know the significance of the square within the map.

4. *Selection*: The listing and naming of the deities registered on Sukul's map basically follow the *Kāśīkhaṇḍa* (72.3–13: Śakti, 72.93: Bhairava; 72.90–91: Durgā, 72.97–100: Vetāla, of which the legend only lists 24 calling the unmentioned *anukta*).

As the electronic Heidelberg index of the map (*www.benares.uni-hd.de/map.htm*) shows, almost each *tīrtha* of the map is listed in the *Kāśīkhaṇḍa*, often in the same sequence. However, it is also important to see what has been selected. Apparently the focus is on śaivite deities, *liṅgas*, goddesses, water places *ghāṭs*.

5. *Coding*: The map is not orientated towards the North but to the East. If you take the colophon as the main indicator of how to hang the map, the Ganges is on the top thus suggesting to the "normal" users of maps that the river flows from "West" to "East". However, at the sides of the map the geomagnetic directions are mentioned, which is somehow a different level of spatial orientation. Only very few topographic indications are registered. Not every holy place has a temple symbol, the architectural pictures mainly concern the *ghāṭs* a few temples and the town hall. There is no legend, i.e. explanation of conventional signs or methods of coding used. What is mentioned is basically written down in words.

Thus, the *Kāśīdarpaṇa* mirrors and reduces a written text, i.e. the *Kāśīkhaṇḍa*. The cartographic representation clearly follows a model derived from the text. In other words, the map does not represent what appears to the eye in "reality" but what is imagined by a reader of the *Kāśīkhaṇḍa*. It is a graphical representation of by and large mythical places, therefore a thematic map, highlighting the Pañcakrośīyātrā and *kāśī-kṣetra* in an idealistic model. The map clearly follows religious ideas and it is certainly not designed for orientation in space.

Kṛṣṇacandra Śarmā's *Kāśīdarpaṇapūrti*

Interestingly, only one year later (1877 A.D.) another map was published by the same publisher which is much more directed to topographical details although the map is similar in its shape, scope and aims: i.e. the map *Kāśīdarpaṇapūrti* designed by Dharmādhikārī Paṇḍita Kṛṣṇacandra Śarmā (Fig. 2–4).[4] Taking again the above mentioned formal criteria, this map is to be characterized as follows:[5]

1. *Scale*: Kṛṣṇacandra's map has three parts (*aṅka*). The first part basically concerns the Pañcakrośīkṣetra and its circumambulation, the second part the Nagarapradakṣiṇā, the third the Antargṛhayātrā. Part 1 uses a grid. The map clearly wants to map what appears to one's eyes in the field. It is not only concerned with mythical imaginations of Kāśī, but also with problems of measuring. It is mentioned in the colophon, that Kṛṣṇacandra Śarmā took the help of a certain Rāmakṛṣṇa Dīkṣita Gor who showed him the various deities at the spot and explained him the various *tīrthas*. Moreover, the processions are indicated in its topographical routes (*mārga*: not *yātrā*!). Finally, it is explicitly said that distances were accurately measured according to English maps (*aṃgrejī nakṣā ke anusār*: legend of *aṅka* 2). Even mistakes are mentioned:

> (Legend on the top of the circle of Part 1:) *So yaha mārga aśuddha hai. kāraṇa kī isa mārga ke anusāra sūtra ghumāvane se kṣetra sīmā devatā Bhīmacaṃḍī kṣetra se bāhara ho jātīṃ haiṃ. isaliye Pāṭhaka Svāmī ne nāpa ke jo mārga nikālā haiṃ vahī śuddha hai.vahī upara likhā hai vahī Madhyameśvara se pāṃca kosa ṭhīka milatā haiṃ. vartamāna mārga kama hotā hai. Mahādeva Bhaṭṭa ne aṭakala se calāyā thā. īskā khulāsā svāmī ne nāpa au śāstrārtha sahita chapavāyā hai so dekha lenā// cha//*

This line of the present route [in the map] is not correct. Because, [if we measure the area] circling with a rope round according to the [distant of the] route [from Madhyameśvara], the area of goddess Bhīmacaṇḍī remains beyond [the area of the measurement]. Therefore, the route which was discovered by Pāṭhaka Svāmī after measuring it, is indeed true. This is exactly five *kosa* (approx. 10 miles) from Madhyameśvara which has been written above. But the present route [indicated by the dots on the map] is shorter [than the Pāṭhaka Svāmī's measurement]. Mādhava Bhaṭṭa [might have] introduced it by his guess. This fact has [already] been published by the Svāmī after measuring and debating (*śāstrārtha*) [on this matter]. [Please] read it.

2. *Frame*: Kṛṣṇacandra's map also uses the mandalic form and he makes it explicit in the legend of Part 1:

(Legend of Part 1, §1:) *śruti smṛti purāṇa ādi me prasiddha kṣetra jo bhavasāgara ke pāra utarane kā sādhana Kāśīnāma se vikhyāta hai | usakā ākāra gola maṃḍalākāra hai |*
The region, famous in *śruti, smṛti* and *purāṇa*, which rescues [any person] from the sea of worldly existence, is quite famous in the name of Kāśī. Its shape is round [and] as a *maṇḍala*.

3. *Centrality*: The map declares the Madhyameśvara as the centre of the map for it continues in the legend:

(cont.): *vaha maṃḍala Madhyameśvara ko kendra mānake pāṃcakosa ke vyāsārddha se rekhā ghumāvane se hotā hai | isa kṣetra ke bhītara eka ke bhītara dūsarā ityādi krama se tīna kṣetra aura haiṃ jina ko nāma Vārāṇasī Avimukta aura Aṃtargṛha haiṃ | ina sava kṣetroṃ ke ākāra aura Varaṇā se Asī taka Gaṃgā ke pāsa ke devatāoṃ ko choḍakara Vārāṇasī pradakṣiṇā ke devatā aura Kāśī pradakṣiṇā ke jisako vaḍī Paṃcakrośī kahate haiṃ sīma grāma aura vāsa sthānoṃ ke sahita yaha gola naksā hai jisapara eka kā 1 aṃka diyā hai ||*
This becomes a *maṇḍala* when a circle of a half the diameter of 5 *kosa* (about 10 miles) is drawn from Madhyameśvara as a centre. Inside this region there are three places one after another; the names [of the places] being Vārāṇasī, Avimukta and Antargṛha. The shapes of all these places, their boundaries, villages, the places for night halts, the names of the gods of Vārāṇasī-pradakṣiṇā and six gods of Kāśī-pradakṣiṇā which is called *baḍā* (grand) *paṃcakrośī* are given in this map under number 1, except the names of the gods which are on the bank of the river Gaṅgā from Varaṇā to Asī.

Kṛṣṇacandra is concerned with myth and "reality". He wants both: Kāśī as a mythical *maṇḍala* and as a circular city. He realises that both need not necessarily go together and that there are other iconic symbols used for Kāśī such as the leaf of the oil plant:

(Legend of Part 1, § 3:) *aura usake bhītara kā kṣetra Avimukta nāmaka Viśveśvara ko kendra mānakara cārasau dhanuṣa ke vyāsārddha se vṛtta ghumāne se Gaṃgā ko aura aṃḍāja chauthāī ke kaṭā huā golākāra hai | aura usake bhītara kā kṣetra Aṃtargṛha nāmaka kucha eraṃḍa patrākāra hai | jisa ke sīmā Deva Varāheśvara, Kaśyapeśvara, Dhruveśvara, Hāṭakeśvara Paśupatīśvara, Pitāmaheśvara, Vīreśvara,*

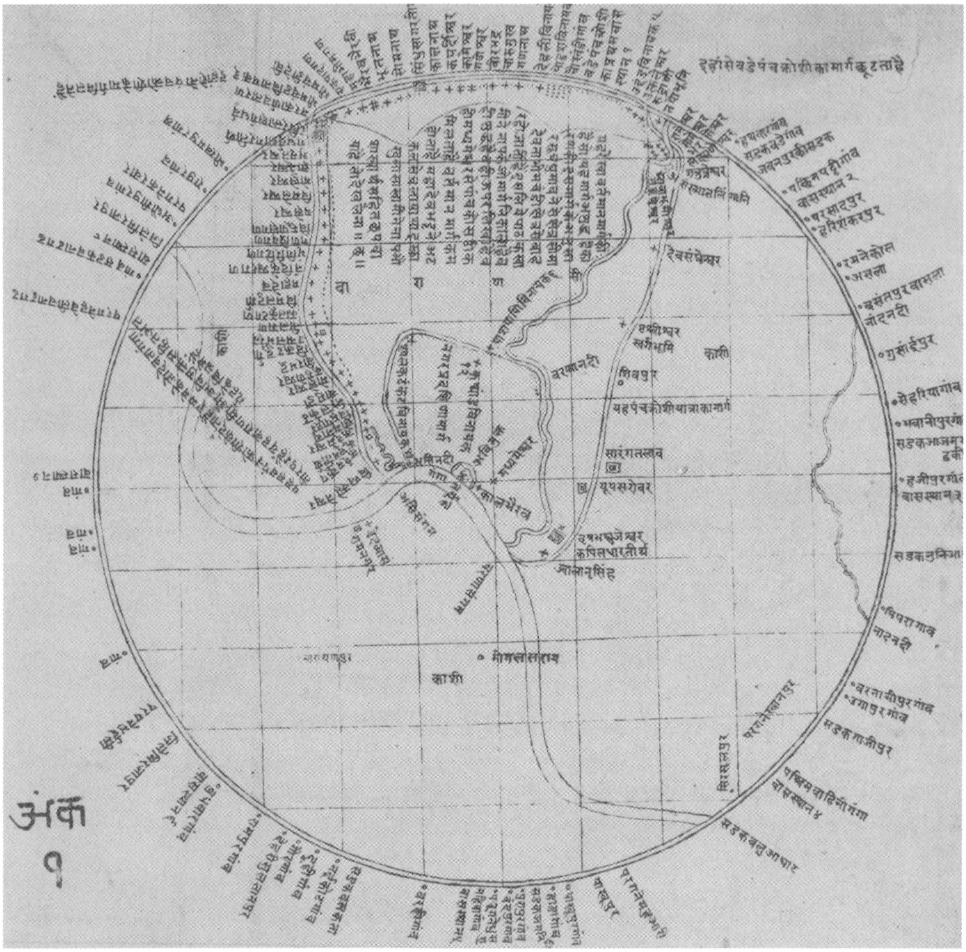

Fig. 2: *Kāśīdarpaṇapūrti*, map of Kṛṣṇacandra Śarmā, part 1.

> *Agnīśvara ye pradakṣiṇa krama se haiṃ ina dono kṣetroṃ ke devatāoṃ ke sthāna aura nāmoṃ ke sahita tīna ke aṃka kā eka naksā alaga dikhāyā hai* |
> And, the intermediate region which is round in its shape with a bit cut towards the Gaṅgā when a circle of a half the diameter of 400 *dhanuṣa* (about 700 m) is drawn from Viśveśvara named Avimukta as a centre. And, the intermediate region of it named Antargṛha is in the shape of a leaf of the costor oil plant (Ricinus Comunis). Its gods in the boundary are Varāheśvara, Paśupatīśvara, Pitāmaheśvara, Vīreśvara, Agnīśvara which are according to the circumambulation. The names and areas of the gods of those both those areas are given in a separate map under number 3 [= Part 3].

Beyond the circle, protective deities are not mentioned as in the map of Kailāsanātha Sukula, but only the names of villages.

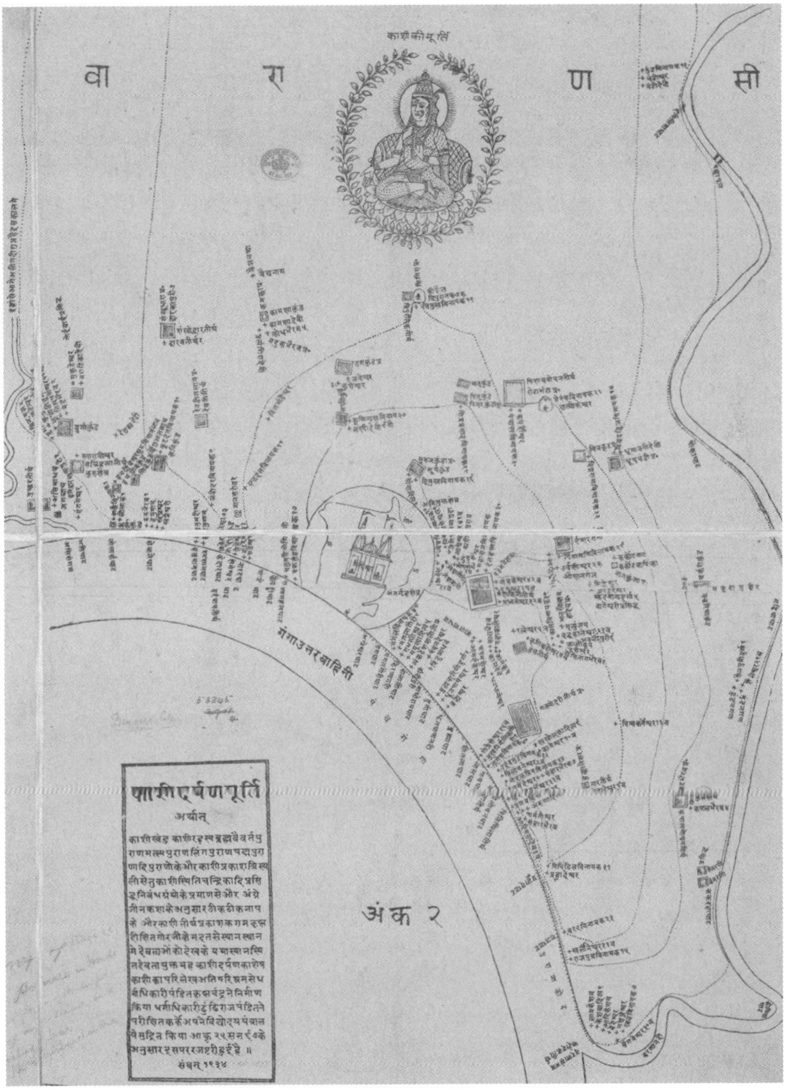

Fig. 3: *Kāśīdarpaṇapūrti*, map of Kṛṣṇacandra Śarmā, part 2.

4. *Selection*: For Kṛṣṇacandra the frame and the scale of the circle creates a problem of selection: the names of the deities of the intermediate region can not be listed because it is too small:

> (Legend of Part 1, § 2:) *isa Kāśī ke avāṃtara kṣetroṃ ke sava devatāoṃ ke sthāna aura nāma isa choṭe gola meṃ nahīṃ āya sakate isaliye kṣetasaṃnyāsiyoṃ kī pradakṣiṇā kā jisako nagara pradakṣiṇā kahate haiṃ mārga ko avadhi rakhakara ādhe Vārāṇasī kā eka alaga naksā jisa para do kā 2 aṃka diyā hai dikhāyā hai* |

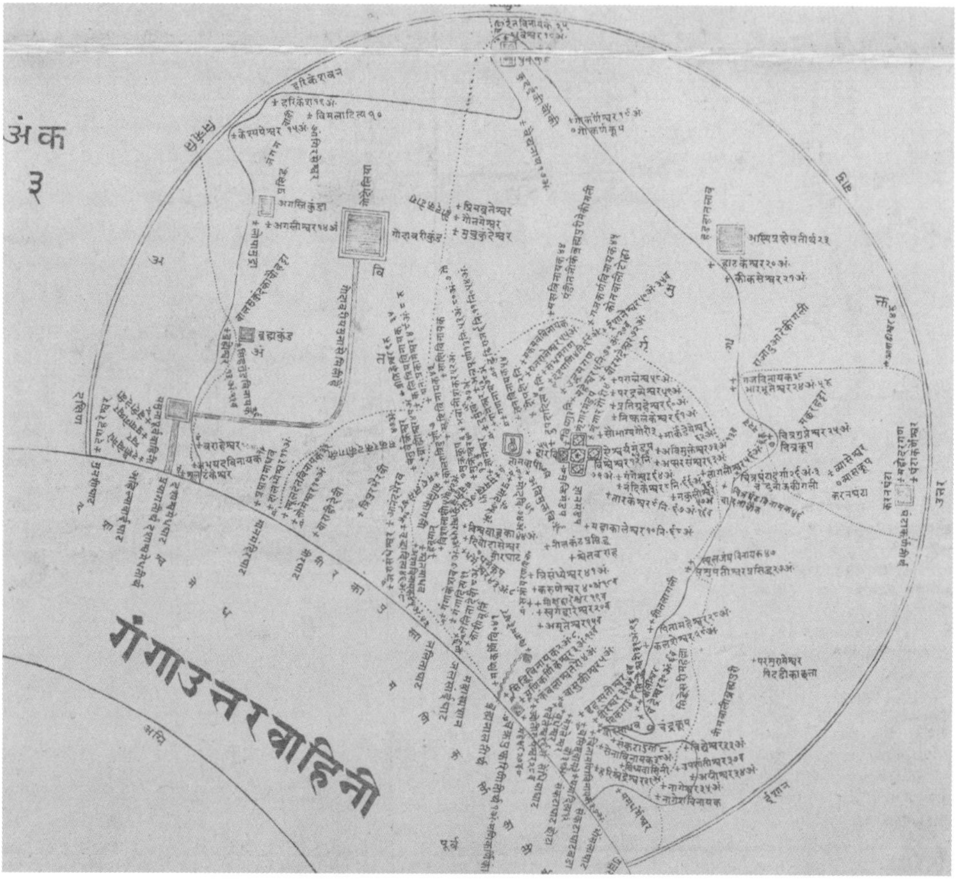

Fig. 4: *Kāśīdarpaṇapūrti*, map of Kṛṣṇacandra Śarmā, part 3.

> The names of all gods and their locations, which are situated in the intermediate region, can not be given in this small circle. Therefore, the route of the *pradakṣiṇā* (circumambulation) of *kṣetra-saṃnyāsis* which is called *nagarapradakṣiṇā* is drawn as it is and a separate map of the half of Vārāṇasī is [also] shown in the number 2 map.

Thus, Kṛṣṇacandra is aware of the selections which are necessary in drawing the map. He knows of its limitations, and he also mentions written texts as more complete sources for studying Kāśī:

> (Legend of Part 1, bottom) *ina tīno naksoṃ ko vyasta samasta rūpa se dekhane se Kāśīyātrā ke aśeṣa devatāoṃ ke sthāna aura nāma sava jāna paḍeṃge | ina sava devatāoṃ kī yātrā kā vidhi aura Kāśī saṃvaṃdhī vahutase viṣaya Paṃḍitavara Śrīyuta Nanda Paṃḍita Dharmādhikārī jinake Keśava Vaijayaṃtī nāmaka Bṛhad-viṣṇusmṛtiṭīkā prabhṛti aneka dharmaśāstra nivaṃdha prasiddha haiṃ unake kie hue*

Kāśīprakāśa nāmaka graṃtha kā bhāṣānuvādarūpa Kāśīdarpaṇa nāmaka hamāre kie hue Hiṃdī Bhāṣā aura Saṃskṛta graṃtha se jāna paḍeṃge | Śivam
(Legend of Part 2:) If you take a careful look at these three maps, you will know all the names and areas of the endless gods of Kāśīyātrā. All the *vidhis* of the *yātrā* of all these gods and more about Kāśī can be known from the book named "Kāśīprakāśa" [originally] written by Paṃḍita Śrīyuta Nanda Paṃḍita Dharmādhikārī whose various books and essays on *dharmaśāstra* including a book named "Keśavavaijayaṃti", a vast *ṭikā* of the famous *Viṣṇusmṛti*, which has been translated by me under the title "Kāśīdarpaṇa". May all be well.
Kāśīkhaṃḍa Kāśīrahasya Brahmavaivartapurāṇa Matsyapurāṇa Liṃgapurāṇa Padmapurāṇādi purāṇoṃ ke aura Kāśīprakāśa Tristhalīsetu Kāśīsthiticandrikādi prasiddha nivaṃdha graṃthoṃ ke pramāṇa se aura aṃgrejī nakaśā ke anusāra ṭhīka ṭhīka nāpake aura Kāśītīrtha prakāśaka Rāmakṛṣṇa Dīkṣita Gorajī ke madata se sthāna sthāna me devatāoṃ ko dekhake yathāsthānasthita devatāyukta yaha Kāśīdarpaṇa kā śeṣa Kāśī kā parilekha ati pariśrama se Dharmmādhikārī Paṃḍita Kṛṣṇacaṃdra ne nirmāṇa kiyā Dharmādhikārī Ḍhuṃḍhirāja Paṃḍita ne parīkṣita karke apane Vidyodaya Yaṃtrālaye mudrita kiyā ā(śvina) kṛ(ṣṇa) 25 sana 67 ke anusāra isapara rajaṣṭarī huī hai || Saṃvat 1934
With much effort, Dharmādhikārī Paṃḍita Kṛṣṇacaṃdra has written this *parilekha* with the deities residing as they are after consulting the evidences of *Kāśīkhaṃḍa, Kāśīrahasya, Brahmavaivartapurāṇa, Matsyapurāṇa, Liṃgapurāṇa, Padmapurāṇa* etc., also the famous books of essays [like] *Kāśīprakāśa, Tristhalīsetu, Kāśīsthiticandrikā* etc.

5. *Coding*: Kṛṣṇacandra's map has a legend that lists various water places (*talāb, kūvā, bāulī, choṭī nadī*) with symbols. It is much more designed to be used as a map in the conventional sense of the term.

(Legend of Part 3:) *viṃduoṃ ke rekhā se 56 vināyaka jānanā. una ke sāta āvaraṇa haiṃ. eka eka āvaraṇa me āṭha āṭha vināyaka hai. vāhara ke āvaraṇa se āraṃbha hai pradakṣiṇa krama hai | 42 liṃga ke yātrā ke aṃkoṃ ke pāsa (va) yaha akṣara likhā hai. aṃtargṛhī yātrā ke aṃkoṃ ke pāsa (aṃ) yaga akṣara hai. durgāyātrā me durgā pada hai (duṃ). gaurīyātrā me gaurī pada hai (gau). nityayātrā ke aṃko me (ni) akṣara hai. ekādaśarudrayātrā ke aṃko para (ru) akṣara. aṣṭabhairavayātrā me bhairava pada hai. dvādaśādityayātrā me arka aura āditya pada hai. paṃcakrośī ke jo devatā gaṃgā tīra me hai vahāṃ (paṃ) yaha akṣara hai. mārga ke to krama hī se haiṃ ||*
Signs of the *yātrā*. The line with dots should be understood as 56 Vināyakas. It has seven circumambulatory paths (*āvaraṇa*). There are 8 Vināyakas in each circumambulation. According to the order of circumambulation, it starts from the outer circumambulation. For 42 Liṃgayātrā "va" letter is written after the serial numbers of the *yātrā*.
For Aṃtargṛhīyātrā "aṃ" letter [is written] after the serial numbers of the *yātrā*.
In Durgāyātrā "duṃ" letter [is given] after the serial numbers of the *yātrā*.
In Gaurīyātrā "gau" letter [is given] after the serial numbers of the *yātrā*.
For Nityayātrā "ni" letter [is given] after the serial numbers of the *yātrā*.

For Ekādaśarudrayātrā "ru" letter [is given] after the serial numbers of the *yātrā*.
In Bhairavayātrā "Bhairava" word [is given].
In Dvādaśādityayātrā "Arka" or "Āditya" word [is given].
The letter "paṃ" is given to the gods of Paṃcakrośī, which are along the side of the river Gaṃgā. There is an order [given] for the route[s].

In the legend of Part 3, the map also uses conventional signs (*yātrā kā saṃketa*) for places of gods, ponds, wells, springs and small rivers.

To sum up: For Kṛṣṇacandra Śarmā, places are not only mythical but also within space. In drawing or designing his map, he is referring to both a religious and a scientific referential system. Although similar in appearance, both maps thus show significant differences regarding the formal features of maps (see table 1):

Table 1: Comparison of K. N. Sukula's and K. C. Śarmā's maps

Formal features of maps	*Kailāsanātha Sukula's Map*	*Kṛṣṇacandra Śarmā's Map*
scale	based on totality	based on appearance
framing	mandalic	mandalic and topographical
centrality	not indicated	indicated and not indicated
selection	mostly religious and text-related items	religious, text-related and physical items
coding	no symbols	symbols and legends

Conclusion: Religious and Scientific Concepts of Space[6]

The religious and the scientific referential systems can not easily be combined. In a religious concept, space and what is in the space form a unit. There is no total space ("Gesamtraum") within which other spaces can be located. Religious spaces are therefore per definitionem omni- and translocal. In a scientific concept of space, however, space is perceived as a medium, in which objects and places are placed. Seen from a religious point of view, there is no difference between Kāśī as a holy field (*kṣetra*) or as a temple, picture, sculpture or map. The mythical Kāśī remains Kāśī no matter how it appears.

Moreover, in a religious context, space is more felt than perceived. It is felt as a sacral power which can manifest itself in various forms: as a disc, conch, leaf, *liṅga* or text. These objects are identical because they share the same sacral power or potency. In a scientific context objects are singularities which are never identical.

In a religious context, Kāśī is always Kāśī independent of its location. Religious spaces are not in relations, but absolute. Uttarakāśī is also Kāśī, not a replica. Centrality is likewise absolute, not dependent on periphery. In a scientific context, however, spaces are defined by three-dimensional coordinates which lead to topographical and metric concepts as well as to norms of longitude and width.

Finally, in a scientific concept, space is dependent on magnetism, the position of the observer, extension, measurements etc. Space, then, is isotropic, i.e. identical with the methods of observation. In a religious concept of space, on the contrary, space is—once again—absolute. There can not be space around space. This is why inner space and outer space can become identical. Religious maps try to represent this idea, but as the case of Kṛṣṇacandra Śarmā's map shows they often have to compromise with scientific concepts of space. Therefore, it is often claimed—as also indirectly by Kṛṣṇacandra—that only religious people can see what is not represented: the Absolute or the totality or the true, inner Kāśī.

In other words, "A compound soul would not be a soul" (Wittgenstein, Tractatus logico-philosophicus 5.5421). The same holds true for religious maps. The composition of many holy places on a map does not necessarily make it a religiously used map. However, the more this aspect is intended—and I do believe that Sukula's and Kṛṣṇacandra's maps partly reflect this intention—the less important are at least three points:

(a) Accuracy: religious maps must not be accurate for they also refer to spaces which are not on the maps but in the maps.

(b) Orientation towards the North (cp. Gordon 1970): more important than an orientation based on magnetism and the polar axes become religious ideas such as the *maṇḍala* type or the orientation towards the sacred East. The religious "East" is always east no matter whether it is in the west or north. Likewise religious maps generally neglect the egocentric orientation, i.e. the position and spatial experience of the observer or pilgrims.

(c) Scale: Religiously seen, any scale or no scale represent Kāśī for it is *in illo locco*, in the heart, beyond space and time. And also beyond maps.

Notes

1 Cosgrove 1999: 2; cp. also Imhof 1972, Nöth 1998 and Schmauks 1998.
2 See also Gengnagel (2000, forthc.) and this volume.
3 Pieper 1979, Gole 1989, Gutschow 1982 and 1994, Gutschow & Michaels 1993, Michaels 2000 and 2003b, Schwartzberg 1992, Singh 1987 and 1988.
4 Thanks are due to Birgit Mayer-König and Nutandhar Sharma for helping in transcribing and studying the map.
5 Since this map has not yet been published, I shall describe it in greater detail than the map of Kailāsanātha Sukula.
6 The following thoughts have been influenced by Cassirer (1924) and Hübner (1985) as well as Eck (1981, 1998) and Panikkar (1991). For a more elaborate analysis see Michaels (2003a, 2003b, 2004: ch. VI.1).

References

Texts and maps

Kāśīkhaṇḍa: *Skandapurāṇa*, Vol. II: Kāśīkhaṇḍa/Āvantyakhaṇḍa, ed. Nag Sharan Singh. Delhi: Nag Publishers, 1982 [reprint, [1]1830].

Kāśīdarpaṇa [map of Kailāsanātha Sukula]. Kāśī: Ḍhuṃḍhirāja Śāstrī/Vidyodaya-Mudrālaya, Saṃvat 1933 (=1876 A.D.). Collection of Niels Gutschow.

Vārāṇasī or *Kāśīdarpaṇapūrti* [map of Dharmādhikārī Paṇḍit Kṛṣṇacandra Śarmā]. Kāśī: Ḍhuṃḍhirāja Śāstrī/Vidyodaya-Mudrālaya, Saṃvat 1934 (=1877). British Library (Nr. 53345 (4.), catalogued under the title "Map of Kashi i.e. Benares showing places of Hindu pilgrimage, shrines and bathing places)".

Secondary literature

Cassirer, Ernst. 1924. *Philosophie der symbolischen Formen*. Part 2: *Das mythische Denken* (Reprint: Darmstadt: Wissenschaftliche Buchgesellschaft, 1973).

Cosgrove, Denis. 1999. Introduction: Mapping Meaning. In: Denis Cosgrove (ed.), *Mappings*. London: Reaktion Books, pp. 1–23.

Eck, Diana L. 1981. Indias *tīrthas*: Crossings in Sacred Geography, *History of Religions* 20, 4: 323–344.

— 1983. *Banaras: City of Light*. London, Melbourne, Henley: Routledge & Kegan Paul.

— 1998. The Imagined Landscape: Patterns in the Construction of Hindu Sacred geography, *Contributions to Indian Sociology* 32, 2: 165–188.

Gengnagel, Jörg. 2000. Visualisierung religiöser Räume–Zur Kartographie von Benares. In: D. Pezzoli-Olgiati & F. Stolz (eds.), *Religiöse Kartographie: Organisation, Darstellung und Symbolik des Raumes in religiösen Symbolsystemen*. Bern: Peter Lang, pp. 217–234.

— forthc. *Visualized Texts–Sacred Spaces, Spatial Texts and the Religious Cartography of Banaras*. Habilitationsschrift, Heidelberg 2006.

Gordon, B. L. 1970. Sacred Directions, Orientation, and the Top of the Map, *History of Religions* 10: 211–227.

Gutschow, Niels. 1982. *Stadtraum und Ritual der newarischen Städte im Kathmandu-Tal - Eine architekturanthropologische Untersuchung*. Stuttgart: W. Kohlhammer.

— 1994. Vārāṇāsī/Benares: The Centre of Hinduism? A Discussion of the Meaning of "Place" and "Space", *Erdkunde* 48: 194–209.

— & A. Michaels. 1993. *Benares. Tempel und religiöses Leben in der heiligen Stadt der Hindus*. Köln: DuMont.

Imhof, Eduard. 1972. *Thematische Kartographie*. Berlin: de Gruyter.

Michaels, Axel. 2000. Konstruktionen von Translokalität und Zentralität. Eine religiöse Karte von Benares. In: D. Pezzoli-Olgiati & F. Stolz (eds.), *Religiöse Kartographie: Organisation, Darstellung und Symbolik des Raumes in religiösen Symbolsystemen*. Bern: Peter Lang, pp. 187–216.

— 2003a. The Sacredness of Himalayan Landscapes (Introduction). In: N. Gutschow, Ch. Ramble, A. Michaels & E. Steinkellner (eds.), *The Sacred Landscape of the Himalaya*. Vienna: Verlag der Österreichischen Akademie der Wissenschaften, pp. 13–18.

— 2003b. Tīrtha—Orte der Transzendenz in hinduistischen Texten, Ritualen und Karten. In: G. Oberhammer & Marcus Schmücker (eds.), *Mythisierung der Transzendenz als Entwurf ihrer Erfahrung*. Wien: Verlag der Österreichischen Akademie der Wissenschaften, pp. 349–380.

— 2004. *Hinduism. Past and Present.* Transl. by Barbara Harshaw. Princeton: Princeton University Press.

Nöth, Winfried. 1998. Kartensemiotik und das kartographische Zeichen, *Zeitschrift für Semiotik* 20: 25–39.

Panikkar, Raimon. 1991. There is No Outer without Inner Space. In: Kapila Vatsyayan (ed.), *Concepts of Space: Ancient and Modern*. New Delhi: Indira Gandhi National Centre for the Arts, pp. 7–38.

Pieper, Jan. 1979. A Pilgrim's Map of Benares: Notes on Codification in Hindu Cartography, *GeoJournal* (Dordrecht) 3, 2: 215–218.

Scafi, Alessandro. 1999. Mapping Eden: Cartographies of the Early Paradise. In: Denis Cosgrove (ed.), *Mappings*. London: Reaktion Books, pp. 50–70.

Schmauks, Dagmar. 1998. Landkarten als synoptisches Medium, *Zeitschrift für Semiotik* 20: 7–24.

Schwartzberg, Joseph. 1992. *The History of Cartography*, Vol. II.1: *Cartography in the Traditional Islamic and South Asian Societies*, University of Chicago Press.

Singh, Rana P. B. 1987. The Pilgrimage Maṇḍala of Vārāṇasī/Kāśī: A Study in Sacred Geography", *The National Geographical Journal of India* 33: 93–114.

— 1988. The Image of Varanasi: Sacrality and Perceptual World in Hindu Codification, *The National Geographical Journal of India* 34: 1–32.

Sukula, Kubera Nātha. 1977. Vārāṇasī Vaibhāva (in Hindī). Patna: Bihār Rāṣṭrabhāṣā Pariṣad.

JÖRG GENGNAGEL

Maps and Processions in Banaras: The Debate Concerning the Pañcakrośīyātrā[1]

The maps of Banaras that have, thus far, been collected and studied by the Vārāṇasī Research Project, under the auspices of the South Asia Institute in Heidelberg, range from pictorial maps, early and modern printed maps to charts, panoramic views, and topographical maps.[2] These maps do not present a uniform visualization of the sacred topography of Banaras. Variation is considerable, although most of the maps are based on the corpus of spatial texts concerning Banaras' sacred topography. In the present volume, Panoramas of Banaras, which "oscillate between a view and a map", are examined by Niels Gutschow, and Axel Michaels deals with aspects of transcendence and translocality in two maps that were lithographed in the latter half of the 19th century. In this article I will offer a contextualisation of these early lithographical maps by examining a debate regarding the performance of the Pañcakrośīyātrā, a topic that has thus far received almost no attention whatsoever.[3]

The earliest printed map of Banaras that we know of is the *Saptapurīyātrādiprakāśapatra* lithographed in 1873 for Jaṅg Bahādur Singh (Citaipur) by the Banaras Akhbar Press.[4] Kailāsanātha Sukula's map *Kāśīdarpaṇa* (1876), and Kṛṣṇacandra Śarmā's *Kāśīdarpaṇapūrti* (1877), were both printed in Banaras at the Vidyodaya Mudrālaya of Ḍhuṃḍhirāja Śāstrī.[5] Factors leading to the production of these early printed maps are multilayered. The emergence of print culture and the spread of lithography in Banaras during the mid-19th century made it possible to produce and distribute printed maps to a wider public. In addition, during the second half of the 19th century the railways of colonial India became an important factor regarding the promotion of pilgrimages.[6] The connection between the promotion of pilgrimage and the distribution of pilgrimage brochures as well as maps is documented by the legends on the map *Saptapurīyātrādiprakāśapatra*, where the station master's bungalow is mentioned as a place where one could obtain a booklet on processions in Banaras (*Kāśīyātrākalpalatā*).[7]

Western cartographical tradition was introduced in Banaras by way of British topographical maps, which made their appearance during the first half of the 19th century. We are aware of only two maps produced prior to the 1870s: James Prinsep's map *The City of Bunarus* (1822) and the map *Cantonments of Sikrol and Pandypoor, also The Civil Station & City of Benares* which was printed in 1869 in Calcutta. It was James Prinsep who first mapped the Antargṛhayātrā in detail (Gengnagel 2003: 262). On the map *Kāśīdarpaṇapūrti* Kṛṣṇacandra Dharmādhikārī explicitly refers to "British maps" (*aṃgrejī nakśā*) as his source material. It is within this context that the debate on the correct performance of the Pañcakrośīyātrā occurred.

Traces of a Debate on the Pañcakrośīyātrā

Although the Pañcakrośīyātrā is the procession in Banaras that has received by far the most attention, references to a conflict concerning the correct path of the Pañcakrośīyātrā are rarely found in written sources. The conflict does not form part of the cultural memory linked to this procession. It was Shashank Singh, art collector at Asī Ghāṭ in Banaras, who first referred me to a document on a "*yātrāśuddhi*" which, apparently, includes a small map that depicts the "corrected" section of the procession. Despite repeated efforts, I did not succeed in tracking down this document, however references to a conflict do, indeed, exist.[8] Kubernātha Sukula, grandson of Kailāsanātha Sukula who produced the map *Kāśīdarpaṇa*, provides in his monograph *Vārāṇasī Vaibhava* (1977) a brief and ambiguous hint by referring to a conflict in Banaras in the mid-19th century concerning the route of the Pañcakrośīyātrā:

> Preceding the seventies or eighties there was confusion about the route in the North of Bhīmacaṇḍī. But through the efforts by the late Paṇḍit Bāpūdeva Śāstrī this has been solved. At this place there exists however even now some confusion about the route of the procession.[9]

Sukula shows awareness of a certain "confusion" about the section of the Pañcakrośī route north of Bhīmacaṇḍī, but considers it unnecessary to go into further detail and describe the causes for the uncertainty.

More explicit still is a printed source published exactly one hundred years earlier. The aforementioned map *Kāśīdarpaṇapūrti*, which was appended by Kṛṣṇacandra Dharmādhikārī to his book *Kāśīdarpaṇam* in 1877, provides us with a very significant reference. This map, which is divided into three parts, depicts in its first section, *aṅka* 1, the Pañcakrośīyātrā in the upper left corner.[10] The sacred territory of Kāśī is encompassed within a full circle which contains a grid, and purports to represent both the small and large versions of the Pañcakrośīyātrā.[11] For our purposes, the upper western section of the map is of primary importance (Fig. 1). The following inscription is linked to a curved line drawn inside the outer circular double line:

> *yaha rekhā vartamāna mārga kī*
> *hai so yaha mārga aśuddha hai kā-*
> *raṇa kī isa mārga ke anusā-*
> *ra sūtra ghumāvane se kṣetrasīmā-*
> *devatā bhīmacaṃḍī kṣetra se bāha-*
> *ra ho jātī haiṃ isa liye pāṭhaka svā-*
> *mī ne nāpake jo mārga nikālā hai va-*
> *hī śuddha hai vahī ūpara likhā hai va-*
> *hī madhyameśvara se pāṃcakosa ṭhīka*
> *milatā hai vartamāna mārga kama*
> *hotā hai mahādeva bhaṭṭa ne aṭa-*

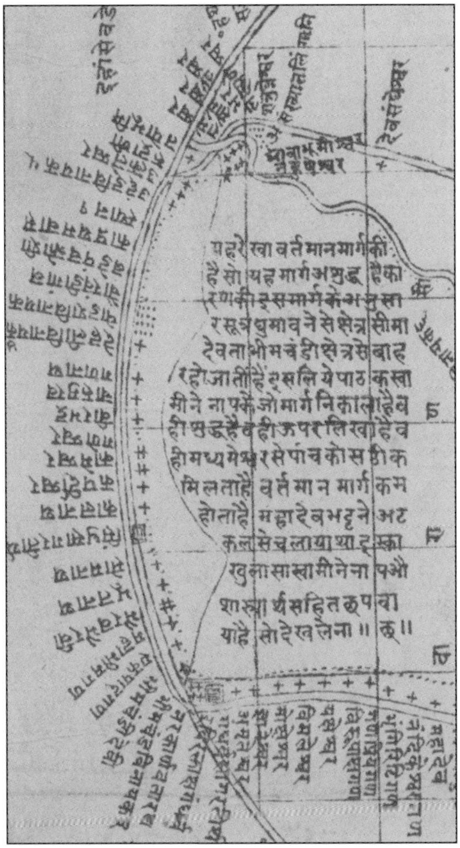

Fig. 1: Detailed view of the *Kāśīdarpaṇapūrti* (1877) with section of the Pañcakrośīyātrā north of Bhīmacaṇḍī.

> *kal se calāyā thā iskā*
> *khulāsā svāmi ne nāpau*
> *śāstrārthasahita chapavā-*
> *yā hai so dekha lenā.*
>
> This is the line of the present route. This route is not correct because by moving a rope according to [the position of] this route the goddess of the border of the sacred field (*kṣetrasīmadevatā*) Bhīmacaṇḍī would be outside the sacred field. Therefore the route that Pāṭhaka Svāmī has deduced (*nikālā*) by measurement is correct. This [route] was drawn above. There [a distance of] exactly five *kosa* from Madhyameśvara is obtained. The present route is less [distant from Madhyameśvara]. Mahādeva Bhaṭṭa has introduced it by conjecture (*aṭakala*). Svāmī has after measurement (*nāpau)* published this fact (*khulāsā*) together with the document of the debate (*śāstrārtha*). Consult it.[12]

This short caption clearly links this section of the Pañcakrośīyātrā, located north of Bhīmacaṇḍī, which is represented by two different lines, to a discussion of an "impure" (*aśuddha*) and "pure" (*śuddha*) route. A Pāṭhaka Svāmī was involved in the endeavour of measurement with a view toward preventing the goddess Bhīmacaṇḍī from being placed outside the limits of the sacred field of Banaras. Since the given information regarding this conflict is rather condensed, the reader of the map is asked to consult a publication about the results of a religious debate (*śāstrārtha*) for more information. The content of this religious debate will be discussed in the following section.

The Booklet *Pañcakrośī ke mārga kā vicāra*

The religious debate and related events are closely linked to the royal family of Banaras, and especially to the period of the reign of Īśvarī Prasāda Nārāyaṇa Singh, Mahārāja of Banaras from 1835 to 1889. His reign is important for our purpose in more than one respect. The Mahārājā founded the Dharma Sabhā in Vārāṇasī around 1870, which included among its members the famous Bhāratendu Hariścandra, who was in close contact with Īśvarī Nārāyaṇa Singh.[13] It was Bhāratendu Hariścandra who published, in 1872, the short booklet *Pañcakrośī ke mārga kā vicāra* ("The examination of the Pañcakrośī road") which deals, over the course of eighteen pages, with the discussion of the correct path of a section of the Pañcakrośīyātrā (Hariścandra 1872).[14] The booklet was reprinted at Rāmnagar in 1997, and appended to the publication of the poem *Pañcakrośa Sudhā*, a metrical description of the Pañcakrośīyātrā composed in 1842 by Kāṣṭhajihva Svāmī, the *guru* of Rājā Īśvarī Prasāda Nārāyaṇa Singh. Kāṣṭhajihva Svāmī was also known as Anantaśrīcaraṇa, Deva Tīrtha as well as Deva Kavi.[15]

In 1872, Hariścandra raised an issue that, in fact, dated back to 1853, when the actual measurement of the Pañcakrośīyātrā took place. Hence, the beginning of the whole debate occurred almost 20 years previously when an assembly of religious specialists signed a document which sanctioned the recently constructed road as the correct path concerning the performance of the procession in question. The entire issue was elucidated by Hariścandra in 1872 in his booklet which included the text of the original document. Finally, in 1997 the Purāṇa department of the library in the Rāmnagar palace drew attention once again to the events of the past by reprinting Hariścandra's booklet in the appendix of a book that praises the Pañcakrośīyātrā and was, subsequently, distributed to pilgrims.[16] The controversy surrounding the route of the Pañcakrośīyātrā was a complex affair that, in spite of repeated efforts to alter existing pilgrimage practice, has passed almost unnoticed.

What then was the key issue of the debate that Hariścandra raised anew in 1872? One basic aspect of the discussion arose from the attempt to apply the statements of spatial texts directly to the city's topography. This was accomplished through the use of modern techniques such as surveying, measurement and the creation of topographical maps. In the case of the Pañcakrośīyātrā, verses attributed to the *Pātālakhaṇḍa* of the *Padmapurāṇa* are quoted as the fundamental scriptural authority. These verses are very significant because they introduce the section on "the dimension of the sacred field of Kāśī and others" (*kāśyādikṣetraparimāṇam*) in Nārāyaṇa Bhaṭṭa's digest *Tristhalīsetu*:

Fig. 2: The circular structure of the *kāśīkṣetra* with the radius between Madhyameśvara and Dehalīvināyaka and the "old" and "new" borderlines (drawing by Niels Gutschow).

> *madhyameśvaram ārabhya yāvad dehalivighnapam*
> *sūtraṃ saṃsthāpya tad dikṣu bhrāmayen maṇḍalākṛti*
> *tatra yā jāyate rekhā tanmadhye kṣetram uttamam*
> *kāśīti yad vidur vedās tatra muktiḥ pratiṣṭhitā*
> *kāśyantaḥ paramaṃ kṣetraṃ viśeṣaphalasādhanam*
> *vārāṇasīti vikhyātam iti*
>
> If one places a cord reaching from Madhyameśvara to Dehalīvighnapa and moves this in the form of a circle to the [four] directions, there within this created line is the excellent *kṣetra* that is Kāśī. The wise know that there liberation is achieved. The inner part of Kāśī (*kāśyanta*) is the most excellent *kṣetra* which is the means (*sādhana*) to special fruits and known as Vārāṇasī.[17]

This description of the circular structure of the *kāśīkṣetra*, with the distance between Madhyameśvara and Dehalīvināyaka defining the radius and extension of the circular field, was then directly applied to the physical space. The distance between the two temples was measured, the length was taken as radius and a circle with the corresponding radius was projected onto the space of Banaras through the use of topographical maps (Fig. 2).[18] This took place in 1853 and an error was discovered with regard to the circular route and, consequently, the route of the Pañcakrośī procession as it was performed at the time. In the terminology of the text: the "real" (*vāstava*) route based on the "reality" of textual evidence and surveying was in conflict with the contemporary or actually used (*pracarita, vartamāna, sāmprata*) path in a small section north of Bhīmacaṇḍī.[19] The ensuing task of identifying the "incorrect" route and reconstructing the "real" route was a complicated process in which many were involved.

Hariścandra's introduction (*bhūmikā*) to the *śāstrārtha* document begins with the remark that the Pañcakrośī procession was neglected, or actually forgotten, under Muslim rule. The procession was re-established when Kāśī was once again ruled by Hindus, but there was uncertainty about the correct path and people had to guess where to go.[20] The Brahman Mahādeva Bhaṭṭa collected money for repairing the road and the construction of bridges etc. In the 1850s, Pāṭhaka Svāmī, who was a pupil of Ketakara Svāmī,[21] undertook a survey of the whole Pañcakrośī road and realized that the road between Bhīmacaṇḍī and Rāmeśvara had been moved and was situated within the *kṣetra*.[22] He considered the location of this section of the road constructed by Mahādeva Bhaṭṭa improper, and in order to rectify this a fund-raising committee with six members, including the District Magistrate of Banaras, F. B. Gavins, was established.[23] A plan of the western section of the Pañcakrośīyātrā was prepared by the members of the committee along with government surveyors, based on research at the location in question.[24] It was discovered that the land needed to construct the new and proper road was owned by the Mahārāja of Banaras, Īśvarī Prasāda Nārāyaṇa Singh. This fact was made known to yet another assembly of influential Banarsī Pandits.[25] They wrote a letter to the Mahārāja, with a document (*śāstrārtha*) presenting the facts.

The donation of land was granted by the Mahārājā and the road was ultimately constructed in accordance with scriptural sources and measurements taken from topographical maps. Along with the construction of the new road, the deities that needed to be moved were installed in their respective places alongside the road.[26] In order to modify the practice of the pilgrims the document of the *śāstrārtha*, which was undersigned by 51 members of the Dharma Sabhā of Banaras, was distributed in Sanskrit as well as in Hindi.[27] In the introduction to this document the reader is reminded that every pilgrim should strictly follow the processional route without the slightest deviation. This point is reinforced by a quotation of the verse *Kāśīrahasya* 10.25: "While performing the circumambulation one should not abandon the *kṣetra* by even one sesame seed" (*kṣetraṃ pradakṣiṇīkurvaṃs tilamātran na santyajet*). The text of the *śāstrārtha* provided a definition regarding the size of the *kāśīkṣetra* and its geometrical structure. The argument is based on the quotation of the above given verses ascribed to the *Padmapurāṇa*. Since the distance between Madhyameśvara and the contemporary Dehalīvināyaka is too short, a circle with this radius would not include the temple of the goddess Bhīmacaṇḍī in the *kāśīkṣetra*. Textual evidence, however, revealed that Bhīmacaṇḍī was one of the deities situated on the border of Vārāṇasī (*vārāṇasīsīmadevatā*). Consequently, it would have been a grave mistake were the goddess to be situated outside the limits of the *kṣetra*.[28] In light of this, the Dehalīvināyaka temple had to be moved to the west so that the circular field did not exclude Bhīmacaṇḍī.

Together with this text, a small and simple map (*nakśā*) of the area between Bhīmacaṇḍī and Rāmeśvara is given (Fig. 3).[29] This sketch shows the corrected route on the outer circle, the path that was actually used which moves inside of the *kṣetra*, and a dotted circular line representing the smaller circle which was calculated by taking the distance between the actual Dehalīvināyaka and Madhyameśvara as radius.[30]

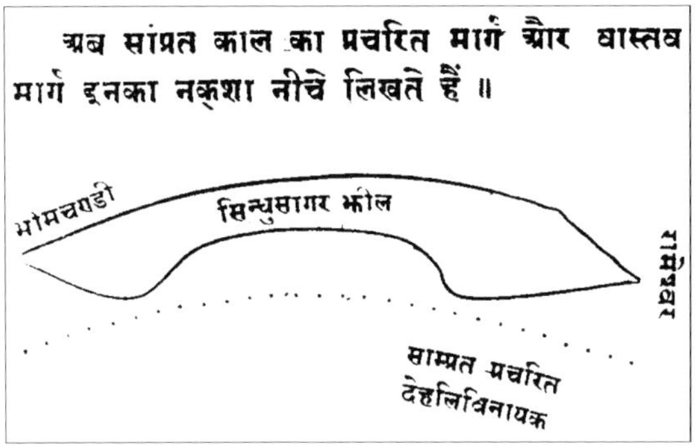

Fig. 3: Map of the section between Bhīmacaṇḍī and Rāmeśvara from the booklet *Pañcakrośī ke mārga kā vicāra* (Hariścandra 1872: 9).

Almost 20 years had passed when Bhāratendu Hariścandra raised the matter again in 1872. He expressed his discontent and astonishment that the pilgrims continued to follow the same route although the corrected route had been constructed in accordance with textual evidence (*śāstra*), logical inference (*yukti*), and correct measurements (*śuddha nāpa*). The pilgrims, however, continued to follow the incorrect route as the corrected route was not sufficiently well-known. Hariścandra complained that the "stubborn people" acted according to the saying "It is better to do something in a deficient way than to do nothing" (*akaraṇān mandakaraṇaṃ śreyaḥ*).[31] Hariścandra strongly objected to this point of view and cited the established practice of circum-ambulations: if a procession road splits and one is uncertain about the correct path, one should always follow the path to the left in order to avoid walking inside the field.[32] He maintained that the common practice followed an incorrect, "impure" (*aśuddha*) route and that the whole procession was therefore imperfect or "destroyed" (*khaṇḍita*).

Several basic questions remain unanswered. Did Hariścandra succeed in this second effort to remind the pilgrims of their duty to take the new path? Is the route in the section of the Pañcakrośīyātrā north of Bhīmacaṇḍī, as it is performed today, based on the revised route? Are there still traces left of both routes? Field research has indeed

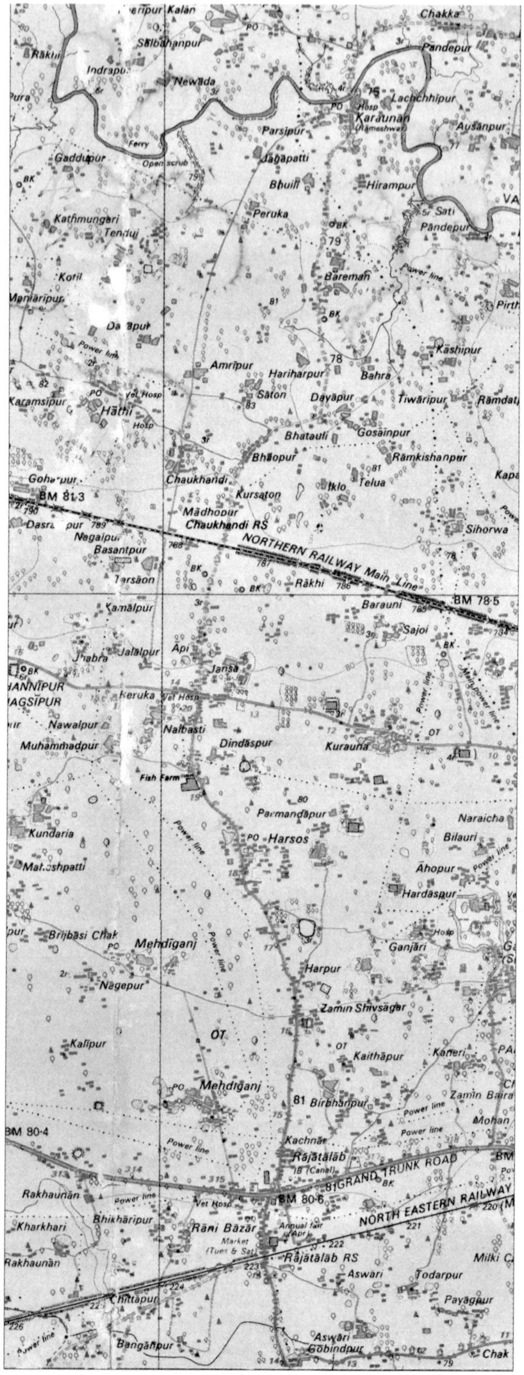

Fig. 4: Area north of Bhīmacaṇḍī, sheet 63 K/15, Survey of India, 1975, 1:50.000.

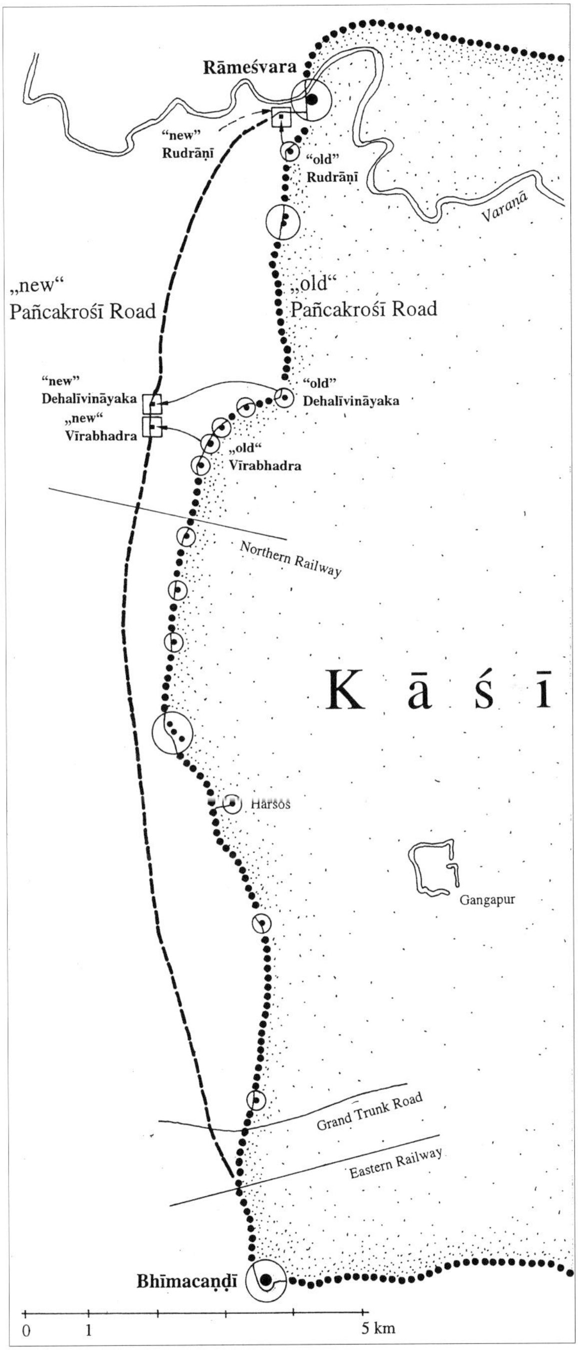

Fig. 5: The section between Bhīmacaṇḍī and Rāmeśvara showing the "new" and "old" Pañcakrośī roads, along with the reinstalled temples (drawing by Niels Gutschow).

demonstrated that traces of the efforts of Hariścandra as well as others can still be found. Nevertheless, their attempts to alter pilgrimage practice "on the ground" were unsuccessful.

The Search for the "Old" and "New" Pañcakrośī Route

If alternative routes existed one would expect to find corresponding evidence in physical space. And there are, in fact, parts of an optional route to be found on topographical maps of the region. One road depicted by a red double line is clearly visible on the 1: 50.000 topographical map dated 1975 (Fig. 4). It runs from south to north, and bends slightly eastward in two places. This section of the route starts at Aswāri Gobindpur in the south–the place where Bhīmacaṇḍī is situated–then cuts across the Grand Trunk Road in Rājātālāb, and passes through the villages Harpur, Harsos, Naibasti, and Jansa. The road then crosses the Northern Railway Main Line at Chaukhandi Railway Station, and moves on to the villages Dayāpur and Bareman. It finally reaches the village Karaunān (Rāmeśvara) in the north. To the west of this road fragments of another path are indeed noticeable. A thin, straight red line starts at Rājātālāb and crosses the village Mehdīganj. The line is then interrupted, and only once again visible at the village Jalālpur where it crosses the Northern Railway Main Line, and extends in a direct line to the villages of Chaukhandi, Amrīpur, Jagāpatti, Parsīpur; before finally reuniting in the north at Karaunān (Rāmeśvara) with the larger road.

Investigation on the spot has revealed that the larger road, which moves somewhat to the east, depicts the Pañcakrośī road that is taken to this day by the pilgrims. The hypothesis that the smaller road was the newly constructed one was confirmed at once when people living along this road were questioned. The fragmentary cartographic rendering of the road proved to be correct. Interviews with elderly inhabitants of the villages Parsipur, Jagāpatti, and Chaukhandi made clear that this very road was, according to their perception, the "old Pañcakrośī Road". Older inhabitants of the village Chaukhandi still retained the memory that the road was used by "holy men and Sants" (*sādhu-sant log*) in their parents' time, and even as recently as their own childhood. In the words of the interviewed persons this small road, along which they live, was clearly the old (*prācīna*, *purāṇa*) one, whereas the road to the east, which the pilgrims actually use, is the new (*nayā*, *adhūnik*) road. For the villagers, the "old" road was the one that was no longer in use while, from the perspective of the ritual and religious experts, the construction of the "new" road merely reestablished what was, in fact, age-old practice. This seemingly paradoxical, and often confusing usage reflects, on a local level, an approach by the religious elites that has been described as "New Antiquarianism" (Dalmia 2001). For convenience sake I will call the route and places along the road constructed in the 1850s "new", while the other road and its respective temples will be referred to as "old".

What remained lacking as final confirmation of this oral identification of the road were the temples which had been reinstalled along the way. In 2001 I found a Dehalīvināyaka temple on the new road, at a crossing just outside the village of Chaukhandi (Fig. 6). It is situated to the west of the well-known Dehalīvināyaka temple located near the villages of Bhatauli and Dayāpur. In addition, a few hundred meters to the south of the "new" Dehalīvināyaka temple, I located a very small shrine with an image of Vīrabhadra (Fig. 8). In the following year, the ruins of a small resthouse for pilgrims (*dharmaśālā*) were discovered in the village Mehdiganj, in the southern section of the "new" road. With respect to the images and temples of Rudrāṇī Devī in Parsipur, which will be discussed later in detail, identification is somewhat ambiguous as they are located in an area where the "old" and "new" roads are almost reunited (Fig. 9). The results of the above findings were incorporated in the drawing of a map of the area between Bhīmacaṇḍī and Rāmeśvara (Fig. 5). The map shows the two roads clearly, and the relocation of the temples Dehalīvināyaka, Vīrabhadra, and Rudrāṇī Devī is indicated as well.

The Two Dehalīvināyakas

Fig. 6: The "old" (left) and "new" (right) Dehalīvināyaka temples located in the village Chaukhandi. On the right, the straight road constructed in the 19th century.

According to inscriptions found at the "old" Dehalīvināyaka temple, the present building was constructed around 1790, and the adjacent well was built a few years later (Singh 2002: 101–103). The *pūjārī*, Raghunātha Dūbe, and his son, Umāśaṃkar Dūbe, live with their family in a building near the temple. According to their statements they have been the Pūjārīs of the Dehalīvināyaka temple for 50 years. Neither of them were aware of a "second" Dehalīvināyaka, nor had they ever heard of a debate concerning the location of the "old" Dehalīvināyaka temple.

The "new" Dehalīvināyaka temple in the village Chaukhandi is relatively small, and the standing, four-armed image is less refined than the one in the "old" temple (Fig. 7). This "new" Dehalīvināyaka shrine shows no signs of regular veneration, and there appears to be no close relationship between the villagers in Chaukhandi and the temple. On the walls of the "new" temple the following inscription was found:

śrī dehalīvināyaka // prācīna sthāna śāstra // pramāṇa se bhī milā hai //
Dehalīvināyaka. The ancient place found also according to scriptural authority.

Fig. 7: Images in the Dehalīvināyaka temples: to the left the "old" image, to the right the image installed in the mid-19th century.

This inscription is crucial regarding the identification of the temple and the statue. In addition, by referring to the scriptural authority (*śāstrapramāṇa*) it relates directly to the quoted verses of the *Padmapurāṇa*, as well as to further textual evidence given in the document of the *śāstrārtha*. The arguments put forth by the members of the Dharma Sabhā were directly inscribed on the walls in order to provide the authority needed for the shifting of the Dehalīvināyaka temple to its "new" location. The inscription confirms that the effort in the 19th century, as described above, did not intend to establish something "new". Rather, it reestablished or reinvented an "ancient" (*prācīna*) spatial reality. In the words of the *śāstrārtha*: the actual pilgrimage practice was an "impure", incorrect (*aśuddha*) deviation and the real (*vāstava*) practice had to be restored.

The Two Shrines of Vīrabhadra

A very small shrine with an image of Vīrabhadra is situated along the "new" Pañcakrośī road in the village of Chaukhandi. Its size and the domination of the small structure by a much higher pile of bricks illustrates the perishable nature of these types of small images installed in the 20th century. An unquestionable identification of this simple shrine with its standing, two-armed white image was only possible because of an inscription on the outer western wall:

Fig. 8: The images of Vīrabhadra on the "old" (left) and the "new" (right) Pañcakrośī roads.

paṃcakrośī meṃ śrīvārāṇasī sīmaga // vīrabhadrāya namaḥ //

On the Pañcakrośī [road], gone to the border of Vārāṇasī. Veneration to Vīrabhadra!

This inscription deals with the issue of the gods situated on the border of Vārāṇasī (*vārāṇasīsīmadevatā*). They define and delineate the borders of the sacred territory of the *kāśīkṣetra*. Their exact location is crucial in terms of the course of the processional road, and therefore decisive in the debate concerning the proper location of this road.

The Two Rudrāṇī Devīs

The matter of the location of Rudrāṇī Devī is somewhat more complicated since the two possible locations are situated in an area near Rāmeśvara where the "old" and "new" roads merge. Both places are known to the pilgrims, and it is impossible to unequivocally prove that the "new" Rudrāṇī was installed in the mid-19th century. No inscriptions whatsoever were found at the two locations. The image that I, personally, would identify as the "old" image is located in a shrine along the "old" route. The place is referred to as *tapobhūmi*, i.e. "site of religious of religious austerity". The "new" image is clearly located on the road constructed in the 19th century.

Fig. 9: The *mūrtis* of the two Rudrāṇī Devīs in the village Karaunān/Rāmeśvara (left), and Parsipur (right).

Conclusion

In the presented case study, references to authoritative spatial texts, as well as the utilization of modern surveying techniques and topographical mapping were important tools used by the local elite in their attempt to change the route of the Pañcakrośīyātrā. In spite of the above and, in addition to the combined efforts of various others including a group of Banarsī Pandits, the Mahārājā of Banaras, the District Magistrate, and, finally, the famous playwright and journalist Bhāratendu Hariścandra; their endeavour did not succeed. The pilgrims follow the same route today as they did prior to when the issue was initially raised in 1853.

The discrepancy between this final result, and the time and energy invested by so many people regarding the shifting of roads, as well as gods, is substantial. Various reasons exist for this evident failure. The contrived construction of a new road and, especially, the relocation of several shrines and temples would have required wide acceptance by both the villagers living along the road, as well as the pilgrims. To this end, a pilgrimage infrastructure with *pūjārīs* for the newly established main temples, such as Dehalīvināyaka, along with caretakers for the smaller shrines and facilities for the pilgrims along the road would certainly be necessary. Such facilities were at least partially provided—ruins of a small rest house were found along the new road—but it seems probable that they catered only to the well-informed, pious religious elite. This appears to be consistent with the memory of the villagers, as they spoke of the "*guru-sant log*" who made use of this road.

Another decision had to be made just a few years after Hariścandra raised the issue again. The *Northern Railway Main Line* was constructed in the 1880s (Köckmann 1982: 34), and traversed the area in question near the village Chaukhandi where, together with a small station, a railway crossing had to be constructed (see Fig. 4). The railway authorities decided to link both places to the "old" road. The new road shows no signs of a crossing and use of the "new" Pañcakrośī road was discontinued at the time of the construction of the railway line. Without further evidence it is impossible to determine the reasons for the decision on the part of the authorities to construct the crossing at the "old" road. It certainly does not, however, suggest that there was broad acceptance of the "new" road at the time of the railway construction, and it has, for practical reasons, contributed to the lack of acceptance of this alternative route.[33]

In the given case, map-making and cartography were extensively utilized in an effort to alter existing pilgrimage practice. As will be shown in the following section, traces of the dispute regarding the correct route of the Pañcakrośīyātrā can also be detected on printed maps published shortly after the activities of Hariścandra. The fact that we know of three maps of Banaras that were lithographed in the 1870s in Banaras is an indication that the debate concerning the correct performance of the Pañcakrośīyātrā increased the interest in promoting pilgrimage practice through the use of maps. In addition to the available spatial texts, the medium of cartography was utilized to plot the corrected route of the Pañcakrośīyātrā. The fact that this cartographic effort was in vain may suggest that map-making is not an effective instrument for changing existing pilgrimage practice. Negotiation and contestation of ritual practice is rarely documented, but occurs frequently. The attempt to reconstruct an ideal practice based on textual evidence, as well as the outcome of surveying and mapping of physical space, has to be reconciled with the reality of the persistence of everyday practice. Based on the analysis of the religious specialists of the 19th century, the present day performance of the Pañcakrośīyātrā would still have to be considered "impure" or "wrong" (*aśuddha*) and "broken" (*khaṇḍita*). This contrasts starkly with the fact that this procession is one of the most famous and popular in Banaras. The spiritual merit one acquires by performing this procession is not questioned. As this example illustrates, sacred spaces are not generated by maps, in spatial texts, or defined by surveying physical space. Rather, they are stamped out by the very feet of the pilgrims over the course of their circumambulations. It is the pilgrims' spatial practice that creates and defines sacred spaces.

Notes

1 This paper is based on research in Banaras in Spring 2001 and 2002 as part of the Vārāṇasī Research Project, Heidelberg funded by the Deutsche Forschungsgemeinschaft. I would like to thank Beat Niederer, Hemant Sarna and Rakesh Kumar Singh for their assistance during my search for old and new roads. Special thanks to Niels Gutschow for many fruitful discussions and for providing me with drawings of maps. My thanks to David Whizin, as well, for revising the English.

2 A list of religious and topographical maps of Banaras is included in Gengnagel (forthc.), Appendix 1. For a catalogue of the Banaras Archive of the South Asia Institute see *www.benares.uni-hd.de/ben-archive.pdf*.

3 See Gengnagel 2005: 69–76 for a short description of the Pañcakrośīyātrā and its textual sources.

4 See Gole (1989: 66) and Gengnagel (2000) for a preliminary description of a coloured version of this map.

5 See Michaels in this volume for the *Kāśīdarpaṇapūrti* (Fig. 2–4). The *Kāśīdarpaṇa* is, to date, the most closely studied map of Banaras. See Gengnagel 2003 for bibliographical details, and the electronic publication *Mirror of Kāśī* (Gengnagel & Michaels 2001).

6 See Kerr 2001 for a study of the effects of railways on pilgrimage in the 19th and 20th centuries.

7 See Gengnagel (forthc.) for details of the legend.

8 This hint marked the beginning of the research on this conflict, but the document itself was never found in Shashank Singh's collection.

9 *sattara yā assī varṣa pahale bhīmacaṇḍī ke uttara bhī kuch mārga kī gaṛabaṛī hone lagī thī, parantu svargīya mahāmahopādhyāya paṇḍita bāpūdeva śāstrī ke prayatna se vaha ṭhīka ho gaī thī, parantu yātrāmārga is sthāna para ab bhī kuch gaṛabaṛa hai.* (Sukula 1977: 217).

10 See Michaels in this volume for a complete reading of the legends of the map and a short description of its structure. Singh (1988: 6, Fig.2) has reproduced a drawing of one version of this map.

11 The large Pañcakrośīyātrā is directly mentioned on the map. At Rāmeśvara an inscription indicates that at this point the path of the longer version diverges from the path of the shorter version: *ihāṃ se vaḍe paṃcakrośī kā mārga chūṭatā hai*.

12 A transcription and draft of a translation of this passage was produced by Birgit Mayer-König and Nutan Dhar Sharma in September, 2000. It has subsequently been revised according to the digitalized reproduction of the map *Kāśīdarpaṇapūrti*.

13 See Upādhyāya: *inko darbār me bhāratendu hasriścandra prāyaḥ āyā-jāyā karate the aur mahārājā inko apne putra ke samān mānate the.* (1994: 122). For an excellent study of Bhāratendu Hariścandra's role in the latter half of the 19th century in Banaras see Dalmia (1997).

14 Vasudha Dalmia has kindly provided a copy of this text.

15 See Upādhyāya 1994: 122, 791–806 and Dalmia 1997: 79–80 on Kāṣṭhajihva Svāmī. Upādhyāya gives Viśveśvara Datta Miśra as his worldly name (*gṛhasthāśrama kā nāma*), and names his *dīkṣāguru* as Vidyāraṇya Tīrtha (Upādhyāya 1994: 793). This might well be the name of the author of the *Pañcakrośa Sudhā*. The colophon of the printed text reads as follows: *iti vidyāraṇyatīrthakṛtā pañcakrośasudhā* (Pañcakrośa Sudhā: 111). Both the publisher of the *Pañcakrośa Sudhā* as well as Upādhyāya (1994: 800), however, give Kāṣṭhajihva Svāmī as the author of the text.

16 The editor Satyanārāyaṇa Khaṇḍūḍī wrote the introduction in the reprint of 1997 and took the same position that Hariścandra did when he introduced the document of the *śāstrārtha* in 1872.

17 The verses are quoted in the *Pañcakrośī ke mārga kā vicāra* with the introduction *tathā ca tristhalīsetudhṛtapātālakhaṇḍīyaṃ pādmaṃ vacanam* (Hariścandra 1872: 11).

18 The Madhyameśvara temple is situated in Dārānagar to the North of Maidāgin. For a photo and a ground plan of this small temple see Gutschow (2006: 76).

19 *saṃvat 1910 meṃ arthāt īsavī san 1853 meṃ vah vāstava pañcakrośī kā mārga baḍe-baḍe pariśrama se nirmāṇa karake prasiddha kiyā hai aur vah mārga bahut saral aur pracarita mārga se lambāī meṃ kuch choṭa hai.* Hariścandra 1872: 3–4.

20 The term applied for the basis of his corrections is *aṭkal se* "by conjecture" which is also used in the legend of the map *Kāśīdarpaṇapūrti*, by Kṛṣṇacandra Dharmādhikārī.

21 The full names of these two Svāmīs are Paramahaṃsa Parivrājaka Svayaṃprakāśāśrama Svāmī and Paramahaṃsa Parivrājaka Sureśvarāśrama Svāmī.

22 *tab jāna paḍā ki bhīmacaṇḍī aur rāmeśvara ke bīc meṃ mārga bahut bhītar se banā hai.* Hariścandra 1872: 5.

23 The following names are given: Devanārāyaṇa Siṃha Bahādura, Bābū Phatahanārāyaṇa Siṃha, Bābū Gurudāsa Mitra, Rāya Nārāyaṇa Dāsa and Bābū Vrajadāsa. Together with F. B. Gavins, a certain Amānī Sāhib Bahādur is also mentioned as having raised funds (Hariścandra 1872: 6). The introduction to the *śāstrārtha* stresses the importance of Gavins' role: *vah vāstava mārga śrīyuta gabins sāhiba bahādura jī kī kṛpā dṛṣṭi se ban gayā hai* (p. 10). For a reference to F.B. Gavins see Motīcandra 1985: 352.

24 These surveyors are called *sarkāra kā bhūmī nāpanevālā* (Hariścandra 1872: 6).

25 *tab pāṭhaśālā ke kāśīnātha śāstrī ādi sab paṇḍita log aur sakhārām bhaṭṭa bhaṭṭa ādi bhaṭṭa log kākārāma paṇḍita gaṅgādhara śāstrī bhāṭavaḍekara ādi nagar ke baḍe-baḍe paṇḍita in sabhoṃ ne dharma kārya samajha ke pañcakrośī ke vāstava mārga ke bhūmī ke liye us mārga ke śāstrārtha ke sameta ek prārthanā patra śrīkāśīrāja ko likhā* (Hariścandra 1872: 7).

26 *phir vah śāstra siddha atiprāktan mārga śuddha prakār se ban gayā aur us par yathāsthān meṃ devatāoṃ kī bhī sthāpanā huī* (Hariścandra 1872: 8).

27 The *śāstrārtha* is undersigned by the following names: Bhaṭṭopākhya Jayarāma Śarmā, Kāśīnātha Śāstrī, Sukulopāhvomārāvaśarma Paṇḍita, Caturveda Paṇḍita, Hīrānanda Śarmā Paṇḍita, Durgādatta Śarmā, Devadatta Śarmā, Gulajāra Śarmā, Bhaṭṭa Prabhākara Śarmā, Bhaṭṭa Vaijanātha Śarmā, Abhayācaraṇadeva Śarmā, Śivanātha Miśra, Pāṇḍeya Īśvaradatta Śarmā, Harinārāyaṇa Śarmā, Kālīprasāda Śarmā, Lajjāśaṅkara Śarmā, Nandarāma Śarmā, Pañcānana Śarmā, Bāpūdeva Śāstrī, Nirupaplava Vecana Śarmā, Nārāyaṇa Śāstrī, Viṭṭhala Śāstrī, Veṅkaṭarāma Śarmā, Gaṅgādhara Śāstrī Thatthe, Nārāyaṇa Meghaṃkara, Bhaṭṭa Rāma Śāstrī, Rāmajasana Paṇḍita, Sevarāma Śarmā, Gopālakṛṣṇayājñika, Śākadvīpīya Rāmanātha Pāṭhaka, Ramāpati Śarmā Tripāṭhī, Kārttika Rāma, Draviḍa Śaṃkara Śāstrī, Jīvanātha Rājaguru, Bhaṭṭa Kamalākarabhaṭṭa Śarmā, Śālagrāma Upādhyāya, Lakṣmaṇa, Chātirāma Śāstrī, Śītala Miśra, Pāṭhaka Vaidyanātha Śarmā, Nāgeśvara Śāstrī, Sundaradīkṣita, Jyotirvid Rāmanātha Śarmā, Paṇḍita Jāgeśvara Upādhyāya, Hanumad Ācārya, Gāḍagila Kṛṣṇarāma Śarmā, Gaṅgārāma Śarmā Tripāṭhī, Gāḍagila Kṛṣṇa Śarmā, Kārlıkara Rājārāma Śāstrī, Devakṛṣṇa Śarmā, and Śıdhora Dādā Śāstrī.

28 The *śāstrārtha* quotes: *tathā ca tāni vacanāni. vārāṇasīsīmadevatā upakramya. nairṛte tu tato vidyād bhīmacaṇḍīvināyakam / āśāpuram iti khyātaṃ yatra tannagaraṃ mahat // kṣetrasya paścime bhāge śrīdehalivināyakaḥ / avatāraḥ sa vijñeyaḥ sākṣad ḍhuṇḍhe svayaṃ prabhoḥ //* (Hariścandra 1872: 12).

29 The map is introduced by the following line: *ab sāṃprat kāla kā pracarita mārga aur vāstava mārga inkā nakśā nīce likhate haiṃ.* In between Bhīmacaṇḍī and Rāmeśvara the pond *sindhusāgara jhīla* is shown. Below the dotted line *sāṃprata pracarita dehalivināyaka* is inscribed.

30 The map is described with the following words: *yahāṃ jo ūpar kī rekhā hai us ke ākār kā vāstava mārga hai. jo bīc ko ḍeḍhī rekhā hai us ke ākāra kā pracarita mārga hai. aur madhyameśvara se sāṃprata prasiddha jo dehalīvināyaka kā sthāna hai vahāṃ tak sūtra dhar ke ghumāne se jo mārga ho yah nīce jo tīsarī biṃduoṃ kī rekhā likhī hai us ke ākāra kā hogā. is nakśe ko dekhane se spaṣṭa hotā hai ki sāṃprata jo dehalivināyaka kā sthāna prasiddha hai us ko vāstava sthāna samajhane se bhīmacaṇḍī, caṇḍagaṇeśa, rāmeśvara ādi sab devatā kṣetra ke bāhar paḍatī haiṃ is liye vah sthāna vāstava mārga ke bhītar hai. arthāt śuddhā nahīṃ hai. is kāraṇa jo vāstava mārga prasiddha kiyā hai usī mārga se log pradakṣiṇā kareṃ nahīṃ to un kī pradakṣiṇā khaṇḍita hogī* (Hariścandra 1872: 9). A more detailed map was probably attached to the original document, and this is most likely the map Shashank Singh referred me to, as mentioned previously, but I was not able to locate it.

31 *aisī sthiti par bhī jo śāstra se, yukti se aur śuddha nāpa se siddha kiyā hai kī bhīmacaṇḍī aur rāmeśvara ke bīca meṃ pañcakrośī kā pracarita bahut bhītar se hai aur isī liye us mārga se pradakṣiṇā karaṇe se vah pañcakrośī khaṇḍita hotī hai is bāt par koi dhyāna nahīṃ dete aur usī aśuddha mārga se pañcakrośī karate haiṃ yah baḍe hi āścarya aur kheda kī bāt hai jo kaho ki pracarita mārga bhītar se hai aur yātrā khaṇḍita hotī hai yah sab ṭhīk hai parantu śuddha mārga prasiddha hī nahīṃ hai tab ānāyatyā lok isī aśuddha mārga se pradakṣiṇā karate hoṃge kyoṃki akaraṇān mandakaraṇaṃ śreyaḥ* (Hariścandra 1872: 3.).

32 *isī liye jahāṃ pradakṣiṇā meṃ do pagdaṇḍī mārga mileṃ aur un meṃ vāstava mārga kā saṃśaya ho to binā pūche niḥsandeha log bāṃe bhāga ke mārga se jāte haiṃ* (Hariścandra 1872: 3).

33 A striking example of changes in the sacred topography of Banaras is the shifting of the course of the Asī rivulet by the City administration during the last decade of the 20th century. As a result, the confluence of Asī and Gaṅgā (*asīsaṃgama*), one of the important landmarks defining the sacred territory of Banaras, was moved southward.

References

Maps

Saptapurīyātrādiprakāśapatra (Banaras Archives no. 4) Vārāṇasī 1873. Printed for Jaṅga Bahādur Siṃha (Citaipur), Banaras Akhbar Press. Lithography on paper. Language: Hindi, script Devanāgarī. Private collection Shashank Singh, Vārāṇasī. 76 x 72 cm.

Kāśīdarpaṇa (Banaras Archives no. 5) Vārāṇasī, 1876. Printed at the Vidyodaya Press for Kailāsanātha Sukula. Lithography on cloth, private collection Niels Gutschow; Lithography on paper, British Library 53345(2). Language: Sanskrit and Hindi, script Devanāgarī. 79 x 92 cm.

Kāśīdarpaṇapūrti (Banaras Archives no. 7) Vārāṇasī 1877. Printed for Kṛṣṇacandra Śarmā at the Vidyodaya Press. Lithography on paper, black & white. London: British Library 53345.(4.) and (3). Language: Hindi, script Devanāgarī. 64 x 94 cm.

Uttar Pradesh, Ghāzīpur and Vārānasi Districts (Banaras Archives T-13), First Edition, 1: 50.000, no. 63 K/15, Surveyed 1972–73, Surveyor General of India Dr. Hari Narain, Government of India 1975, Survey of India.

Texts

Anantaśrīcaraṇa Devatīrtha (Kāṣṭhajihva Svāmī) 1997. *Paṃcakrośa sudhā. anuvādak evaṃ sampādak: Ācārya Paṃ. Satyanārāyaṇa Śāstrī Khaṇḍūrī*. Vārāṇasī: Sarvabhāratīya Kāśīrāja Nyāsa.

Hariścandra 1872. *Pañcakrośī ke mārga kā vicāra. śrīyuta bābūhariścandra kī ājñā se chāpā gayā. saṃvat 1926. Banāras. Meḍikal hāl ke chāpekhāne meṃ chāpā gayā. San 1872.*

Kāśīdarpaṇam, nānāpurāṇebhyaḥ saṃgṛhītam, hindibhāṣayā anuvāditañ ca. Saṃgrāhakaḥ prakāśakaś ca śrīvārāṇasīnivāsī Śrīkṛṣṇacandradharmādhikārī Bhāratānunvādakaḥ. Kalikātā, Kāvyaprakāśa-nāmadheya-yantrālaye, sāma-bhāgavatānuvādaka Śrībrahma-vrata-sāmādhyāyi-Bhaṭṭācāryeṇa saṃśodhya mudritam. Saṃvat 1932 [=1875].

Kāśīrahasyam, Brahmavaivartapurāṇapariśiṣṭam saṭīkam kāśīrahasyam, Rādhākṛṣṇa Mor (ed.), Gurumaṇḍalagranthamālā No. 14, Vol. 3, Calcutta: 1957.

Tristhalīsetu of Nārāyaṇa Bhaṭṭa. Gaṇeśa Śāstrī Gokhale (ed.), Poona: Ānandāśrama-mudraṇālaya, 1915 (Ānandāśrama Sanskrit Series 78).

Secondary Literature

Dalmia, Vasudha. 1997. *The Nationalization of Hindu Traditions. Bhāratendu Hariśchandra and Nineteenth-century Banaras*. New Delhi: Oxford University Press.

Gengnagel, Jörg. 2000. Visualisierung religiöser Räume—Zur Kartographie von Benares. In: Daria Pezzoli-Olgiati & Fritz Stolz (eds.), *Religiöse Kartographie. Organisation, Darstellung und Symbolik des Raumes in religiösen Symbolsystemen*. Bern: Peter Lang, pp. 213–230.

— 2003. Mapping Sacred Spaces—Aspects of Cartography in 19th Century Banaras. In: Vera Dorofeeva-Lichtmann & Daniela Moritz (eds.), *Creating and Representing Sacred Spaces*. Göttinger Beiträge zur Asienforschung. Göttingen: Peust&Gutschmid, pp. 247–263.

— 2005. *Kāśīkhaṇḍokta*: On Texts and Processions in Vārāṇasī. In: Jörg Gengnagel. & Ute Hüsken & Srilata Raman (eds.), *Words and Deeds. Hindu and Buddhist Rituals in South Asia*. Wiesbaden: Harrassowitz, pp. 65–89. (Ethno-Indology. Heidelberg Studies in South Asian Rituals 1).

— forthc. *Visualized Texts—Sacred Spaces, Spatial Texts and the Religious Cartography of Banaras*. Habilitationsschrift, Heidelberg 2006.

— & Axel Michaels. 2001. *Kailāsanātha Sukula's Mirror of Kāśī (Kāśīdarpaṇa, 1876). A historical map of Vārāṇasī based on the Kāśīkhaṇḍa and related texts*. Electronic Publications of the Vārāṇasī Research Project, South Asia Institute, Heidelberg. (*www.benares.uni-hd.de*).

— & Birgit Mayer-König. 2003. *Descriptive Catalogue of mainly Visual Material on Varanasi kept in the Banaras Archive of the South Asia Institute*. Vārāṇasī Research Project, South Asia Institute, Heidelberg. (*www.benares.uni-hd.de/ben-archive.pdf*).

Gole, Susan. 1989. *Indian Maps and Plans: From Earliest Times to the Advent of European Surveys*. New Delhi, Manohar.

Gutschow, Niels. 2006. *Benares–The Sacred Landscape of Vārāṇasī*. Stuttgart/London: Edition Axel Menges.

Kerr, Ian J. 2001. Reworking a Popular Religious Practice: The Effects of Railways on Pilgrimage in 19th and 20th Century South Asia. In: Ian J. Kerr (ed.), *Railways in Modern India*. Delhi: Oxford University Press, pp. 304–27.

Köckmann, Uwe. 1982. *Hindutradition und Stadtentwicklung. Varanasi: Analyse einer gewachsenen ungeplanten Pilgerstadt am Ganges*. Bochum.

Singh, Rana P. B. 1988. The Image of Varanasi: Sacrality and Perceptual World, *The National Geographical Journal of India* 34: 1–32.

— 2002. *Towards the Pilgrimage Archetype. The Pañcakrośī Yātrā of Banāras*. Varanasi: Indica Books.

Sukula, Kubernāth. 1977. *Vārāṇasī Vaibhava*. Patna: Bihār Rāṣṭrabhāṣā Pariṣad.

Upādhyāya, Baldeva. ²1994. *Kāśī kī pāṇḍitya paramparā (kāśīstha saṃskṛta vidvānoṃ jīvancarit evaṃ sāhityak avadānoṃ kā prāmāṇik vivaraṇ) [1200–1980]*. Vārāṇasī: Viśvavidyālaya Prakāśan.

SUMATHI RAMASWAMY

Enshrining the Map of India: Cartography, Nationalism, and the Politics of Deity in Varanasi[1]

On the 25th of October 1936, a very unusual shrine opened its doors to the public in Varanasi (Fig. 1). Named "Bharat Mata Mandir", "Mother India Temple", it was, in the words of one contemporary visitor, "a new kind of shrine which would call forth the devotion and service and sacrifice" of every Indian, "without distinction of caste or colour or creed, race or sex".[2] Its opening was attended by every major figure associated with the Indian National Congress.[3] And no less a personage than Gandhi presided over the inauguration ceremony. Those who spoke at the opening ceremony noted that it was a shrine which did not call forth "blind worship, but a dedication to an idea".[4] That idea was nationalism, and it called for "a new kind of worship and a new kind of shrine".[5] In the words of one speaker, "It called forth all the children of the land to the worship of the Mother, forgetful of divisions and differences (...). The Father of us all had created us and given us the land to live on and food to eat. We, His children, often quarreled amongst ourselves, but the Mother called [us] together to forget [our] bickerings at her peace-giving feet".[6] The temple was, indeed, variously characterized in contemporary reports as "a new House of God",[7] "a shrine of Love", "a place to forget communal differences",[8] and in Gandhi's words, "a cosmopolitan platform for people of all religions, castes, and creeds", that would promote "religious unity, peace, and love in the country" (Gandhi 1958: 388).

But perhaps the most unusual feature of the temple, frequently noted by contemporaries as well, was not so much the idea that it embodied as what it enshrined—or did not (Fig. 2). As Gandhi noted in his very opening words, "the temple contains no image of any god or goddess. It has only a relief map of India made of marble stone (...)" (Gandhi 1958: 388). In the words of Mahadeo Desai, one of Gandhi's aides, "There is no image of an imaginary God or Goddess in the shrine, though Babu Shivprasad [Gupta] is a devout believer in India's Gods and Goddesses, inasmuch as he knows that there is the grand imagination of Indians behind those images. But he saw that the new times demanded a new kind of worship and a new kind of shrine which would rally India's sons and daughters. A relief map of India in marble was what he conceived of (...)".[9] Both the story of what led the Banaras philanthropist named Babu Shivprasad Gupta (1883–1944)[10] who financed the building of the temple to enshrine this cartographic image of India, and the details of the map itself, are inscribed in marble on the inside walls of the temple.[11] The idea of using the map of India to represent "Bharat Mata" dawned upon Gupta in 1913 when he visited a school founded for young widows in Pune by the Marathi social reformer, D. K. Karve (1858–1962). He saw etched there on the ground a relief map of India fashioned out of mud

Fig. 1: Bharat Mata Temple. Benares. Built by B. Shiva Prasad Gupta. Postcard, c. 1936 (courtesy Joachim Bautze, Heidelberg).

and clay. Subsequently, on a visit to the British Museum in London, he marveled at the many printed maps, big and small, he saw displayed there. Returning to Banaras, he commissioned in 1918 his friend Durga Prasad to design a relief map of India, made to scale and laid out neatly on a grid.[12] The foundation stone for the temple was laid in April 1927,[13] and work began on the temple with the help of 30 sculptors and 27 masons (all of whose names are inscribed on one of the inside walls of the temple). These local sculptors, most of who were natives of Banaras, had to be taught

> the new art of making mountains and plains, and rivers and lakes, of the right proportion, from slabs of marble. Seven hundred sixty-two square pieces (11" x 11") of marble (with smaller bits here and there) have gone to the making of this relief map, 31 feet and 2 inches long, and 30 feet and 2 inches broad. Every one of the physical features shown is to scale, a surface inch being equal to 6 miles and 704 yards, and an inch of height meaning 2,000 feet. Thus Mt. Everest has been cut out of one piece of marble 15 inches high. The height of all of India's plains and mountains from the sea level has been accurately shown and gradation of heights from 500 feet to 30,000 feet have been shown in figures to draw the spectator's attention to relative heights. Over 400 peaks of India's mountains have been shown, all her rivers with their relative

length and breadth and depth have been shown, as all of India's principal lakes. All the latest geometrical knowledge has been utilized for making the relief map as accurate as possible, and all important cities and towns and villages have been shown.

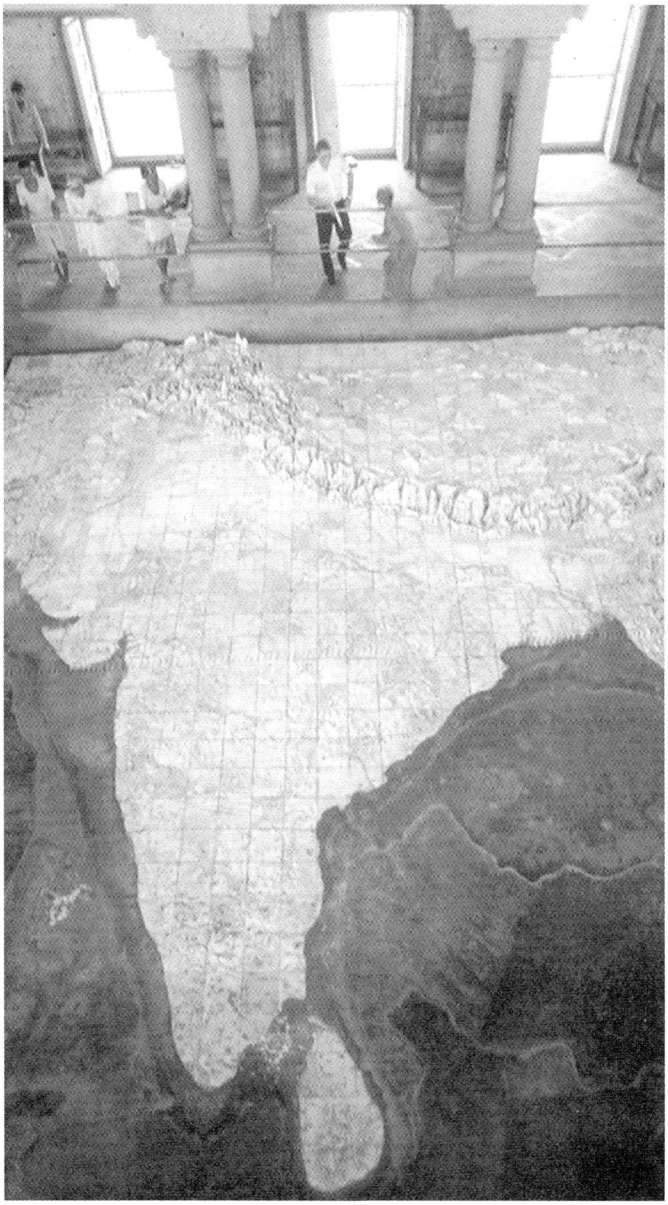

Fig. 2: Map of India, Bharat Mata Mandir, Varanasi. Contemporary post card.

> Twenty artisans were at work cutting, shaping and putting together these marble pieces, from day to day during five long years. The walls of the temple have maps in colour showing the different periods of Indian History as also various features of physical geography.

So marveled Desai, concluding his ecstatic comments with the declaration that "the relief map is a poem in marble, and the whole shrine is a temple not only of art but of instruction and knowledge".[14]

Monumentalizing the Map of India

My own interest in this unusual temple and its quixotic object of enshrinement flows from the very idea of creating a monument to the nation that is centred on a relief map of its territory.[15] This is all the more unusual because I have suggested elsewhere that Indian modernity is even today not particularly map-minded (Ramaswamy 2001, 2003). And this was especially true in the 1930s when this temple to Bharat Mata opened its doors to the public.

As a monument to the nation and its territory, the temple memorializes the nation-space in a particular way, as extending from Afghanistan in the west into Burma in the east, and from the Himalayas in the north to the ocean to the south, incorporating in the process the island of Sri Lanka. This is "akhand Bharat", "undivided India", a geographical notion that first found voice in ancient Sanskrit texts and then got a new lease of life under *pax Britannica* (albeit in a somewhat truncated form), but no longer prevails after the end of empire, the Partition, and the creation of numerous independent nation-states in the subcontinent. Yet this memorial to the nation built prior to its catastrophic partition freezes a moment in time—under British rule—when much of this space was seemingly unitary and singular, and operated under the sign of "India".

As such, the map-as-monument attempts to produce what I would characterize as visual piety for the nation.[16] Undivided India is enshrined and presented to the viewer-patriot in much the same way that in the generic Hindu temple, gods and goddesses are consecrated and displayed. Diana Eck notes that "the central act of Hindu worship, from the point of view of the lay person, is to stand in the presence of the deity and to behold the image with one's own eyes, to see and be seen by the deity" (Eck 1981: 3). Through such acts of seeing and being seen—as embodied in the practice of *darśan* (lit. seeing)—Hindu devotees do not just see the deity enshrined in the image, but see themselves being seen by the enshrined image. So, Hindu worship is intensely ocular, predicated as it is on a series of visual interactions and exchanges between god and devotee (Babb 1981). Similarly, through the device of the map enshrined in the Bharat Mata Mandir, the nation's territory is presented to the patriot for visual contemplation and consumption. What makes this act of enshrinement particularly intriguing is that the map functions both as an iconic object as well as an aniconic artifact in this temple. From the perspective of territorial realism, the relief map of India is iconic in the Piercian sense of the term, predicated as it is on a formal correspondence or resemblance between sign and referent. As a representation of Bharat Mata in this first temple built to memorialize her, however, the map is aniconic, replacing Mother India's familiar anthropomorphic presence with a disembodied cartographic form.[17]

Yet, because neither as iconic representation of Indian territory nor as aniconic replacement for Bharat Mata does the map of India have a deep and visible purchase in the popular and public culture of its putative audience, the capacity of the temple of Mother India to function effectively as an institution through which visual piety for the nation could be productively generated amongst its citizenry was compromised from the start, a point to which I will return at the end of this essay.

I forefront this question of the temple's capacity to mobilize patriotic sentiments because the map of the India was clearly monumentalized at the Bharat Mata Mandir in order to generate what the geographer Yi-Fu Tuan once called "geo-piety", which he defined as passionate longing for and attachment to a piece of land, in this case, the nation's territory (Tuan 1976). This much is apparent from Gupta's own rendition of why he built the temple (Gupta 1983b) as well as from the statements of those who were present at its inauguration. The temple with its enshrined map was presented as a memorial to the nation, a shrine which unlike other sectarian ones, would bring together Indians of all creeds and walks of life, instead of separating them from one other. So defending his effort from criticisms, Shivprasad Gupta announced on the eve of its inauguration, "The Mandir means merely a house in which a large relief map of India in marble is placed. The image is merely this map, and on the walls are painted astronomical and historical charts. The temple is open to all and every religion, creed, and caste".[18] What makes this particular attempt at producing geo-piety particularly ironic is that the object used to produce reverence for and attachment to the idea of India was itself a product of colonial science. In fact, it is noteworthy that when Gupta decided to build a "temple to nationalism", dedicated to the "worship" of "Bharat Mata", he turned, not to the rich iconographic and architectural traditions of Hinduism, but to the knowledge practices of colonial and scientific modernity. As historians of cartography like Matthew Edney write in great detail, the topographic mapping of India, its coastlines, its mountain ranges, its river courses, and so on, was one of the earliest projects that the British colonial state and the Survey of India took on from the closing decades of the 18th century. The conversion of irrational Hindu space into the precisely surveyed and measured Euclidean place of modern scientific cartography was the most important consequence of this act of mapping India, from the colonial perspective. The resulting idea of a bounded and circumscribed peninsular place named "India", was then circulated through geography textbooks and manuals, and cartographic devices like maps and globes (Edney 1997; Ramaswamy 2003).

Product though it may be of colonial cartography, the topographic map of India came in useful for Shivprasad Gupta and his fellow nationalists for it shows the nation-space as a singular territory, not divided and carved up into various pieces.[19] The topographic map also naturalizes India, presenting it—in marble, no less, at the temple—as something that exists naturally and empirically like the mountains, rivers and valleys that appear on its surface. One may see at a glance what kind of terrain the nation possesses, for the relief map serves as "a condensed image of the state's configurated landscape", bringing into the patriot's vision "the drama of depth". (Casey 2002: 148,

150). So, "India" is no fiction of imagination, but as real and living as the very soil that constitutes it. Not least, the topographic map with its emphasis on India's natural features also allows the nation's territory to be perceived as timeless as the mountains to the north, the rivers that make their way patiently across its landscape, and as imperishable as the marble used to memorialize it within the walls of the Bharat Mata Mandir.

Somaticizing India

Yet the colonial geographic project, driven as it was by the imperatives of the modern state, made virtually no attempt to accommodate an alternate enchanted imagination of the same territory that was now named "India", as sentient, sacred, and female.[20] This latter imagination, that I have elsewhere characterized as "somatic", first surfaced in the 1870s and 1880s amongst Bengali nationalists—long after the colonial geographic and cartographic projects had taken off—when a new conception of India as a goddess and as a mother named Bharat Mata, took shape (Ramaswamy 2001; see also Ramaswamy 2002). As others have demonstrated, in one of her earliest visual representations as it finds expression in a well-known painting by Abanindranath Tagore, produced around 1905, she appears as a young, desexualized, ascetic figure, her four arms bearing the blessings of food, clothing, learning, and salvation (Sen 2002: 17–30). Soon after, from around 1909, she comes to be visually presented to the Indian public, occupying the map of India. In another essay, I detail the manner in which the mother's body and the map of India come together within the nationalist cartographic practices of the past century (Ramaswamy 2001), but here I will limit myself to noting that there are at least four different ways in which Bharat Mata is made to relate to the map of India in the various print-visual productions of the emergent nation that I refer to as "bodyscapes", as these find expression in calendar art, textbook illustrations, jacket covers of nationalist pamphlets and books, newspaper cartoons, greetings cards, and the occasional painting. The historian of art from Bengal, Pratapaditya Pal recalls "how such pictures would excite us when the atmosphere was charged with the electrifying sparks of the freedom movement" (Pal 1997), and other contemporaries note the power of such visual images to move patriots to heroism and martyrdom. It is this power that the burgeoning printing industry of colonial India deployed in the service of generating visual piety for the nation and geo-piety for its territory. First and most commonly, Bharat Mata literally occupies the map of India, filling up what would otherwise be empty cartographic space with her body (Fig. 3). There are many ingenious ways in which her body blurs the neatly-drawn boundaries of state cartography which virtually disappear as her head is made to occupy Kashmir; a scarf streams out to conceal Pakistan, a toe or flag-staff reaches out to claim Sri Lanka, and so on. In a few examples, her body shares the cartographic space of India which it occupies with the bodies of well-known "Indians" and nationalists, almost always male (see for example, Ramaswamy 2003: Fig. 4). In a second type of bodyscape, rather than her body occupying the cartographic space of India, Bharat Mata stands (or sits) on a map of India whose outlines are etched on a terrestrial globe. In such visualizations which increase in popularity from the 1940s, Bharat Mata literally stands on top of the world, her head reaching out into the very cosmos itself, signifying her liberation from British rule (Fig. 4).

Fig. 3: Bharat Mata, artist unknown, 1958. Frontispiece, V. Lakshmanan, *Putiya Aarampakalvi Tamil (Moonraam Puttakam)*, Mannargudi: Shri Shanmugha Publishing House (courtesy Tamilnadu State Archives, Chennai).

Less common, but quite intriguing, are those bodyscapes in which Bharat Mata appears to springs out of the map of India, the lower part of her body disappearing into its cartographic contours (Fig. 5). Such prints most clearly capture an important suggestion made by many nationalists that there is no distinction between the territory named

"India" by colonial geography, and Bharat Mata who resides in all its constituent parts. As the nationalist Aurobindo wrote in 1905, "What is a nation? What is our mother country? It is not a piece of earth, nor a figure of speech, nor a fiction of the mind. It is a mighty *Shakti* [force, power], composed of all the Shaktis of all the millions of units that make up the nation (...)" (Ker 1973: 37).

Fig. 4: Roop Kishore?, "Shaheed Bhagat Singh", Rising Art Cottage, Calcutta, circa. 1940–45 (courtesy Christopher Pinney Collection).

Fig. 5: "Ma ka Aahvaan" (Call of the mother), artist unknown, 1993 (courtesy Christiane Brosius).

Finally, and most pertinent to the concerns of this essay, are those bodyscapes in which the map as an identity prop is dispensed with, and the body of Bharat Mata literally stands in for the map of India, outlining in this process the cartographic shape of India. An early emergent example is a 1923 bodyscape that appeared on the cover of a San Francisco-based journal of the expatriate Ghadar Party in which Bharat Mata stands on a globe on which is etched the cartographic outline of India out of which she emerges; her sari and her hair map out the cartographic shape of India (this is reprinted in Ramaswamy 2003: Fig. 1). A more striking example is a 1937-poster from Coimbatore in which Bharat Mata appears clad in the Indian tricolour. The contours of her body echo the cartographic outline of India, as her tricolour sari billows out laying claim to the different territorial spaces of the emergent nation (Fig. 6). In striking contrast to the Banaras temple where the anthropomorphic image of Bharat Mata is dispensed with to be replaced by the aniconic map of India, here, the map of India is displaced by the image of Bharat Mata.

Fig. 6: P.S. Ramachandra Rao, "Vande Mataram" (Homage to Mother), Rao Brothers, Coimbatore, 1937, colour lithograph (courtesy Robert del Bonta collection).

The Politics of Representation: The Map versus the Mother

When I first started this project a few years ago, I was convinced that visual productions of Bharat Mata needed the map of India as a prop for establishing her identity as the deified embodiment of the emergent nation, as a goddess of polity and territory. How else could her devotees visually set her apart from the innumerable other Hindu goddesses on whose iconographic traditions she was parasitic, and with whom she could be readily confused if it had not been for the map of India? Here, to some extent I was lulled into thinking thus by Benedict Anderson's insightful comments on the place of maps in modernity, and especially by his observation that the logo map of the nation "penetrated deep into the popular imagination, forming a powerful emblem for the anti-colonial nationalisms being born" everywhere in the colonies in the first half of the twentieth century (Anderson 1991: 175). Today, I am no longer convinced of the salience of maps in India's nationalist struggles or even for Indian modernity. To be sure, there are scattered examples of the symbolic use of the map of India by various individual patriots and nationalist organizations from the early years of this century which I have documented elsewhere (Ramaswamy 2001; 2003). In fact, a map of undivided India showing major mountain ranges and rivers is the logo for the Kashi Vidyapeeth (a nationalist university started in Banaras by Shivprasad Gupta in 1921), on whose campus the Bharat Mata Mandir stands. This logo continues to be used today on the university's letter-heads and publications. Maps of India were also used in nationalist posters and cartoons from the 1930s which cartographically depict the economic consequences of colonial rule, and which, not surprisingly, were proscribed and put out of circulation by the colonial state. In these instances, the cartographic space of India is not occupied by the fulsome and prosperous Bharat Mata, but by the emaciated body of the Indian peasant, near-naked, starving, or dead. And many posters published in the 1940s and 1950s showed the outline map of India occupied by the bodies of well-known male nationalists like Gandhi, Nehru or Bose.

In many such patriotic deployments, the map of India was frequently extracted from the disenchanted geographic imaginary in which it circulated in colonial discourse, and re-embedded in the somatic imagination of India as the sentient, divine, and female Bharat Mata. That is to say, almost from the start, the map of India, as it was used in patriotic visual practices, is associated with the vocabulary and imagery of Bharat Mata, as in an early 1907 print which shows the goddess sitting with her arms resting on a globe on which is etched the cartographic outline of India (Ramaswamy 2003: 156). So much so that I now suspect that it is not so much Bharat Mata who needs the map to prop up her identity, but the map that needs the familiar image of the goddess to legitimize its claim to (cartographically) represent "India". This is also the reason that I do not agree with Matthew Edney's assertion in his important study of colonial cartography in British India that "the colonial geographic conception of India was adopted *without question* (...) by Indian nationalists" (Edney 1997: 15–16, emphasis mine). The bodyscapes produced by visual patriotism are a striking reminder of the strategic disjuncture between colonial cartographic practices and at least one strand of nationalist cartography in modern India. This disjuncture—and difference—emerges from the insertion of the deified and feminized image of Bharat Mata into the disenchanted geographic conception of the colonial state. So, we learn that around 1903, in the context of the nationalist fervor that had spread through Bengal Presidency, a group of young Bengali men resolved to sacrifice their lives fighting for independence from British rule by taking a pledge in front of a map of India but revealingly, in the name of Bharat Mata (McKean 1996: 252–253). Similarly, the logo of the so-called "Anti-Circular Society" (started in November 1905 in Calcutta) was a circle with an outline map of India; inscribed on the map are the words, "United Bengal 80000000, United India 300000000", while around it are written the phrases, "For God and Motherland" and "Liberty, Equality, Fraternity" in English, and most strikingly in Devanagari, the words "Vande Mataram", "Homage to the Mother".

It is worth dwelling on the association between the map of India and the phrase "Vande Mataram" in this early 1907 hybrid logo, especially because the same phrase is inscribed in numerous places on the walls of the Bharat Mata Temple in Banaras as well. As many scholars of Indian nationalism have noted, "Vande Mataram" became a key nationalist slogan from the early years of this century, as the hymn to Bharat Mata from which it was derived became the signature anthem of the nascent nationalist movement. Although the hymn itself is not explicitly cartographic in its celebration of Bharat Mata, it is interesting to note that in the 1970s, during the centenary celebrations commemorating its original composition around 1875, numerous publications on the hymn featured maps of India as well as bodyscapes of Bharat Mata (Ramaswamy 2003: 171). I suggest that they do this because it is through the somatic imagination of national territory as a sentient, deified and gendered being in the form of Bharat Mata that the map of India has been made familiar to the national subject during the past century or so of patriotic activity and mobilization. In other words, its circulation through textbooks and other productions of the colonial and post-colonial state notwithstanding, the map of India gains visibility amongst patriotic Indians not just as a disenchanted representation of the territorial space of India, but also as an artifact that commands their reverence and worship, even sacrifice of body and self. This reverential and worshipful attitude is facilitated by the association of Bharat Mata with the map. Early in the century, Rabindranath Tagore had one of his patriotic fictional heroes insist that "no one can give up his life for a map" (Tagore 1985: 90–91). I want to suggest that Indian nationalism "solved" the problem of getting Indians to die for an abstract territorial space named "India" by inserting the image of Bharat Mata into the empty social space of the map of India. Such bodyscapes of Bharat Mata which were circulated through the various visual activities of nationalism, converted the citizen-subject from a neutral observer of the cartographic image of India into its dedicated patriot, so that the map was not just a map but a "loving" homeland and motherland (Ramaswamy 2003). So, in 1905, the nationalist Aurobindo Ghoshe, a teacher in Baroda at that time, insisted that the map of India hanging on his office wall was "not a map, but the portrait of Bharat Mata; its cities and mountains, rivers and jungles form her physical body; all her children are her nerves, large and small. Their literature is her memory and speech; their vitality is her life...Concentrate on India as a living mother, and worship her with nine-fold devotion", he commanded (Dave et al. n.d: 38).

It is because Mother India's visual image has played such an important role in this crucial conversion of the citizen into patriot that its absence in the Bharat Mata Mandir in Banaras, the first temple to the goddess to be built, is striking. To be sure, the absent mother is seemingly present everywhere in the temple: in the very name of the temple to start with, and in the inscription of her slogan Vande Mataram, both on the portals as well as on inside walls. The inauguration ceremony in 1936 began with a hymn to Bharat Mata; numerous speakers at the ceremony invoked her name, and the famous Hindi poet Mythili Sharan Gupta composed and recited a poem in honour of the temple which ended with reminding all those who had assembled in "Bharat Mata's courtyard" of the "fraternal bonds" that tied them together; the poet also recalled the Mother's "labor pains" in giving birth to her "sons". The ceremony closed with a recital of the

Vande Mataram hymn.[21] Shivprasad Gupta himself, in an essay that he published on the temple, repeatedly referred to the map itself as "matramurti", "mother's image"; as "mata ki murti", "mother's image", and even explicitly as "Bharata Mata ki murti", "Bharat Mata's image".[22] At the same time, as I noted earlier in the paper, contemporaries were quick to observe—*and* appreciate—that the image of Bharat Mata was nowhere to be seen in the temple, and that *only* a map may be found inside its walls. They were also keen to reiterate that no rituals of worship would be performed, and even today, visitors are asked not to offer fruits or flowers to "the image".[23]

It is also important to underscore that everyone present at the inaugural ceremony was quick to point to the spirit of inclusivism that permeated this new temple. Newspapers reported that close to 25,000 people from *all* of India's communities—Hindus, Muslims, Christians, Jains, Parsees, Buddhists and the so-called "Untouchables", Gandhi's Harijans—attended the inauguration; prayers from various scriptural traditions were recited at the ceremony, and the temple was declared open to all, regardless of religious affiliation. As one contemporary reporter who was present at the inaugural ceremony wrote,

> The temple of Bharat Mata will be the first of its kind in India and has behind it a principle so very essential to the freedom of our country. The Muslims, the Christians, the Jews, the Parsis, and the Hindus need not now go to separate places of worship to realize the highest in them [*sic*]. Here is a temple dedicated to the most sacred goddess of Hindustan [India] where all without distinction of caste and creed can offer worship and realize the highest moral and spiritual beatitude.[24]

The same spirit of inclusivism finds a place in a tourist guide-book published as recently as 2000 in which the Bharat Mata Mandir is introduced thus:

> A unique temple dedicated to Mother India. The presiding "deity" of this temple is a relief map of India, engraved on a marble slab. By avoiding the use of customary idols and images that go into the making of a Hindu shrine, this temple stresses the oneness of religion and seeks to foster unity among different communities of India (Srinivasan 2000: 19).

Yet, the undertow of Hinduism was apparent here, as indeed in so many other public and "secular" events in colonial India: the foundation stone for the temple was laid in April 1927 with the recitation of hymns from Hindu scriptures alone and to the accompaniment of Hindu rituals of consecration; the inauguration ceremony took place on Vijaya Dashmi, one of the holiest days of the Hindu ritual calendar; contemporaries frequently referred to the map enshrined in the building as the "*murti*", "image;" and the building that houses the map, although not resembling a conventional temple (see Fig. 1), is after all referred to by the generic Hindi term for a Hindu house of worship as *mandir*.[25] Most tellingly, since at least the mid-1950s, the Bharat Mata Mandir has been incorporated into the popular pilgrimage maps of Hindu Banaras that many scholars have written about (Eck 1996; Pieper 1979; see also Gengnagel this volume). Almost invariably, the icon of an anthropomorphic Bharat Mata, sometimes occupying a map of India, marks the spot where the *mandir* to the goddess stands on the pilgrimage route (see, for example, Fig. 7–8).[26] Shivprasad Gupta and his colleagues may have banished

Bharat Mata's anthropomorphic image from the inner precincts of the temple to her that they fashioned, but popular Hindu imagination has re-asserted itself and brought the mother back in (sometimes with and sometimes *sans* the map). The anthropomorphic appears to have triumphed over the abstract when it comes to the matter of representing Indian national territory, at least in popular pictorial representations of the Mandir.

The Politics of Deity: The Map as *murti*

What does remain striking about the Bharat Mata Mandir, however, is the absence of the anthropomorphic *murti* of Bharat Mata, rendered familiar through decades of visual patriotism. What accounts for this absence from a site that is ostensibly meant to commemorate her? Why did the makers of the temple banish her familiar bodily presence, replacing it with the impersonal cartographic map of India? And how successfully does the map perform its new role as *murti,* iconic of India's territory even while aniconically representing Bharat Mata?

I answer such questions by considering the contentious survival of Hindu modes of worship within India's tortured modernity. On the one hand, as the proliferation of the visual and verbal work done around Bharat Mata from the 1870s suggests, India's nationalists, many of whom had undoubtedly developed a very ambivalent public attitude to mainstream Hinduism in the context of trying to forge a "secular" nationalism, came to increasingly depend on one of that religion's diacritical features, namely, image worship, in their quest to enliven and sacralize the otherwise disenchanted abstract territorial space named "India". Hence the frequent invocation of Bharat Mata in nationalist discourses of both the religious and secular hues.

On the other hand, the worship of images by Hindus had become by the early twentieth century a cultural and political embarrassment not just to those "secular" nationalists who came to believe that such a practice alienated non-idolatrous citizens of India; but even to Hindu reformers and intellectuals who at least since the 1830s had responded to a scathing colonial and missionary critique of Hinduism by rejecting the worship of images in temples on various grounds. As Ashis Nandy notes in a recent insightful essay, for these men,

> the country's main problem was its idolatry, and the rather poor personal quality of its gods and goddesses. These reformers wanted Indians to get rid of their superfluous deities and either live in a fully secularized, sanitized world in which rationality and scientific truth would prevail or, alternatively, set up a regular monotheistic God, as "proper" Christians and Muslims had done. (Nandy 2001: 132).

Many such reformers who by the closing decades of the 19th century began to participate in an emergent Hindu nationalism, even questioned the temple itself as institution which divided Hindu from Hindu. So, in 1923, one such reformer Dr. Balkrishna Moonje, a leading figure from Nagpur of the Hindu Sabha, commented on

> the absence of a common meeting place in the Hindu polity for the castes from the highest to the lowest [to meet] on perfectly equal terms, just as the Mohammedan has

> in his masjid (...). Cannot our temples be made to serve the purpose which the masjid does for the Mohammedan? (Quoted in Jaffrelot 1996: 22).

Almost as if in response to Moonje's rhetorical question, the Punjab-based Swami Shraddhananda, a follower, like Gupta, of the Arya Samaj drew up an agenda in 1926 for "saving" what he characterized as "the dying Hindu race". This agenda included as its very first step the building of

> one Hindu Rashtra Mandir [temple of the Hindu nation] in every city and important town, with a compound which could contain an audience of 25,000, and a hall in which *Katha* [prose selections] from Bhagavad Gita, the Upanishads, and the great epics Ramayan and Mahabharata could be daily recited. The Rashtra Mandir will be in charge of the local Hindu Sabha which will manage to have Akharas [places for physical exercise] for wrestling and *gatka*, etc. in the same compound. While the sectarian Hindu temples are dominated by their own individual deities, the Catholic Hindu Mandir would be devoted to the worship of three mother-spirits: the Gau mata [the cow-mother], the Saraswati-mata [goddess of knowledge]; and the Bhumi-mata [motherland]. [....] *Let a lifelike map of Mother-Bharat [India] be constructed in a prominent place, giving all its characteristics in vivid colours so that every child of the Matri-bhumi [earth mother] may daily bow before the Mother and renew his pledge to restore her to the ancient pinnacle of glory from which she had fallen.* (Quoted in Jaffrelot 1996: 22, emphasis mine).[27]

Even as Swami Shraddhananda was writing thus, his fellow Arya Samajist Shivprasad Gupta had already attempted to bring life to such a vision in the Bharat Mata Mandir in Banaras. In Gupta's vision, as we have seen, the map of India operates as a material sign of Bharat Mata, aniconically conjuring up for the benefit of the patriotic viewer "her" vast expanse, the beauty of "her" landscape, the majesty of "her" hills to the north, the fertility of "her" rivers, and so on.[28] The map of India, that colonial artifact produced to control and domesticate the irrational inchoate Indian space, and to iconically represent Indian territory, was thus appropriated by Indian nationalists and aniconically deployed by them in their quest to create the new "religion" of nationalism centred around the worship of Bharat Mata. In this new religion, indebted though it was to its Hindu moorings in various ways, the embarrassment posed by the very Hindu body of Bharat Mata was dealt with by aniconically displacing the anthropomorphic *murti* of the deity with the abstract map of the nation. Patriotic visual practices from the early decades of the century had established an intimate connection between Mother India's body and the map of India. This meant that when the deification of India's territory as an anthropomorphic Hindu mother goddess became problematic, as it increasingly did in the context of the nationalist mobilization of non-Hindus, the map could aniconically stand in for her idolatrous body as it did in the Bharat Mata Mandir.

Fig. 7: Section of *Asalī kāśī kṣetra kā naksā Map of Kashi*, Puja Publications, Varanasi, n. d. pilgrimage map. Archives Vārāṇasī Research Project, South Asia Institute, Heidelberg. The Bharat Mata Mandir appears on the upper right corner of the map, just south of the railway line that snakes its way down the eastern edge of the map. The Mandir is designated by a building, and the image of Bharat Mata (*sans* map). The building is named "Bharat Mandir" in Hindi, and Bharat Temple in English.

But the question we are left with is, could it? Soon after the temple was opened, the Bombay-based *Indian Social Reformer* expressed its doubt thus:

> We view with some anxiety the prospect of the growth of new cults based not on religious but territorial idealism. The relief map of India, according to the *Guardian* of Madras, comprises the whole of India with its sister countries, like Afghanistan, Baluchistan, Tibet, Burma, part (why part?) of the Malaya Peninsula and Ceylon. All these are to be worshipped as Bharat Mata which, according to the ancient Hindu geographers, was the land bounded by the Himalayas and the three seas. The temple, no doubt, will have its daily puja and hierophants to perform it. *The Indian has a passion for worship but we much doubt if his sentiments of reverence will be aroused by a map, even though it is done in mekran marble.*[29]

The Bombay Chronicle was even more skeptical. In an editorial entitled "Cosmopolitan Temples", it noted, commenting on Gandhi's approval of the temple that

> we share the hope, if only because India surely needs religious unity and peace. But we feel at the same time that a great deal more than building cosmopolitan prayer houses is necessary if that hope is to be fulfilled in any ample measure and in the near future. We do not see why all the existing temples, mosques, and churches should not be open to all persons seeking them. In any case religious unity is possible only to the extent that every religion is interpreted in a thoroughly rational manner and

> consistently with the dictates of humanity. In the absence of such an interpretation of religion, temples, mosques, and churches will be, as they have been in the past, a source of religious conflict rather than of religious unity. Bigotry glorified as religion has come to stink in the nostrils of all right-thinking men.[30]

Today, the Bharat Mata Mandir wears an air of dismal abandon as it stands amidst the hustle and bustle of down-town Banaras. The typical visitor to the temple today is more likely to be a foreign tourist rather than the Indian patriot, although local schools occasionally bring the young citizen to view the map and learn the country.[31] The capacity of the temple to forge the new "religion" of nationalism was severely compromised both because of the inability of the map of India to serve as either iconic symbol of Indian territory or as aniconic incarnation of Bharat Mata, and also because of the failure to evolve effective "nationalist" rituals that would have rendered the map-as-*murti* into an object of patriotic reverence. The temple does not even attract patrons of the older religion of Hinduism that the new religion of nationalism sought to supplant. On the face of it, the fact that the temple was built in Banaras was a historical accident, the man behind the enterprise being a wealthy native of the city. Nonetheless, this was an historical accident that could have been very productive, given Kashi's reputation as the spiritual capital of Hindu India (Eck 1982). One would have thought that the millions of devout Hindus who annually visit the city and its numerous temples would have also repaired to the Bharat Mata Mandir, especially since it also appears in the popular pilgrimage maps in the company of the more well-known temples as well as finds mention in various guide-books. Geographically distant though Shivprasad Gupta's edifice might be from the hallowed temples along the river Ganga, it still falls within the limits of the sacred geography of the holy city (Gutschow 1997).[32] But yet it is largely ignored by Varanasi's faithful pilgrims.[33] In the opinion of its manager Shyam Das Singh with whom I spoke first in 1998 and who made much the same point to me when I saw him again in 2001, if only there had been an anthropomorphic *murti* in the temple, and not a mere map, then millions of Indians would come through its doors. And indeed, a newspaper feature on the temple that appeared a few years ago noted, "Many pilgrims from rural areas who come to Varanasi, go to the Bharat Mata Mandir with flowers, Ganga water and other "puja articles", *only to find that there is no idol there*". (Bhattacharya 1993: 21, emphasis mine). And indeed, it is quite possible that today's Hindu nationalists have learnt from Shivprasad Gupta's (costly) "mistake" for in the multi-storied temple to Bharat Mata that opened in the city of Haridwar in 1981 under the sponsorship of the Vishwa Hindu Parishad (VHP), both a map of India *and* an anthropomorphic image of Bharat Mata are enshrined (McKean 1996). Hindu nationalist cartography of recent years has made prolific use of both maps of India and bodyscapes of Bharat Mata, but it has also attempted to deal with modern Hinduism's ambivalence over idolatry by anthropomorphizing the map itself in an ingenious new way. In a recent VHP greeting card featuring the map of India, Bharat Mata is absent even while metonymically—and aniconically—signaling "her" presence with her crown, one disembodied hand holding aloft the saffron flag, the other in a gesture of blessing (Brosius 1997)!

Fig. 8: Anon. "Bharat Mata", Contemporary picture postcard. Rasthriya Swayamsevak Sangh, Karnatak.

What we might facetiously characterize as the mystery of the missing mother in Varanasi's Bharat Mata Mandir offers a very telling lesson about the dilemmas of colonial and national modernity in India. This is a modernity where older practices of reverence have been rendered increasingly problematic, even an embarrassment, but new ones wrought to replace these ancient ways have proven to be inadequate. The enshrined map of India ought to have attracted millions of reverential citizens, but it fails to do so in a context where its purchase on the popular imagination, either as an iconic representation of Indian territory or as an aniconic stand-in for the familiar anthropomorphic form of Bharat Mata, is weak at best despite a century of nationalist cartographic activity and visual patriotism. At the beginning of the previous century, one of Tagore's fictional heroes had wondered whether the map-form had the power to move Indians to die for the country; in other words, to generate geo-piety. A couple decades later, when the *Indian Social Reformer* questioned whether "sentiments of reverence will be aroused by a map", it questioned the ability of the map to produce visual piety for the territory it iconically represented. The flagging fortunes of the Bharat Mata Mandir in Hinduism's paradigmatic sacred city are a striking reminder that in the subcontinent, the map of India is an indispensable but nonetheless inadequate sign—neither having purchase as iconic of the nation's territory nor succeeding in aniconically summoning into presence the hallowed figure of the nation's embodiment as Bharat Mata.[34]

Post-script

On a recent visit to the temple in October 2004, the first thing I noticed as soon as I entered its doors was a giant painting of Bharat Mata occupying a map of India. This painting has been modeled on an image that has been widely used by Hindu nationalist parties like the VHP, the RSS, and the BJP in their visual activities (Fig. 8), and was done by a local painter "a couple years ago", according to the manager Shyamdas Singh who commissioned it. With the introduction of this painting into the temple's precincts, the anthropomorphic appears to have certainly triumphed over the abstract, despite the original intentions of Shivprasad Gupta. The mother is no longer really missing in this first temple built in her honour.

Notes

1 Acknowledgements: Versions of this paper were presented at seminars in Ann Arbor, San Diego, Göttingen, Heidelberg, and Delhi. I thank members of the audience in all these venues for their comments. In particular, I am very grateful to Richard Davis, Jörg Gengnagel, and Ajay Sinha for their insightful suggestions. In Varanasi, Sharda Shankar Dwivedi, Anand Krishna, R. B. P. Rana, and Arjun Tiwari were generous with their time in providing me information and helpful leads. Funding for my work in Varanasi was provided by the American Philosophical Society and the American Institute of Indian Studies.

2 *Hindustan Times* (Delhi) 1936, October 26, p. 7.

3 In addition to Gandhi, the luminaries present included Sardar Patel, Khan Abdul Ghaffar Khan, Jawaharlal Nehru, Babu Rajendra Prasad, Bhulabhai Desai, Prakash Narain Sapru, Gurumukh Singh, Acharya Kriplani, G. B. Pant, Mrs. Vijayalakashmi Pandit, Purushottam Das Tandon, and Jamnalal Bajaj, all of whom had arrived in Banaras for the All India Congress Parliamentary Committee meeting.

4 *Hindustan Times* (Delhi) 1936, October 26, p. 7.

5 *Indian Social Reformer* 1936, November 7, Vol. XLVII (10), p. 154.

6 *Hindustan Times* (Delhi) 1936, October 26, p. 7.

7 *Hindustan Times* (Delhi) 1936, October 26, p. 7.

8 *Hindustan Times* (Delhi) 1936, October 25, p. 1.

9 *Hindustan Times* (Delhi) 1936, October 26, p. 7.

10 I have seen no critical studies, especially in English, of this fascinating figure, although I have benefited from reading an unpublished essay titled "Babu Shiva Prasad Gupta and Benoy Kumar Sarkar" by Giuseppe Flora (I thank Sharda Shankar Dwivedi for sharing this paper with me). Gupta hailed from a family of wealthy Banaras landed magnates and British loyalists. In his reminiscences, Gupta himself writes of the fact that as a young adult he was exposed to the Arya Samaj and his traditional religious views underwent a radical transformation, and that consequently his "mind moved away from idol worship" (Gupta 1983a: 88). Soon after, he also began to support the activities of Bengali nationalists. Around 1914, Gupta set out on a world tour that took him first to Egypt, and then to London where he met the fiery Arya Samaj leader Lajpati Rai with whom he subsequently linked up again in New York. By October 1915, when he was homeward bound via Japan and Singapore, Gupta was arrested on the charge of consorting with seditious conspirators against the British Raj, but was released soon after on the grounds that his uncle Raja Sri Moti Chand was a loyalist. A 1916-government report on him characterizes him as "an active nationalist and Arya", and noted his tendency to seek out the company of "known revolutionaries". On his return to Banaras in early 1916 after his travels, he started in 1920 the Hindi daily *Aaj* (which continues to be published till this day), and also donated one million rupees towards the establishment of the Kashi Vidyapith in 1921, a centre for national education following Gandhian principles.

11 Shiv Prasad Gupta himself wrote in detail (in Hindi) about his reasons for building the temple, the sources of his inspiration, and his hopes for the memorial (see Gupta 1983b)

12 Although identified by some contemporaries as "Professor of Science", Durga Prasad (1875–1952) appears to have been a local antiquarian and numismatist, and an amateur geographer (Singh & Rana 2001: 159). He taught at the Kashi Vidyapith (the nationalist college also established by Gupta), and was a member as well of the Executive Committee of the Banaras Hindu University. It appears that Durga Prasad made extensive use of the detailed relief maps of the subcontinent produced by the colonial Survey of India (Gupta 1983b: 391).

13 For details of the *shilanyas* ceremony presided over by the eminent Bhagwan Das, see (Dwivedi and Singh, forthc.).

14 *Hindustan Times* (Delhi) 1936, October 26, p. 7. For details regarding the construction of the map, see also Gupta 1983b.

15 Contemporary estimates suggested that the entire project had cost Shivprasad Gupta a lakh of rupees, the marble map alone a hefty Rs. 25,000. When Gandhi was asked to comment on why he was lending his approval and blessing to "this waste when the poor of the land needed this money badly", he is said to have responded that "the poor of the land realize more than he imagines that man does not live by bread alone". Gandhi's aide Mahadev Desai also insisted that "the shrine is a worthy embodiment of a worthy idea", and that Gupta's "lakh of rupees could not have been better spent". (*Hindustan Times* (Delhi) 1936, October 26, p. 7). Monumentalizing a map at such cost was rather a singular event, the only other comparable example Mussolini's grand attempt in Rome around the same time (Minor 1999).

16 I adapt this notion from Victor Morgan who suggests that visual piety is "the visual formation and practice of religious belief" (Morgan 1997: 1). Following Morgan (1997: 9), I suggest that like religious art, certain types of maps, especially those incorporating images of Bharat Mata or of the martyrdom of nationalist leaders, bring to the fore "the ontological presence" of someone or something sacred. It is the sacralisation of national territory attempted by such cartographic "pictures" that I am interested in underscoring.

17 Vidya Dehejia notes that "The term 'aniconic' carries the dictionary meaning of 'symbolizing without aiming at resemblance,' and 'aniconism' is defined as 'worship or veneration of an object

that represents a god without being an image of him'". (Dehejia 1991: 45). Her essay demonstrates, for example, that a variety of emblems such as footprints or the parasol were deployed in Buddhist art from very early on to invoke the person of the Buddha. Similarly, Geeti Sen reminds us that in India "there had always existed the predilection to worship through sign/symbol in place of iconic representation. Recall some of these: the Bodhi tree, the *dharma chakra*, and the *stupa* to represent the Buddha; calligraphy as the noblest medium to invoke the name of Allah; the *Guru Granth Sahib* to invoke the Sikh Gurus; the *tulsi* plant for Vishnu, the *linga* as the purest of the 'signs' to represent Shiva, and the *Shri Yantra* to invoke the Devi. And in certain senses (...) was the map not a *yantra* to invoke and represent the Motherland [Bharat Mata]?" (Sen 2002–2003: 167–68).

18 *Hindustan Times* (Delhi) 1936, October 25, p. 2.

19 Interestingly though, Indian territory is shown broken up into numerous coloured patches in the map of India painted on the sign board outside the temple announcing the Bharat Mata Mandir. The sign board is obviously a much later addition to the premises.

20 See in this connection the comment by Geoff King that "the power to map or to narrate, or to keep other forms of mapping at bay, is a key element in the [state's] ability to claim a territory (...).The western colonial map is an abstraction that tends to extinguish other dimensions of reality in an act of violent appropriation". (King 1996: 139).

21 *Leader* (Allahabad), October 28, 1936, p. 12.

22 Gupta 1983b, p. 389, p. 391.

23 The failure—or is it refusal?—to create a set of (secular) rituals that would have increased the capacity of the monumentalized map to serve effectively as a patriotic symbol points in turn to the compromised character of the Bharat Mata Mandir, fatally suspended between the older familiar religion of Hinduism and the new "religion" of nationalism.

24 *Leader* (Allahabad), October 26, 1936, p. 10.

25 Note the following comment in an inexpensive guide book published in the 1950s: "(...) we call this building a temple and we Indians pay our deep regard and respect to it in the same devotional way which we do to other temples in general. But why? It is enjoined in our Vedas..that the MOTHER AND THE MOTHERLAND ARE GREATER THAN HEAVEN ITSELF [sic] (...). And it is in the same way that our early Indian thought deified our MOTHER-INDIA and elevated patriotism into a religion. And accordingly since the earlier days of our Indian culture every Indian meditate [sic] BHARAT-MATA (Mother India) as a physical and visible form in her various aspects. This deity of secular democracy, the deity of the Cosmic Mind, the deity of the common, collective, and the national mind, which claims and commands universal homage and worship from each and every Indian was enshrined in this building (...)." (Anon. 1957: 29).

26 So in *Asli Bada Naksha Kashi Kshetra Ka: Map of Kashi* which was available in the bazaars of Banaras around 1953, the Bharat Mata Mandir is designated by the figure of the goddess, holding the Indian flag and offering her blessing. The temple itself is identified as Bharat Mandir, "India temple" (Lannoy 1999: 518). The same pilgrimage map was available into at least the 1970s. In another recent map (which circulates in a wide range of sizes) with a similar title, the temple appears on the upper right hand corner of the map, just south of the railway line that snakes its way down the eastern edge of the picture. The Mandir is designated by a building and the icon of Bharat Mata (*sans* map) holding a trident in one hand, the other in a gesture of blessing. A flag (not truly the tricolour, but close enough) is also drawn on the building. The building is called "Bharat Mandir" in Hindi, and Bharat Temple in English. A car is parked in front of it (see Fig. 7). Most strikingly, in a map catalogued by the Archives Varanasi Research Project, South Asia Institute, Heidelberg, the temple is designated by a bodyscape in which Bharat Mata stands as a 2-

armed goddess holding a trident and occupying the map of India. In a four-inch Banaras Guide Map published by the colonial Survey of India offices in 1933 and currently in the British Library, the site for the Bharat Mata Mandir is clearly labeled but no pictorial depiction accompanies this (British Library Cat. No. 53345 (7)). By contrast—and very tellingly—a recent tourist guide map of the city uses the same icon to designate the site as it does to designate all Hindu temples (Sreenivasan 2000). I am immensely grateful to Jörg Gengnagel who urged me to consider these examples, and shared many of them with me. In a larger study on the Bharat Mata Mandir, I hope to do more with such maps and their representations of this unusual memorial to the nation.

27 Jaffrelot is quoting here from Shraddhananda's *Hindu Sangathan—Savior of the dying race*, published in 1926. Around the same time, Subramania Siva unsuccessfully sought to build a "temple" to Bharat Mata in the Madras Presidency in 1921. It is not clear from the information we have on his plans whether he planned to install a map of India in this temple. Of course, prior to all this, Aurobindo Ghosh had already issued the call around 1905 for the construction of a temple to "Bhawani Bharati, Mother of India" in his pamphlet *Bhawani Mandir*. He insisted that the temple should be built "in a place far from the contamination of modern cities and as yet little trodden by men, in a high and pure air steeped in calm and energy. This temple will be the centre from which Her worship is to flow over the whole country (...)". (Ker 1973: 34–43).

28 See the following comment in a Varanasi guide-book, "The Himalayan range with its high pointed peaks, the plateau of Tibet, commonly known as the roof of this world behind them, the rivers with their rise and falls, the seas and oceans with their islands, the variations of the heights of the different parts of the country from the sea-level, the famous KHAIBER-PASS and the BULAN-PASS—through which invaders entered the country—are all vividly displayed in this extraordinarily beautiful and most natural relief map of the vast land." (Anon. 1957: 31). Or, here is a more recent response to the enshrined map: "Every mountain range has been shown with its individual features. The valleys, ravines, and passes through the range have been shown true to their depth. Four hundred odd peaks of the Himalayan range recreated in such a magnificent way that their height can be visualized merely by looking at them. For a visitor, following the range of Kailash mountain ranges with all their magnificent peaks and glaciers on the map will be an experience" (Bhattacharya 1993: 21).

29 *Indian Social Reformer* 1936, October 31, Vol. XLVII (9), p. 131, emphasis mine.

30 *Bombay Chronicle* 1936, October 27, p. 6.

31 Personal communication Nita Kumar. Interview with students of the South Point School, Nagwa, Varanasi. Interestingly, these students also told me that one of the principal reason that school kids go to visit the temple is to buy printed maps of India for their school geography lessons. It should also be mentioned that members of the local RSS are known to frequent the temple and salute the map (personal communication, Ramu Pandit, December 2001).

32 I am thankful to Jörg Gengnagel for urging me to stress this.

33 Diana Eck, however, writes of the Bharat Mata Mandir that "It is a popular temple with today's pilgrims, who circumambulate the whole map and then climb to the second-floor balcony for the *darshana* of the whole. Looking down at this map, they can see at a glance the great distances many of their ancestors traveled on foot" (Eck 1982: 39).

34 My conclusion obviously differs from that of historians of Hinduism like Diana Eck and David Kinsley who seem to suggest that the map functions adequately as a *murti* (see, for example, Eck 1999: 34–35; Kinsley 1987: 183–84). See also Geeti Sen's recent observation, "To this day homage continues to be paid by the ritual of circumambulation (*parikrama*) of a vast map of the country, laid on the ground with its rich undulating topography of mountains, rivers, seas" (Sen 2002: 36).

References

Anderson, Benedict. 1991. *Imagined Communities: Reflections on the Origin and Spread of Nationalism*. London: Verso.

Anon. 1957. *Banaras*. Banaras: Oriental Publishers.

Babb, Lawrence A. 1981. Glancing: Visual Interaction in Hinduism, *Journal of Anthropological Research* 37: 387–401.

Bhattacharya, Bharathi. 1993. Once Deified, Now..., *The Hindu*, March 5, p. 21.

Brosius, Christiane. 1997. Motherland in Hindutva Iconography, *The India Magazine of Her People and Culture* 17, 12: 22–28.

Casey, Edward S. 2002. *Representing Place: Landscape Painting & Maps*. Minnesota: University of Minneapolis Press.

Dave, J. H. et al. n.d. *Munshi: His Art and Work, Volume 1*. Bombay: Shri Munshi Seventieth Birthday Citizens' Celebration Committee.

Dehejia, Vidya. 1991. Aniconism and the Multivalence of Emblems, *Ars Orientalis* 21: 45–66.

Dwivedi, Sharda Shankar & Direndranath Singh. forthcoming. Shri Bharatmata Mandir. In: *Rastranirmaata Shri Shivprasad Gupta*. Varanasi: Gyan Mandal Limited, pp. 1063–1090.

Eck, Diana. 1981. *Darśan: Seeing the Divine Image in India*. Chambersburg, PA: Anima Books.

— 1982. *Banaras: City of Light*. Princeton: Princeton University Press.

— 1996. Rose-Apple Island: The Myths that made up India, *The India Magazine* 16 (3/4): 7–16.

Edney, Matthew H. 1997. *Mapping An Empire: The Geographical Construction of British India, 1765–1843*. Chicago: University of Chicago Press.

Gandhi, Mohandas. 1958. Speech at Bharat Mata Mandir, Banaras. In: *The Collected Works*, Volume 63. Delhi: Publications Division, Government of India, pp. 388–89.

Gutschow, Niels. 1997. Varanasi/Benares: The Centre of Hinduism? A Discussion of the Meaning of "Place" and "Space", *Erdkunde* 48: 194–209.

Gupta, Shivprasad. 1983a. Sankshipth Atma-Katha. In: *Kashi Vidyapith Heerak Jayanthi: Abhinandan Granth*. Varanasi: Kashi Vidyapith, pp. 85–95.

— 1983b. Shri Bharatmata Mandir. In: *Kashi Vidyapith Heerak Jayanthi: Abhinandan Granth*. Varanasi: Kashi Vidyapith.

Jaffrelot, Christophe. 1996. *The Hindu Nationalist Movement in India and Indian Politics, 1925 to the 1990s*. New Delhi: Penguin India.

Ker, James C. 1973. *Political Trouble in India (1907–1917)*. Reprint ed. Delhi: Oriental Publishers.

King, Goeff. 1996. *Mapping Reality: An Exploration of Cultural Cartographies*. New York: St. Martin's Press.

Kinsley, David. 1987. *Hindu Goddesses: Visions of the Divine Feminine in the Hindu Religious Tradition*. Delhi: Motilal Banarsidass.

Lannoy, Richard. 1999. *Benares Seen from Within*. Varanasi: Indica Books.

McKean, Lise. 1996. Bharat Mata: Mother India and Her Militant Matriots. In: J. S. Hawley & D. M. Wulff (eds.), *Devi: Goddesses of India*. Berkeley: University of California Press, pp. 250–80.

Minor, Heather Hyde. 1999. Mapping Mussolini: Ritual and Cartography in Public Art during the Second Roman Empire. *Imago Mundi* 51: 147–162.

Morgan, David. 1997. *Visual Piety: A History and Theory of Popular Religious Images*. Berkeley: University of California Press.

Nandy, Asish. 2001. A Report on the Present State Health of the Gods and Goddesses in South Asia. In: *Time Warps: The Insistent Politics of Silent and Evasive Pasts*. New Delhi: Permanent Black, pp. 129–156

Pal, Pratapaditya. 1997. The Printed Image: An Iconographic Excuse. In: G. J. Larson, P. Pal & H. Daniel Smith (eds.). *Changing Myths and Images: Twentieth-century Popular Art in India*. Bloomington: University Art Museum, University of Indiana, pp. 29–38.

Pieper, J. 1979. A Pilgrim's Map of Benares: Notes on Codification in Hindu Cartography. *GeoJournal (Dordrecht)* 3, 2: 215–218.

Ramaswamy, Sumathi. 2001. Maps and Mother Goddesses in Modern India. *Imago Mundi* 53: 97–113.

— 2002. The Goddess and the Nation: Subterfuges of Antiquity, the Cunning of Modernity. In: Gavin Flood (ed.), *The Blackwell Companion to Hinduism*. Oxford: Blackwell, pp. 549–566.

— 2003. Visualizing India's Geo-body: Globes, Maps, Bodyscapes. In: Sumathi Ramaswamy (ed.). *Beyond Appearances? Visual Practices and Ideologies in Modern India*. New Delhi: Sage, pp. 157–195.

Sen, Geeti. 2002. *Feminine Fables: Imaging the Indian Woman in Painting, Photography, and Cinema*. Ahmedabad: Mapin Publishing.

— 2002–2003. Iconising the Nation: Political Agendas. *India International Quarterly* 29, 3&4: 155–175.

Singh, Rana P. B. & Pravin S. Rana. 2001. *Banaras Region: A Spiritual and Cultural Guide*. Varanasi: Sacred Books.

Srinivasan, P. (ed.) 2000. *Varanasi*. Chennai: TTK Healthcare Limited—Printing Division.

Tagore, Rabindranath. 1985. *The Home and the World*. Translated by Surendranath Tagore; reprint ed. Madras: Macmillan.

Tuan, Yi-Fu. 1976. Geopiety: A Theme in Man's Attachment to Nature and to Place. In: D. Lowenthal & M. Bowden (eds.), *Geographies of the Mind*. New York: Oxford University Press, pp. 11–39.

Images

NIELS GUTSCHOW

Panoramas of Banaras

Introduction[1]

Views of the city of Banaras came to be known from 1775 A.D. onwards, after Joseph Tieffenthaler[2], a German apostolic missionary, visited the city. His view of the city as seen from the east across the Gaṅgā was also the first panoramic view. More picturesque views followed when William Hodges visited the city in August 1781 as a member of Warren Hasting's retinue (Hodges 1785–88). The gifted brothers, Thomas and William Daniell, followed in November and December 1786. Their presentation of India and Banaras in particular contributed decisively[3] to the European view of the subcontinent for decades (Archer 1980, Martinelli & Michell 1998). A milestone was set a generation later by James Prinsep, who was the Assay Master of the Mint in Banaras from 1819 to 1830. His lithographic views (Prinsep 1831) and even more so his map of the *City of Bunarus* constructed an image of the city that continued to be confirmed through countless reprints.

Photographs became prominent after 1858. The network established by the famous British photographer Samuel Bourne made sure that his photographs taken in Banaras in 1865/66 could be ordered in all the major cities of the world. When he came first to Banaras in April 1863 he already noted down that "the place deserves the attention of the photographer". Especially because "from the opposite side it appears with its temples and mosques and their minarets and domes shining in the sun entirely magnificent and impressive" (Pohlmann & Siegert 2000: 14). The French photographer Louis Rousselet followed in April 1868. Bourne's and Rousselet's photographs were transformed into wood-cuts and wood as well as steel engravings and published worldwide without referring to the original authors. By the end of the 19th century Banaras was thus a familiar place, with Maṇikarṇikā Ghāṭ and Pañcagaṅgā Ghāṭ figuring prominently.

While an extensive literature has developed on the 18th century picturesque views and 19th century photography (Pohlmann & Siegert 2000: 264–265), little is known about panoramic views which probably were produced from the 1830s onwards as paintings. Only at the beginning of the 20th century were they printed as sheets to be pasted together at home. The sheer length of these products meant they never were included in publications like the *Illustrated London News*, which played an important role in bringing the oriental world to the homes of those who never travelled.

From Painted Panoramas to Panoramic Photography

The term "panorama" is an artificial expression derived from the Greek word *pān* = "all" and *hórama* = "to see". It was probably invented by Robert Baker, an artist who produced a painting of London in 1788 which, as the artist claimed, came close to reality "in size as well as in all other respects" (Oettermann 1980: 9). When he secured his invention as a patent in 1787 he first named it "la nature à coup d'oeil", and "an entire view of any country or situation, as it appears to an observer quite round" (Dewitz & Nekes 2002: 283). Within a few years the news of Barker's invention spread across Europe and the world. In Germany a popular magazine told the readers in 1795: "Man kennt die malerischen Zaubereyen (...) die vor einigen Jahren unter dem prunkenden Namen Panorama (Allansicht) in England so großen Beyfall erhielten" (Oettermann 1980: 8).

The first exhibited panorama of London covered 1479 square meters and depicted the city in full circular view of 360 degrees. The display was so perfect that the spectator felt "as in a dream", or "like flying" (Oettermann 1980: 18). Within a decade the experience of the horizon turned into a form of art. The unusual nature of the experience was greeted with great exitement. The yearning for the fantastic which was beyond reach was met with ever larger and more spectacular circular views. At the same time the display of a panorama in a specially designed structure was regarded as a technical-scientific innovation. The first building that allowed the spectator to enjoy a circular view was constructed by Barker in 1793 at Leicester Square in London. It had two floors to allow the visitor two experiences on the same day. In the early 19th century the subjet of battle scenes enjoyed increasing popularity. Illusionary elements like fire and the ocean wind were created. At the end of the 1820s, Paris was in the grip of "oramas" fever: on display were cosmoramas, georamas, dioramas, uranoramas and neoramas. Whereas a neorama placed the spectator in the middle of a room, preferably of a church, the diorama presented half of a circle in such a way that the spectator was outside the presented space.

Since the 1830s, any wide angle presentation of an urban environment or a landscape was labelled a panorama. Folded as a concertina booklet or kept rolled in a tube, panoramas that did not create a circular view but one in length started to become fashionable in the 1820s. The *Description of the Most Remarkable Places between London and Richmond* was presented by John Clark with a length of 1815 cm and a height of just 21.5 cm. A decade later, *A Panorama of the right Bank of the Elbe from Hamburg to Blankenese*[4] was printed in 18 sheets and traded as a portfolio with bilingual title. Such portfolios with riverfront panoramas must have reached India in the 1820s or 1830s. It is not unlikely that the earliest panoramas of Banaras were odered on the basis of such prototypes.

The earliest known panorama of Banaras, published by Rai Anand Krishna in 2003, is made up of 12 sheets and covers the entire stretch from Assi to Ādikeśava, similar to those of the early 20th century, but with greater accuracy in the distances between the individual places. Rai Anand Krishna suggests that the Company School of painting flourished as early as around 1800 to serve a growing market created by travelling

Europeans. He dates the panorama to "the beginning of the 18th century, for the simple reason that the grand pylons of the Bhonsale temple-palace (...) are yet to be completed" (Krishna 2003: 5). Likewise, no traces of the Amṛta Vināyaka at Naya Ghāṭ, established by Amrit Rao Peshwa in 1807, can be found. Whereas prominent structures such as the Chet Singh Ghāṭ or Rāṇāmahal Ghāṭ are easily identifiable, less dominant features along the riverfront with its 26 stepped embankments are only roughly indicated or simply forgotten. We must take into consideration that the painter intended to present a colourful scene of which the details roughly depicted the riverfront, which he may have sketched from a boat. Equally important is the hustle and bustle on the river with a variety of boats carrying dignitaries, pilgrims, musicians, dancers, and of course goods. As on the panorama of Tieffenthaler, the painter chose a view that incorporates scenes on the opposite river bank. It looks straight towards the high waterfront with its wooded hinterland. While the sky with its stylized clouds obtains ample space, no depth is indicated. Only decades later, the foreground was cut to raise the position of the spectator. More and more depth was created until, finally, the panoramas of the early 20th century appear as maps with distorted dimensions.

By the end of the 20th century, panorama-photography developed hitherto unknown standards. The photographs by Jaroslav Poncar, taken in India between 1976 and 1996, are based on a horizontal angle of 120 degrees. Published in 1998 on the occasion of the 50th anniversary of India's independence, he named his collection *Panorama of India*, probably not knowing of the *Diorama of Upper India* on display in London almost 150 years earlier. Poncar decidedly avoided the plural as he wanted to transcend the confines of single photographs. His aim is to present India in its entirety, much as the local vendor these days at Maṇikarṇikā Ghāṭ in Banaras presents the *Map of Kashi*, saying in English "this all Benares", emphasizing "all". In Poncar's collection Benares (Poncar 1998: 15) is presented with a view across the Gaṅgā, covering Daśāśvamedha Ghāṭ and the entire bend of the river south towards Bhadainī. Poncar adds a poem by Mirza Ghalib: "The Ganges shows its longing ardently / and makes its waves into embracing snares (...)".

In recent years, a revival of true panoramas in full circle has been observed. In 1995, small circular structures were set up at crucial places in the urban landscape of Berlin to present visions of the future (2005) by full circular views, and in summer 2003 the author of these blown-up paintings, Yadegar Asisi, created an imaginary mountain scene in a large gasometer in Leipzig, Germany. The panorama of Mount Everest could be experienced as a continuous 36 metre high picture from an elevated platform. In line with the riverfront panoramas of London, Hamburg or Banaras, panoramas of major German cities with a circular view of more than 360 degrees went on sale in 2003 as postcards under the label of "Funorama". In search of new labels with traditional associations, a panorama-map of Frankfurt in the shape of an axonometric view is currently distributed under the label "Emporama".

Banaras Panoramas 1786 to 1910

None of the pictures presented in the following chapters were known under the term panorama. But all of them try to transcend too narrow confines of that genre, oscillating between a view and a map.

The Panorama of "Banares" by Joseph Tieffenthaler, published in 1786

Following the definitions of the early 19th century, the view presented by Tieffenthaler (Fig. 1) would be a diorama. The viewer stands on the eastern side of the "Ganges" behind a nondescript foreground which is framed architecturally on the left. Clearly identifiable is the Alamgir Mosque on the right, while one can only suggest that the dominant structure towards the left, with corner towers and a tiered centre depicts Chet Singh's palace, which must just have been completed when Tieffenthaler visited the place in 1765. Not a single towering *śikhara* structure which characterizes North Indian temples, is visible. Instead, we see domes that are peaked by crosses that remind us of the Christian background of those who might have produced this view back home in Berlin. It is unknown what kind of sketch Tieffenthaler made on the spot, or whether he took local versions of such a view along with him. The apostolic missionary may even have felt obliged to present the heathen environment with symbols of his own religion. The single tower on the left cannot be explained, nor the hilly landscape with a fort-like structure behind. The structures between the ten stepped *ghāṭs* form a kind of backdrop to create an urban atmosphere. Thus the picture presents a true panorama as it does not intend to present more than what can be seen from the opposite side of the river.

The Panorama of Benares at the Great Globe, Leicester Square, London (1860s)

Already in 1858 the Great Globe at Leicester Square, which had been taken over by Robert Burton in 1826, presented *The Fall of Delhi* and the *City of Lucknow*. Burton tended to mix touristic sites with presentations of heroic battles. This type of mixture was presented in the early 1860s under the title *Diorama of Upper India*. On display was *The Ganges, the cities of Calcutta, Benares, Agra, with the rock temples and the temple of Juggernaut*, and finally also *Scenes of the Military Insurrection*. While *Upper India* could be seen daily at 12 am and 6 pm, *The Diorama of Russia* could be seen at the same place at 3 and 8 pm. *The Panorama of Benares* at the Great Globe (Fig. 2) was also sold in folded form, packed in a square folder measuring 15 by 15 cm. Printed probably as a wood engraving, the view of Benares as seen across the Ganges was only 4,5 cm high. The structures covering the high bank of the river have been drawn schematically, while picturesque boats populate the water. As is true of all the other panoramas, prominent structures like the Alamgir Mosque, the Bhoṅsalā palace and the impressive structures at Rājārājeśvarī Ghāṭ can be identified, while the remaining buildings seem to serve as a kind of backdrop. No specific shapes allow the spectator to discern or recognize individual buildings. In contrast to the early 19th century full circle panoramas of London and Paris, a strict horizon is presented. The panorama wanted to present nothing more than the view from the opposite bank.

Fig. 1: Banaras—panorama between Assi and Pañcagaṅgāghāṭ with the imposing Alamgir Masjid to the right, published by Joseph Tieffenthaler, who visited Vārāṇasī in 1765, printed in Berlin 1786.

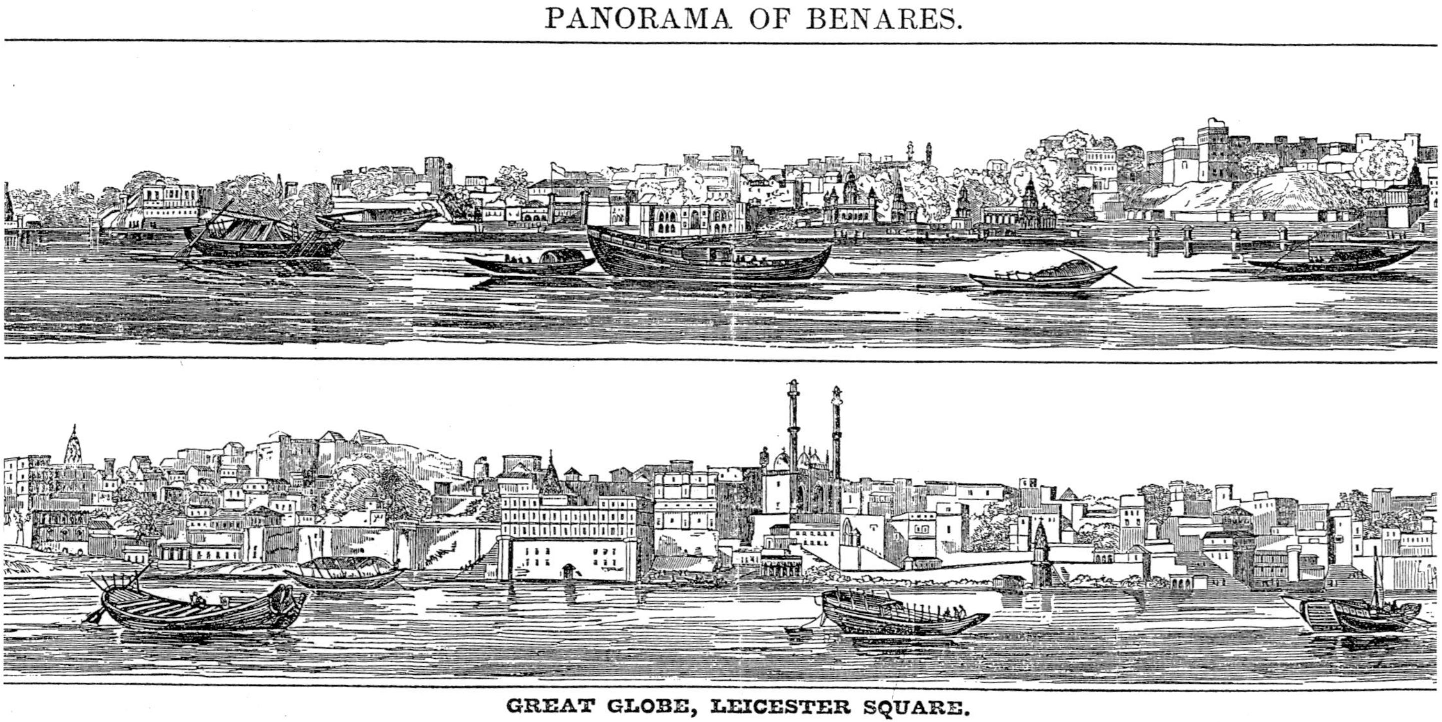

Fig. 2: Detail of the panorama of Benares between Daśāśvamedhaghāṭ to Boṅṣalāghāṭ (above) and Nayaghāṭ to Rājmandirghāṭ (below), displayed at the Great Globe at Leicester Square in London in the 1860s. (Source: Joachim Bautze).

Panoramic Elements of 19th Century Picture Maps

Two late 18th or early 19th century picture maps are kept in the Victoria & Albert Museum in London and in the National Museum of New Delhi. Painted on cloth, both pictures intend to map the sacred topography of Vārāṇasī/Kāśī and above all its processional routes. To illustrate the multitudinous nature of the religious infrastructure, hundreds of temples are depicted in front view. Their view is either presented in variation, or, as in late 19th century versions, stylized and reduced to a symbol. The more geometrical version at the Victoria & Albert Museum takes care to create a strip along the Gaṅgā that can clearly be identified as the *ghāṭs*, but not more than six buildings are attached to this area. Among these is the dominant Alamgir Mosque with its towering minarets. On the second painting, from Delhi's National Museum, the *ghāṭs* are depicted in a faithfulness that is never achieved again in similar attempts to present the architectural framing of the river. It starts with the *liṅga* of Assisaṅgameśvara in a domed structure, turns to the palace of Chet Singh at Śivāla Ghāṭ and presents Hanumān in his temple high above a flight of steps. Likewise, the *liṅga* of Kedāreśvara can be seen at the end of a long flight of steps that originate at Gaurīkuṇḍa, a small water tank located on a platform above the Gaṅgā. Rāṇāmahal Ghāṭ follows with its characteristic three towers, Bhoṅsalā Ghāṭ and—less dominant than in the earlier examples—the Alamgir Mosque. Both paintings are in a way illustrated maps. Indicated processional routes are punctuated by temples which are visited by pilgrims in a prescribed sequence. The presentation of the *ghāṭs* alone turns away from this purpose to add an additional aspect, to turn the painting into an informative and at the same time literally picturesque product.

The third picture map was produced by one Kailāsanātha Sukula in 1876. Under the title of *Kāśīdarpaṇa*, literally "Mirror of Kāśī", he presented a new mixture of text, symbols and topographical as well as architectural elements. Jörg Gengnagel has called it a "word-picture" (Gengnagel 2003: 249), while Axel Michaels said that religious maps like the *Kāśīdarpaṇa* must be understood as "texts of religious spaces".[5] Sukula's *Mirror of Kāśī* has been widely published.[6] In the present context I shall deal exclusively with its depiction of the panorama that lines the Gaṅgā.

The *Mirror of Kāśī* faces East because its title is placed in the upper left corner. The text on the map, however, turns in all directions. The small detail of the map presented here already demonstrates that one would have to turn it many times if one wanted to read the "word-picture". Viewing the picture-map would be no less confusing. The *ghāṭ* and the prominent architectural features behind the *ghāṭs*, however, can fairly well be viewed on turning west (as presented in Fig. 3). Only four identifiable structures can be seen along the entire stretch between the tributaries of the Gaṅgā, the Assi in the south (left) and the Varaṇā in the north (right, Fig.4). As on the earlier paintings and panoramas, the palace of the Bhoṅsalā with its Lakṣmī-Nārāyaṇa temple, built in 1795 by the Maratha commanders from Nagpur, and the Alamgir Mosque, built more than one hundred years earlier above Pañcagaṅgā Ghāṭ on the site of a temple on the order of Aurangzeb, who was also called "Emperor of the World" (= Alamgir) figured prominently.

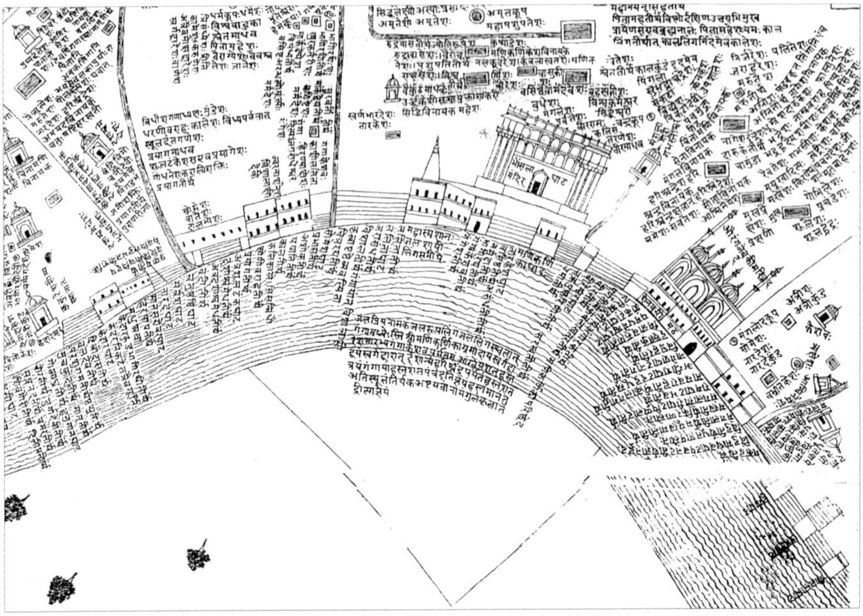

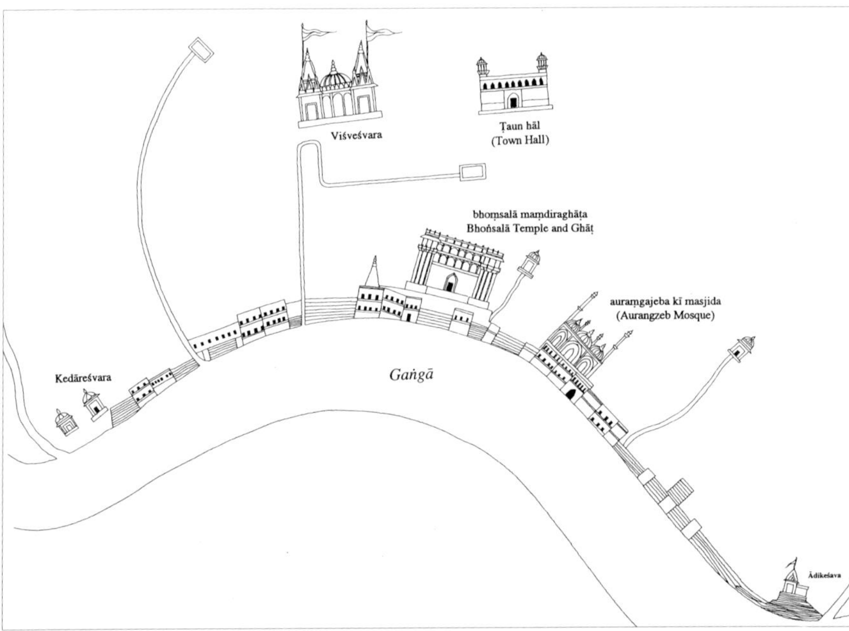

Fig. 3 & 4: *Kāśīdarpaṇa*—detail of the picture map that was published by Kailāsanātha Sukula in 1876 under the title "Mirror of Kāśī". Along the Gaṅgā a number of *tīrthas* and *ghāṭs* are identified in written form, while only very few built structures are named. Size 79 by 92 cm. (Source: author).

Two temples towards left and right, Kedāreśvara and Ādikeśava add to a scene that is dominated by stylized steps which stretch from south to north, interrupted only by some 13 structures marked with black holes, indicating any kind of built environment.

Sukula certainly did not intend to present a panoramic view of Kāśī. The depiction of the *ghāṭs* can be seen rather as an illustration to the map, which presents Kāśīkṣetra with more than 1200 written entries. Worth mentioning, however, are the two dominating figures which are placed almost in the middle of the picture, the temple of Viśveśvara and the Town Hall. The temple mirrors the reality of the 1870s as does the Town Hall. The domed central space of the temple is dedicated to Vaikuṇṭheśvara, Viśveśvara is found under a *śikhara*-tower towards the right and Daṇḍapāṇīśvara under an identical tower towards the left. The representation of the Town Hall (*ṭaun hāl*) offers a rare clue to the motivation behind compiling the *Mirror of Kāśī*. The building commemorates the visit of Prince Alfred to Vārāṇasī in January 1870. Construction started in January 1873 and was completed in December 1875. The festive opening by the Prince of Wales was enacted in January 1876, little more than six months before Sukula had his picture map lithographed using five stones on both paper and cloth. The depiction of the newly built hall on the *Mirror of Kāśī* documents the quest to put a contemorary stamp on his product. The proud structure represented the colonial rule as Aurangzeb's Mosque represented Muslim rule. These two are the only structures that reach beyond the obvious purpose of the map, i.e. to present an inventory of the sacred space of Kāśī.

The Panorama at the Museum der Kulturen, Basel[7]

In contrast to the 19th century picture maps, the panorama kept in the collection of the Museum der Kulturen in Basel intends to be a view (Fig. 5). Behind the view of structures that form the riverfront, one can see stylized houses and temples with domes or *śikhara*-towers in three steps behind the *ghāṭs* which emphasize the impression of the bend of the river as an amphitheatre with ascending steps. To present the bend in full view, the *ghāṭs* have been folded to form a continuous scene. Tieffenthaler tried to communicate the bend of the river when he chose to look across the entire river from a slightly raised vantage point. In his view the opposite river bank bends more decisively towards the left as it touches the frame. The Basel panorama presents the Gaṅgā in front, populated with 19 boats manned by pilgrims, and more boats at the banks of Hanumāna Ghāṭ and Gau Ghāṭ/Gaī Ghāṭ. Some 80 people are seen taking a ritual bath or even swimming. Washermen can be seen at work in Bhadaini at Vacarāya Ghāṭ/Vaccarāja Ghāṭ and Maṃnamandīra Ghāṭa/Mānmandir Ghāṭ and at Phuṭai Ghāṭa/ Phūṭā Ghāṭ.

Along the Gaṅgā a strip of almost equal breadth is reserved for the depiction of the 50 *ghāṭs*, of which 15 extend down to the river by a flight of steps, supported by some 60 structures, most of which seem to have an interior chamber. The middle register depicting the built environment is only little wider than the Gaṅgā and rests on a firm baseline. Only at Manasaravāra Ghāṭa, Pāne Ghāṭa/Causaṭhī Ghāṭa/Munasīyā Ghāṭa, Jalasahī Ghāṭa/Jalaṣāyī Ghāṭ, and Manakankā Ghāṭa/Maṇikarṇikā Ghāṭ do a few structures (like Rāṇāmahal and Tārakeśvara) break this line. The two tanks above the Gaṅgā, Gaurīkuṇḍa (at Kedāra Ghāṭ) and Cakrapuṣkariṇīkuṇḍa (at Maṇikarṇikā Ghāṭ), presented almost in top view, form landmarks which help to locate specific places among the otherwise nondescript depiction of the urban landscape. Cremations at Hariścandra Ghāṭ and at Jalaṣāyī Ghāṭ and the depiction of the *liṅga* of Śmaśāneśvara on a high platform at Hariścandra Ghāṭ add to the recognizability of various places.

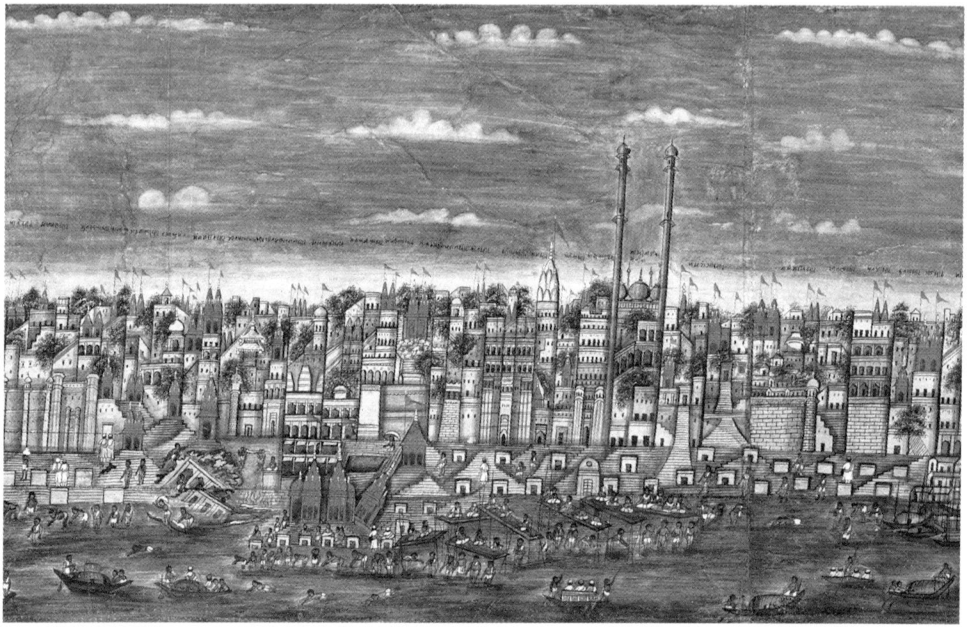

Fig. 5: Detail of *Panorama of Banaras/Vārāṇasī*, dated to the second half of the 19th century. 59 places are identified in a row (from left/South to right/North) above the horizon: besides *ghāṭs* one temple, two palaces, and the imposing Ālamgīr Mosque, Mādhorāja kā dharārā. (Source: Museum der Kulturen, Basel, IIa 9939, ca. 4 m long).

Of the dominant built structures, the palace of Chet Singh (named Pātasākāhavelī, lit. "palace of the emperor"), the Bhoṅṣālā palace (named Ghosalā Ghāṭ) and the Alamgir Mosque (named Mādhorāja kā dharārā) stands out, as in all panoramic views up until the beginning of the 20th century. The Siddhagirimaṭha can clearly be seen at Lalitā Ghāṭ (named Lalatā Ghāṭa), as can the Nepalese Samrājyeśvara temple, established in 1843. This date can certainly serve as a point of reference. If we identify the impressive structure (Rīvakoṭhi) at Assi Ghāṭ as the building that was completed by Lālā Miśra, a minister of the king of Rewa in June 1879, we must assume that the panorama was produced in the 1880s, as the palace of the king of Vizianagar at Kedār Ghāṭ, built in 1890, cannot be seen on the panorama. Also the bridge at Rāj Ghāṭ, which was completed in 1887 and which became a major landmark in all later picture maps and panoramas, is not yet in place.

The entire scheme of the panorama is in a way surprising because it leaves almost half of the space for the sky. The silhouette of the urban scene with its 85 red fluttering flags on top of the almost invariably red *śikhara*-towers from the many temples is underlined by a light horizon suggesting the setting sun against a blue sky with schematic clouds.

What is equally surprising is the fact that except for the *kuṇḍas*, all architectural components are strictly facing the viewer. The tributaries, the Assi to the left, a small rivulet at Hariścandra Ghāṭ, Telianālā beyond Prahlād Ghāṭ and the Varaṇā on the right edge of the panorama create, however, a feeling of depth. The temple of Jagannātha on a high supporting platform to the left and that of Ādikeśava to the right frame the panorama—in complete disregard of the considerable distance of these landmarks to Assi and Rāj Ghāṭ.

A Complete View of the Benares City, dated about 1911[8]

The Complete View is printed on six sheets, which are pasted together and coloured by hand (Fig. 6 & 7). It was published by Beni Singh Bookseller. In the lower register of sheets one and four the printer is mentioned in Urdu: Matbūa(-i) Sulaimānī Press, Banaras. Stamps on sheets three and six record the 3rd January 1913 as the acquisition date of the British Museum. As the shrine (*samādhi*) in memory of Svāmī Bhaṣkarānanda, built in 1910, is already documented on the panorama, so we must assume that the panorama was printed in an updated form in 1911 or 1912.

Notes about the Entries on the Map in Hindi and English

Altogether 81 objects presented on the picture scroll are identified, either in both Hindi and English or—in 14 cases—only in Hindi (Fig. 8). The Hindu version has a strong influence of the local idiom Bhojpuri. It is not clear whether the person who produced the original drawing after which the lithograph was printed also put the identifications into script. The pictorial presentation, however, displays a naivity and amateurship that equals the style of writing, especially the mode of transcription.

To start with the title of the picture scroll, *ACOMRLEVIEWOFTHEBENARESCITY*, it is presented like a Bhojpuri version without identifying single words. The word "complete" alone has a mistake in spelling and remains incomplete. In the same way, most of the identifications display inconsistencies. In the beginning, the designation "*ghāṭ*" is separated from its name, like "Assi Ghat", or it uses the genitive, saying "Ghat of Lallamisser", although the Hindi version correctly states Lalā Miśrā, who as the secretary of the king of Rīvā initiated the building of the *ghāṭ* and on its bank "Rewa Kothi", inaugurated, as an insciption on the building says, in June 1879. As the scroll develops, it says Golaghat or Pahladghat. It seemed most difficult for the writer not to loose the "r" as in Tilochanghat, Pahladghat or Lallamisser, but vowels get also lost, as in Hanman or Hinducolleg. The basic mistake appears in Pahladghat, when the foot of the "L" turns left instead of right. The most incomplete representation turns out

Fig. 6: *A Complete View of the Benares City* (*Śrī Kāśīviśvanāthajī ka sampūrṇa naṃkaśa*)—published circa 1911 by Beni Singh Bookseller at Rāj Ghāṭ, Kashi, size 17 by 305 cm. Detail covering the southernmost *ghāṭs* (*Lalāmiśra kā ghāṭa*) as compared with the situation in November 2000.

Fig. 7: *Complete View of the Benares City* (*Śrī Kāśīviśvanāthajī ka sampūrṇa naṃkaśa*)—published circa 1911 by Beni Singh Bookseller at Rāj Ghāṭ, Kashi, size 17 by 305 cm. Detail covering Śivāla-Ghāṭ/Chet Singh Ghāṭ (no. 12) and Hanman Ghat/Hanumān Ghāṭ as compared with the situation in November 2000.

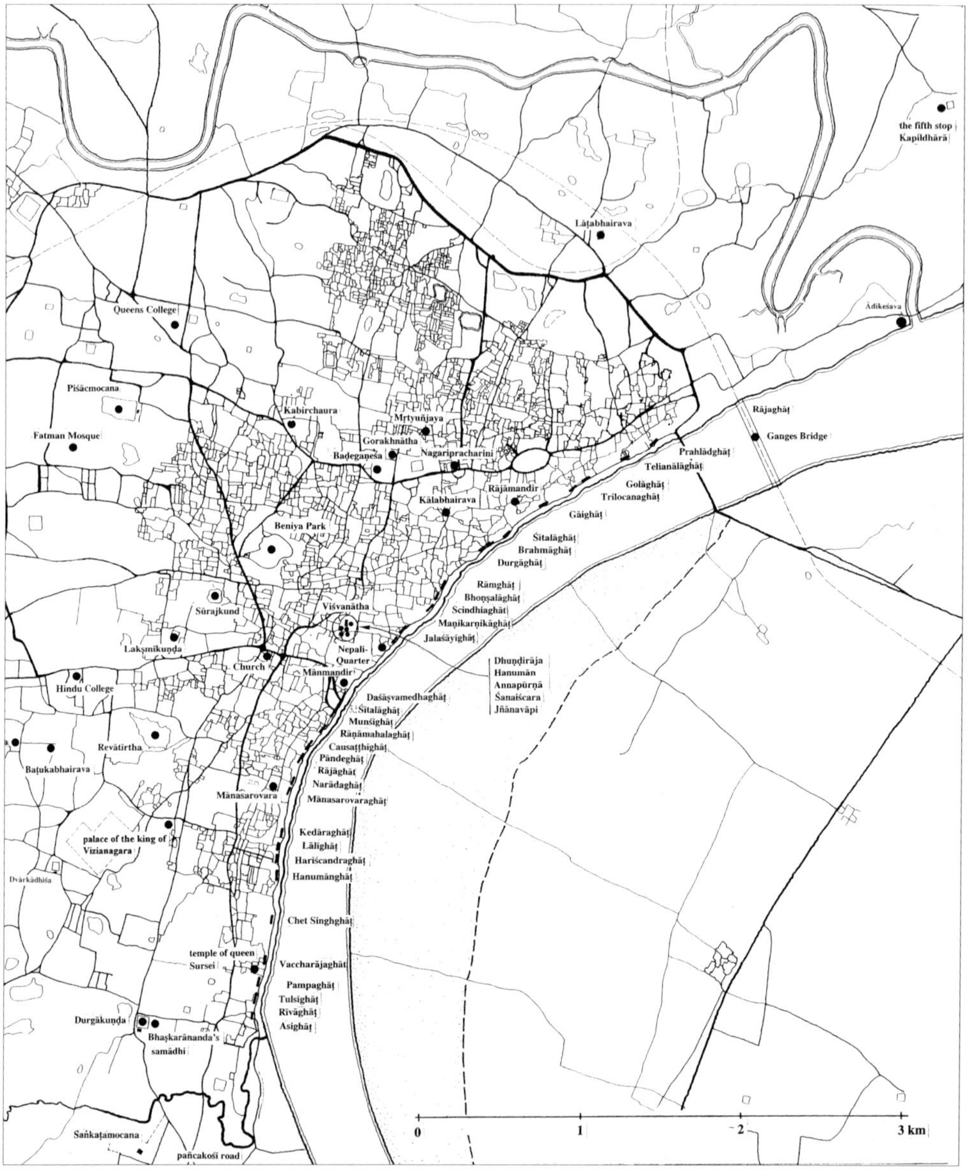

Fig. 8: Location of 76 of altogether 81 places that are identified on *A Complete View of the Benares City* (*Śrī Kāśīviśvanāthajī ka sampūrṇa naṃkaśa*). Apart from 33 *ghāṭs* along the Gaṅgā, temples, palaces, six water bodies and educational institutions are marked. The second to fourth stations along the Pañcakrośī route, as well as two places at Sarnāth, are not seen on this map.

to be "Mankanka Ghat" (probably by coincidence the same term is used on the Basel Panorama, probably 50 years earlier), referring to Maṇikarṇikā Ghāṭ.

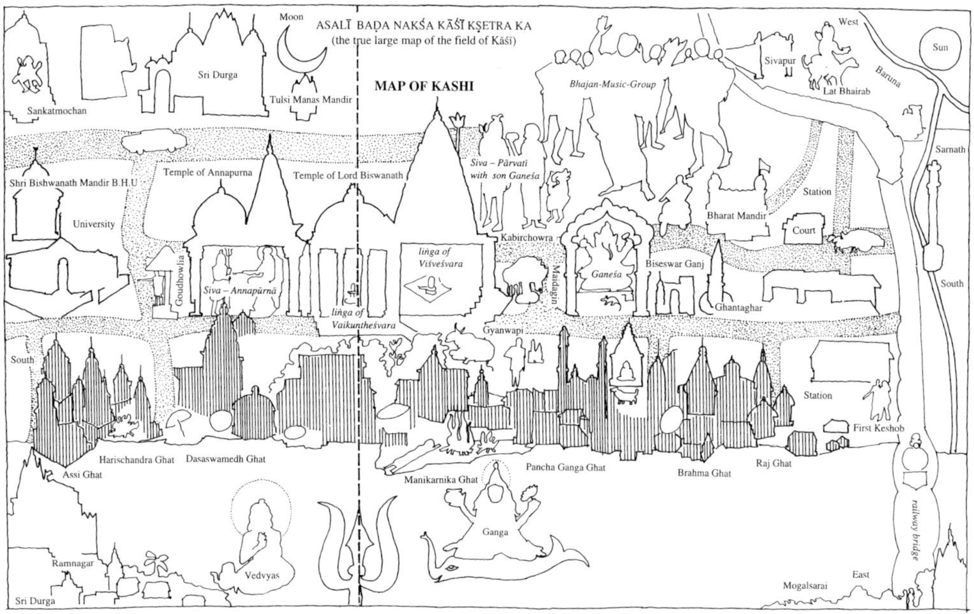

Fig. 9: Tracing of the *Map of Kashi–Asalī baḍa nakśā kāśī kṣetra ka* ("the exact large map of the sacred realm of Kāśī"), published by Thakur Prasad & Sons–Book Seller; Rajadarwaja-Branch: Kachanrigalli, Varanasi, printed circa 1975. The Gaṅgā (also personified as a goddess), with only six *ghāṭs* mentioned, dominates the picture, which depicts traces of a topographical map: a grid of stylized roads, populated by a car, a scooter and an ox cart suggests an urban background. (Source: author).

At the periphery in the background, five places are indicated where *dharamśālās* allow the pilgrims to stop overnight while performing the Pañcakrośīyātrā. The first entry is more informative, saying "firstplaceof panchkosikandwa", later it simply says "2nd place" or "3rd place".

Type of Place Presented on the Scroll

Benares, presented here as an imposing urban identity parallel to a river whose name is not mentioned, is mainly seen as a continuous development of *ghāṭs*, stretching from Assi Ghāṭ to Rāj Ghāṭ. Altogether 32 *ghāṭs* or 40 percent of all entries designate *ghāṭs*, not many more names existed at that time because many of those in use at present originated only in course of the 20th century. One more *ghāṭ* at that time was simply known as Telianālā. Eight *tīrthas* (tanks/*kuṇḍa* or *tālāb*, well/*vāpī*, pond/*tīrtha*) are added to the *ghāṭs*: Durgākuṇḍa, Mānasarovara, Lakṣmīkuṇḍa, Piśācmocana, Jñānavāpī, Surajkuṇḍ and also Revātīrtha, a large pond which already in the 1930s was drained and opened up for urban development. Asthitakṣepatīrtha, which still figures as such at the Pañcakrośī Mandir (established in the 1870s), had already been drained at the end of the 19th century and converted into Beniya Park. Located on the periphery along the Pañcakrośīyātrā is Rājātālāb.

Surprisingly, only very few temples are named on the scroll. It starts with the temple of the Queen of Sursand, a principality along Nepal's border near Janakpur. The Mulk Rani Kuwar built the temple above Jānakī Ghāṭ which was named after the mother of Sītā, whose birthplace is in Janakpur. The temple has four shrines, of which two are dedicated to Śiva and two to Rām and Jānakī. They are placed in a square courtyard, but the scroll somehow wanted to document them all by placing them in a row above the *ghāṭ* at which an undated inscription mentions the queen. As this must have been done by the end of the 19th century, a *zamindar* of the principality renovated the *ghāṭ* after a generation in 1935. Similarly, the 16th century palace of King Savāī Mān Singh of Amber figures prominently as Mānmandir above the *ghāṭ*.

The temple of Śrīviśveśvaranāthjī or "Golden Temple" as the English version says, does not figure prominently in the second line behind the *ghāṭs*. The temples of Annapūrṇā and Ḍhuṇḍirāja are seen beside the Golden Temple, while Hanumān and Śanaiścara are represented by simple shrine blocks. Jñānavāpī is separated from the central ensemble, represented by the Aurangzeb Mosque and placed beside Maṇikarṇikā Ghāṭ. The temples of Kālabhairava, Mṛtyuñjaya, Baḍegaṇeśa and Baṭukbhairava are represented as simple blocks, and of Vaidyanātha as a wide quadrangle, while Gorakhnāth figures as the usual image, a temple with a tower in *śikhara* fashion. For the shrine of Lāṭabhairava (also named Kapālībhairava) not the stele carrying Bhairava's face is represented but the adjoining tank, Kālamocana. The most impressive temple is presented at the end of the scroll as the abode of Ādikeśava, built in 1790 by the *dīwān* of the Scindhias of Gwalior.

Surprisingly, only two palaces are represented on the scroll. Above Hariścandra Ghāṭ is the palace of the king of Vizianagar, who in 1890 also added a *ghāṭ* next to Kedār Ghāṭ. which is given a prominent role within the overall panorama. The other palace is Rājmandir above Brahma Ghāṭ, established in the 16th century by the king of Būṇdī.

The most interesting category which also sheds considerable light on the intentions of the publisher of the scroll is represented by modern institutions. The most recent establishment that figures on the scroll is probably the memorial (*samādhi*) of Bhaṣkarānanda, built east of Durgākuṇḍa in 1908, only a year or two before the scroll was designed. Similarly, the Hindu College, established in 1897 by Annie Besant, the church of the Missionary Society established in 1888, the Queens College and the bridge across the Ganges, inaugurated in 1887, contribute to the obvious intention to document the recent development of the city.

Compared to the achievements of the colonial rule, those of the Muslim past which remains represented by one third of the city's population is clearly underrepresented. The minarets of the mosque at Pañcgaṅgā Ghāṭ are seen towering up but they are not identified. The Fatman Mosque alone, found behind Jalaśāyī Ghāṭ, represents the only identified Muslim item on the scroll.

Besides those 81 identified *ghāṭs* and sites, 84 other unnamed structures are scattered across the entire scroll: 59 houses of differing shape, with or without roof, all rendered in ochre colour. Only one of these represents a monumental structure which looks almost like a church. Moreover 13 temple-structures, easily identifiable by a *śikhāra* tower, are rendered red, as are four *kuṇḍas*. Three parks rendered in green have also not been identified, although the Maidagin park with its rectangular *kuṇḍa* in the middle is topographically placed correctly between Gorakhnāth and the office that propagates the use of Devanāgarī script—a building that is architecturally well presented.

The *Map of Kashi—Asalī Baḍā Nakśā Kāśī Kṣetra Kā* ("The Exact Map of The Sacred Territory of Kāśī"), Printed circa 1975

The *Map of Kashi* (Fig. 9) has to be seen in the tradition of panoramic views that appeared in the middle of the 19th century. It does not locate places in a defined sacred territory as the 19th century maps did, but presents highly stylized images combined with an explanatory text. The format is determined by the dominating element, the Gaṅgā, which covers the entire foreground. All text is provided in Bengali, English and Hindi.

The frame of the map with its 42 vignettes presents the subcontinent and the realm of the gods. The lower register documents the 12 Jyotirliṅgas, the unbounded *liṅga* of light, "the form assumed by Śiva which compelled both Brahma and Viṣṇu to acknowledge his supremacy" (Stutley 1986: 130), with Tarakeswar and Bhubaneswar (the English version follows the orthography presented on the map) having been added. The upper register and the sides present only one more place, Paśupatināth in Nepal. The remaining vignettes depict five forms of Durgā and Lakṣmī, forms of Śiva such as Sadasiva, Nataraj and Mrityunjoy (Mṛtyuñjaya—Śiva in his capacity to overcome Death), and scenes relating to Śiva or Viṣṇu.

Within the frame three layers can be seen: the Gaṅgā and the riverfront with its *ghāṭs* in the foreground, the temples of Viśvanātha and Annapūrṇā in the ideal centre, and the contemporary urban scene populated by both devotees and gods.

No doubt, the Gaṅgā, the sacred river, forms the base: "seeing, touching or drinking the water removes all sin" as the *Agnipurāṇa* says. Personified as a goddess—seated on her mount (*makara*) she holds a lotus and a water pot—she is almost placed on the central axis. Behind the goddess, burning fires and a corpse indicate the place of Maṇikarṇīkā, where the genesis of the universe occurred at the beginning of time. Viśvanāth, the Lord of All, is placed above the cremation ground as the metaphysical centre of Kāśī in the eastern *garbagṛha* of the temple, below the most dominant architectural element of the entire picture. In the central chamber, below a domed structure, we see Vaikuṇṭheśvara, and left of that the temple of Annapūrṇā, the goddess of food and plenty. Below Vaikuṇṭheśvara is a trident rising out of the Gaṅgā—not only as the emblem of Śiva denoting his function as creator, but as the creative tool that supports the entire realm of Kāśī, lifting it above the rest of the world. Axel Michaels (2000: 194/195) has pointed out the absence of centrality in a similar map and in pilgrim's maps in general. Jonathan Parry (1981: 337) identifies Maṇikarṇikā as the navel of Kāśī, but the place does not occupy or even form the centre of the map. It appears as an illustration, one among many along the river. Other, equally important places like the well of knowledge, Jñānavāpī, in which Viśveśvara is said to have taken refuge when his temple was destroyed in 1667, appears only in the form of a text without any image.

The *Map of Kashi* represents, like its 19th century predecessors, first and foremost a *tīrtha*, a sacred place imbued with the quality of translocality, a place of purity where one obtains merit or attains liberation, *mukti*. At the same time the map is something like an enlarged postcard, a souvenir as it were. It depicts recent achievements, landmarks in the 20th century development of Vārāṇasī such as the university (built in 1852), the clocktower (*ghaṭīyantra*, built in 1867), two railway stations, and the bridge, constructed in 1887. Even the Bhārata Mātā Mandir, founded by Shivprasad Gupta in 1936, is depicted as a temple which houses "Mother India", represented by a large map of the entire subcontinent.

As prominent as in earlier panoramas, the Alamgir Masjid figures at Pañcagaṅgā Ghāṭ as Madhav Rao ka Dharahara. There is no valid explanation for why the mosque is no longer identified after its creator (in Sukula's map one could read: *auraṃgajeba kī masjid*). Even the official plate at the site says "Dharhara Mosque—Alamgir Masjid". The term *dharahara* denotes a monument, preferably in the shape of a tower. Notwithstanding the loss of the minarets half a century ago, the *Map of Kashi* follows earlier prototypes with tall minarets. But who is Madhav Rao or Madho Das—a variant named in Murray's handbook since 1892? Most sources agree that the name recalls a "mistri", an architect or builder or even a *diwān* of Aurangzeb who had guided the construction of the mosque.

Behind the image that illustrates Kashi as a *tīrtha* and some of its hundreds and thousands *tīrthas* in concrete shape appears the pattern of a topographical map in the shape of a grid of roads. The grid is again of a highly stylized nature, because it does not represent reality but transforms it into a pattern. The depiction of traffic police at Godaulia, the crossing of the city with a high sense of centrality, contributes to the contemporary flavour. An ox-cart, a scooter and a car stand for traffic, transportation of goods and people.

Contemporaneity is achieved by including a small image of the Tulsi Manas Mandir; in reality a large and demanding structure built in 1966. The Town Hall in Sukula's *Mirror of Kāśī* of 1876, and Svāmī Bhāskarānaṇḍa's Samādhi at Durgākuṇḍa on the panorama dated circa 1910, figure in a similar manner. Obviously, the *Map of Kashi* offers multiple perspectives: it wants to present Kāśī as a *tīrtha*, a sacred place embedded in India, and as an urban environment with a modern infrastructure—a multivalent environment.

The latest version of the *Map of Kashi* is published by Sobhit Publications and is on regular sale at stalls near Daśāśvamedha Ghāṭ and Viśveśvara, and at Maṇikarṇikā Ghāṭ. The scenes along the *ghāṭs* are very simplified. The Bharat Mandir now appears as "Bharatmāta" as well as in built form. The Swami Narayan Majhodri Temple beside the clocktower, has been newly added.

Totally new also is the addition of four pictorial representations of prominent temple interiors—*Śrī Kāśīviśvanāth darśana* and *Śrī Kāśīviśvanāth śṛṇgara darśana* to the left, and to the right *Śrī Kāśīviśalakśi darśana* and *Śrī Kāśīsvarnamayi Annapūrṇā darśana*. In the lower register seven photographs have been added depicting the Golden Śikhara of Viśvanātha's temple, Daśāśvamedha Ghāṭ, Kālabhairava, Maṇikarṇikā Ghāṭ, Ḍhuṇḍhirāja, Hariścandra Ghāṭ, and the Viśvanāth temple of the Benares Hindu University. The reduction of the map in favour of pictorial representations adds to the flavour of a touristic souvenir. The vendor at Maṇikarṇikā Ghāṭ, however, still guides the pilgrims through Kāśī with his finger, explaining the individual items on the map.

Notes

1 Research in Vārāṇasī started in 1987. In 1991 I started to perform pilgrimages in the company of Rana P. B. Singh, guided by Kedarnāth Vyās. From 1999 to 2003 my fieldwork was funded by the German Research Council. The account of panoramas will also be part of an extensive publication on place and territory in Kāśī.

2 Joseph Tieffenthaler had been in India since 1743 and is credited with making the first geographical survey of the course of the Gaṅgā. His account of India was first published in German in 1785, the French edition followed in 1786: Description Historique Géographique de l'inde, qui présente en trois Volumes, enrichis de 964 Cartes et autres Planches. Time I: Contenant la Géographie de l'Indoustan, avec 30 Planches, Berlin 1786, planche VII (opposite page 228) "Banares".

3 I owe this insight to Joachim K. Bautze, who compiled some 400 views published from 1786 to 1911.

4 Panorama des Rechten Elb-Ufers von Hamburg bis Blankenese in achtzehn aneinander gereihten Blättern, nach der Natur aufgenommen von T. B. Wilms, die Schiffe von E. Hasse, auf Stein gezeichnet von I. W. Vos, gedruckt und Verlegt im lithographischen Institute von Charles Fuchs in Hamburg (no date).

5 Michaels (2000: 187): "Religiöse Karten sind Texte von religiösen Räumen. Sie bilden religiöses Wissen ab und sind daher auch Bilder. Der Unterschied zwischen Bildern, einschließlich Karten, und Texten ist ein konventioneller, wie ich zu zeigen versuche, denn ich lese Raumtexte als Karten und Karten als Texte von religiösen Räumen".

6 Three copies of the *Kāśīdarpaṇa* printed on cloth were handed over to the author, Jan Pieper and Rana P. B. Singh in 1975, 1976 and 1978 by Kubernāth Sukul. Copies printed on paper are preserved in the Bharat Kalā Bhavan in Vārāṇasī and the British Library (no. 53345.(2.)). It is widely published but has never been sufficiently analyzed: Pieper 1979: 212; Gutschow 1977a, 1977b: 73; 1978: 292; 1994: 196; Gutschow & Michaels 1994: 32–33; Lannoy 1999: 441; Michaels 2000: 216; Schwartzberg 1992: 454; Singh 1988: Fig. 5; 1993: 6; 2002, 44. The cloth print has been exhibited on the occasion of the exhibition "Mandala—Der Heilige Kreis im tantrischen Buddhismus" at the Völkerkunde Museum in Zürich in 1992 and at the exhibition of Vārāṇasī maps at the Museum für Völkerkunde in Heidelberg in summer 2002. Since 2001 the *Kāśīdarpaṇa* is accesible on the internet and as a CD (Gengnagel & Michaels 2001).

7 Two more panoramas of the kind discussed below are known to me: one is kept in the Museum der Kulturen in Basel, No. IIa 10970. In this version the architectural details of the various structures along the *ghāṭs* are depicted with great realism, although only structures facing the *ghāṭs* are depicted; no text. The second is kept at the Völkerkunde Museum in Heidelberg, size 22

by 400 cm, No. 38349. Both must date from to the second half of the 19th century, as the Nepalese Samrājyeśvarī temple is depicted (1841), but not the bridge (1887).

8 The "complete view" is kept at the British Museum, No. 53345(5), size 17 by 305 cm. It was published by Beni Singh Bookseller at Rajghat, Kashi, and printed at Matbūa (-i) Sulaimāni Press, Banāras in 1911 or 1912 and aquired by the British Library in January 1913. Two similar panoramas are known to me. One is kept in the British Library, No. P 761, published in a detail by Richard Lannoy (1999: 513). It is a coloured lithograph, printed by Durga Pershad in Allahabad 1901 and as such represents a refined predecessor of the one printed in Banāras. A second version is kept at the Bharat Kala Bhavan in Vārāṇasī.

References

Archer, Mildred. 1980. *Early Views of India. The Picturesque Journeys of Thomas and William Daniell 1786–1794. The Complete Aquatints.* London: Thames and Hudson Limited.

Gengnagel, Jörg. 2003. Mapping Sacred Spaces—Aspects of Cartography in 19th century Banaras. In: Vera Dorofeeva-Lichtmann & Daniela Moritz (eds.), *Creating and Representing Sacred Spaces.* Göttinger Beiträge zur Asienforschung. Göttingen: Peust&Gutschmid, pp. 247–263.

— & Axel Michaels. 2001. *Kailāsanātha Sukula's Mirror of Kāśī (Kāśīdarpaṇa, 1876). A historical map of Vārāṇasī based on the Kāśīkhaṇḍa and related texts.* Electronic Publications of the Vārāṇasī Research Project, South Asia Institute, Heidelberg. (*www. benares.uni-hd.de*).

Gutschow, Niels. 2006. *Benares. The Sacred Landscape of Vārāṇasī.* Stuttgart: Edition Menges.

— & Axel Michaels. 1993. *Benares. Tempel und religiöses Leben in der heiligen Stadt der Hindus.* Köln: DuMont Buchverlag.

Dewitz, Bodo von & Werner Nekes (eds.). 2002. *Ich sehe was, was Du nicht siehst! Sehmaschinen und Bilderwelten. Die Sammlung Hermann Nekes, Köln.* Göttingen: Steidl.

Hodges, William. 1785–88. *Select Views in India, drawn on the spot, in the years 1780, 1781, 1782, and executed in aqua tinta.* London: printed for the author.

Krishna, Rai Anand. 2003. *Banaras in the early 19th century. Riverfront Panorama. Panorama of Banaras viewed from across the Ganges by 'a native artist'.* Varanasi: Indica.

Lannoy, Richard. 1999. *Benares Seen from Within.* Bath: Callisto Books.

Michaels, Axel. 2000. Konstruktionen von Translokalität und Zentralität. Eine religiöse Karte von Benares. In: Daria Pezzoli-Olgiati & Fritz Stolz (eds.), *Cartografia Religiosa: organizzazione, codificazione e simbologia dello spazio nei sistemi religiosi.* Bern etc.: Peter Lang SA, pp. 187–216.

Oettermann, Stephan. 1980. *Das Panorama. Die Geschichte eines Massenmediums.* Frankfurt/Main: Syndicat.

Pieper, Jan. 1979. A Pilgrim's Map of Benares, *GeoJournal* 3: 215–218.

Pohlmann, Ulrich & Dietmar Siegert. 2000. *Samuel Bourne. Sieben Jahre Indien. Photographien und Reiseberichte 1863–1870.* München: Schirmer/Mosel.

Poncar, Jaroslav. 1998. *Panorama of India.* New Delhi: Timeless Books.

Prinsep, James. 1831, 1833. *Benares Illustrated in A Series of Drawings. Lithographed in England by Eminent Artists.* Calcutta: Printed at the Baptist Mission Press. [Reprint, with an introduction by O. P. Kejariwal: Varanasi: Vishwavidyalaya Prakashan, 1996.]

Renié, Pierre-Lin & Sabine du Vigneau. 1992. *L'Inde. Photographies de Louis Rousselet 1865–1868.* Bordeaux: Musée Goupil.

Singh, Rana P. B. 1988. The Image of Varanasi: Sacrality and Perceptual World in Hindu Codification, *The National Geographical Journal of India* 34: 1–32.

JOACHIM K. BAUTZE

Examples of Unlicensed Copies and Versions of Views from Benares: Their Authorship and Identification[1]

This paper considers mainly published views of Benares from the 18th century to 1911, the year in which, from December 7th onwards, the last Imperial Darbar was held in Delhi. During the festivities of this "Imperial Assemblage"—as the Darbar was officially called—movie-cameras and an endless variety of hand-held box-cameras were brought to India. As these new cameras enabled virtually anybody to take photographs, the huge amount of photographic views of Benares taken ever since escapes any serious attempt of cataloguing them, even when the cataloguing is restricted to published views only. The amount of views from Benares produced and reproduced during the remaining 134 years[2] is still considerably large.

Copper Engravings, Aquatints, Lithographs and Woodcuts

One of the earliest panoramic views of the city as seen across the Ganga is a copper engraving in Tieffenthaler 1786 (see Gutschow Fig. 1).[3] A very closely related view is reproduced in Bhattacharya 1999.[4] As it is more likely that the artist of the painting saw a copy of Tieffenthaler's book rather than Tieffenthaler coming across this individual painting in the bazars of Benares we presume that the painting reproduced by Bhattacharya is an extended copy of Tieffenthaler's engraving. In this case, the copy is unique, with no further circulation through contemporary prints made on the basis of this painting.

With most views, however, it is the other way round: An artist produces a view on the spot which becomes the basis for the engraver in Europe for the production of a reproducable form of the view, mostly in the shape of a lithograph, acquatint or engraving. As a rule the actual view was reproduced only after the invention of successfully printing photographs through a lithographic process, some time after 1895.

William Hodges (1744–99) painted his view of Aurangzeb's mosque[5] in about 1781.[6] An aquatint based on this work was published by Hodges himself as early as 1787.[7] The same mosque was painted by Thomas Daniell (1749–1840) in about 1813 based on a much earlier watercolour drawn on the spot,[8] to be published as a small sized engraving in 1834.[9] Both the Daniells, Thomas and William (1769–1837), created the most famous and largest published views of the city, based on their watercolours made when in India. Several of these watercolours were published as small-sized steel-engravings, decades after their visit to India. Another one of those, the *Shuwallah-Gaut,*

Bernares,[10] was drawn as early as 1788 or 1789 only to be published in 1834 in the *Oriental Annual*.[11] An oil-painting of this Ghat also exists, it was, however, published only very recently.[12]

The most influential publication in terms of views from the sacred city is *Benares Illustrated, in A Series of Drawings. Lithographed in England by Eminent Artists*, by James Prinsep (1799–1840), published in different parts from 1831 onwards. Some of Prinsep's original watercolours surfaced only very recently and it seems, that Prinsep made many more watercolours which were not reproduced in his book.[13] As no copyright then existed, Prinsep's lithographed watercolours were the most influential in illustrating the Western view of Benares. We offer here a selection of his most "successful" illustrations.

His lithograph titled "A PREACHER EXPOUNDING THE POORANS In the Temple of Unna-Poorna, Benares"[14] was copied twice. The first version was published as "Un Pundit expliquant les Saintes Ecritures" in De Jancigny & Raymond 1845,[15] the second appeared as *A Brahmin reading the Shasters* in Wright 1853.[16] This latter publication, brought out first in 1848, saw many editions until about 1860. It is the most influential American publication in the mid 19th century with regards to views of India. Not one of the many woodcuts contained in that book is authentic: All illustrations were copied from earlier woodcuts, lithographs or engravings. To quote another example: "Specimen of Architecture at Benares. Two of the eight columns which support the vestibule of a Temple, represented by engraving (...)"[17] is simply a duplicated version of Hodges' copper-engraving "A COLUMN taken from the TEMPLE of VIS VISHA at Benares", published in 1793.[18]

Prinsep's lithograph "A THAKOOR-DWAREE / Interior of Lala Kashmeree Mul's House, Benares / Dii Penates and Sacra in the Sanctum"[19] re-appears in the same work by Wright as "Interior of a House in the City of Benares".[20] His "VIEW OF THE GYAN BAPEE WELL, BENARES"[21] became *A View in the City of Benares* in Wright's book.[22] The version published by Wright not only shows significant differences in the composition but also a different interpretation of the monuments shown. Prinsep's lithograph "KUPULDHARA TULAO, BENARES"[23] was also used by Wright.[24] Prinsep's "ELEVATION OF A HINDOO TEMPLE"[25] reappears in small-sized engraving in De Jancigny & Raymond's book of 1845 as "Temple Hindoo".[26] The same authors also used Prinsep's "MUNIKURNIKA GHAT, BENARES / from the river"[27] for their "Munikarnika Ghât à Bénarès".[28] Prinsep's lithograph of "BENARES / A Brahmin placing a Garland on the holiest Spot in the sacred City"[29] reappears as rather small-sized wood-engraving in what appears to be the earliest book on Benares in the German language, Albert Ostertag's *Leupolt's Erinnerungen an das Missionswerk in Benares*, frontispiece.[30] Ostertag's work also contains several other woodcuts which were copied from earlier publications.[31] Prinsep's lithograph "EVE OF AN ECLIPSE OF THE MOON / 25th November 1825"[32] became "HINDOUSTAN. - Bénarès la veille d'une fête religieuse" as late as 1888.[33] Of this lithograph, Prinsep's original watercolour recently appeared in the art-market.[34] Also the original watercolour of "VIEW from UGNESWAR GHAT, BENARES"[35] was offered for sale (cf. Fig. 1).[36]

Fig. 1: "View of the Ugneswar Ghat, Benares" (James Prinsep).

Photographic Views: Their Possible Earlier Versions and Later Copies

Samuel Bourne

After 1850, more and more views of Benares were photographed rather than drawn or painted. The most successful commercial photographer in those days was Samuel Bourne (1834–1912), who travelled through India between 1863 and 1869. Bourne took upwards of two-thousand photographs in India: monuments, landscapes, historical sites, group- as well as individual portraits, in short: everything that might sell. He had ateliers in Calcutta, Simla and later also in various other Indian cities. As he formed a business with another photographer, Mr. Shepherd, his firm became known as Bourne and Shepherd even after Bourne's return to England in 1870. The company of Bourne and Shepherd exists to this day in Calcutta (Kolkata).

The company of Bourne and Shepherd was not to remain the only photographic company in India. Others like that of E. Saché, Frith, G. Lawrie, Clifton, T. Rust, etc. followed, often copying Bourne's composition. As the exposure time in those days exceeded a few seconds, every photograph had to be composed, the position of the "natives", their distribution in the view to be photographed and thus made permanent, was usually arranged by the photographer.

By the mid 19th century, a photograph was to be reproduced by the photographic process alone. A satisfactory method of reproducing photographs by a more economic lithographic process was only introduced during the late 19th century. Before that, photographs were usually engraved and printed as woodcut. A writer who wanted to illustrate her or his book on India no longer had to go to an art-school before leaving for India, nor was she or he to be accompanied by an artist. She or he would visit one of the photographic studios and select the photographic views which she or he thought would be most suitable for her or his book. Back home, local artists would be requested to produce blocks for printing these photographs. As a result, most illustrated books published between about 1860 and 1900 are illustrated with either woodcuts or steel-engravings based on original albumen photographs. By the time the books were published, the actual source of the illustrations was by no means evident. The authors gave generally rather vague information like: "from a photograph" without mentioning the name of the studio and/or the photographer. This very often led to the assumption that the writer had taken the photographs himself, which was mostly not the case, a fact that was revealed only recently, when the same photographs were published together with the name of the original photographer and/or studio.

Bourne's photographs could be ordered according to size and catalogue number. In about 1869 a catalogue by Bourne and Shepherd was published, which lists 20 different photographs of Benares, Ramnagar and Sarnath in the most often used size of 12 x 10 inches.[37] 14 more photographs are offered under the smaller size of 4 x 4 ½.[38] Most of these smaller sized photographs are very closely related versions of the larger sized views. This was done in order to enable less wealthy clients to afford such photographic views. The price for a larger view by Bourne and Shepherd is given as Rs.4, the smaller ones cost Rs.1 and 8 Annas each. This system of photographing more interesting views twice, viz. with a smaller and larger glass-plate was also used by one of the greatest 19th century Indian photographers, Lala Din Dayal (1844–1905).

As Bourne was aware of photographs of Indian views taken before he left for India, we must assume that he was also aware of views done in other mediums than photography, such as lithographs or aquatints. To give an example: item 1170 in the Bourne and Shepherd Catalogue of c.1869, titled "'Vishnú Pud' and other Temples near the Burning Ghât"[39] is a well-known photograph today.[40] Its composition strongly resembles a lithograph of almost the same scene by Prinsep, mentioned before.[41] Bourne considered this view so important, that he took *two* different photographs of this composition,[42] to which he assigned the same, i.e. identical negative number, viz. 1170! After the photographers were influenced by engraved and lithographed views such as the one by Prinsep, the engravers and lithographers of the second half of the 19th century in turn were to engrave photographs. Bourne's "first" version of his "1170" was turned into a woodcut published as double-page plate in Eugene Goblet d'Alviella's *Inde et Himalaya. Souvenirs de Voyage* of 1880.[43]

The above mentioned problem of identical Bourne & Shepherd catalogue numbers for two similar but in the end different photographs is not restricted to 1170 of that company. One encounters a similar problem when considering one of the previous numbers within the Bourne & Shepherd catalogue of about 1869, viz. no.1168, showing "The Great Mosque of Aurangzebe, and adjoining Ghats" (Fig. 2).[44] Of this often published photograph[45] a second, lesser known version with the same number exists (Fig. 3).[46] It is inscribed in the negative: "The Great Mosque of Aurangzebe. Benares. 1168" (all in capitals). This "second" version of 1168 captures almost the same angle

Fig. 2: "The Great Mosque of Aurangzebe, and adjoining Ghats" (Bourne, original).

Fig. 3: "The Great Mosque of Aurangzebe. Benares" (second version).

Fig. 4: "Raja Jey Singh's Observatory, and Dasaswa Médh Ghât" (Bourne & Shepherd Catalogue).

as Bourne's signed original, its composition, however, is different. The attention of the viewer of Bourne's original view (Fig. 2) is taken by the boat contrasting against the bright surface of the river in the right hand part of the photograph. This focal point is entirely missing in the "second" version from the atelier of Bourne & Shepherd. It is, however, present in the very first version of this view, the lithograph by Prinsep (cf. our Fig. 1 for the watercolour of the same).[47] Bourne either had Prinsep's earlier view in mind when he photographed the "first" version of 1168 or, and this possibility must also be taken into consideration, he had an eye for such scenes like Prinsep. Why then was the "second" and compositionally weaker version of 1168 produced? We do not know for certain but we presume that the glass-negative of the original 1168 broke some time during the early 1880s and had to be replaced by another, naturally later version of the same or similar view, as copies of the original version of 1168 were probably in high demand. Like 1170, Bourne's original view of 1168 was also engraved, as a large-sized woodcut as early as 1875.[48] Many other photographers produced views from the stairs of the Bhonsla Ghat towards the Panchganga Ghat.[49] As similar views exist in the medium of watercolour[50] and oils[51] as late as 1888, the related photographic views were apparently as successful as the "first" version of Bourne's 1168.

No. 1171 in the Bourne & Shepherd Catalogue of about 1869 lists "Raja Jey Singh's Observatory, and Dasaswa Médh Ghât".[52] The photograph (Fig. 4)[53] is one of the earliest photographic views of this monument taken from that angle. It influenced most subsequent photographic views of this monument.[54]

Louis Rousselet

Initially, Louis Rousselet (1845–1929) was not a photographer like Bourne, when he set out for India in June 1863. He learned photography when in India, and after his return to France he published his diary which he kept in India along with illustrations in the well-known French weekly journal *Le Tour du Monde*. One issue in those days only cost 50 Centimes and contained about 15 pages with engravings, maps etc. The woodcuts that accompany Rousselet's text were either done after native drawings, Rousselet's own sketches, Rousselet's own photographs or photographs by other photographers, such as Shepherd & Robertson or Bourne & Shepherd. The latter sources, however, are never mentioned and it is only today that we are in the position to determine the sources of many engravings published in Rousselet's work.

Rousselet is important because his illustrations are probably the most often copied visual sources on 19th century India. His long series of articles brought out in *Le Tour du Monde* was followed by a bulky, lavishly produced publication *L'Inde des Rajahs. Voyage dans l'Inde Centrale, dans les Présidences de Bombay et du Bengale*. It saw its first French edition in 1875, followed by a second edition in 1877. There were several English editions after 1876 and even an Italian edition. All editions, with the exception of the English edition of 1882[55] retain Rousselet's illustrations, most of which were already published in *Le Tour du Monde* from 1871 to 1874.

Almost all illustrations on Benares were first reproduced in No.685 of *Le Tour du Monde*, published on February 21st, 1874 (Rousselet 1874). All subsequent illustrations were probably done from the same block, so there is but little variety. Revealing, however, is a comparison of the different captions of an otherwise identical illustration. "La fête de Ganésa, à Bénarès", to give an example, is based on a sketch by Rousselet in 1874,[56] whereas it was done after unspecified "Indian paintings" in the edition of 1877.[57]

The photographs after which the woodcuts were produced, are credited in two different ways. They are either by Rousselet himself or from an unmentioned source ("d'après une photographie"). This anonymous source, however, is mostly Bourne & Shepherd, in some cases also Shepherd & Robertson. All of Rousselet's own photographs were recently published (Renié & Vignau 1992) and it must be noted here that a good number of his pictures was not turned into woodcuts for the publication of his *opus magnum*. That Rousselet knew the stock of photographs of Bourne & Shepherd very well is shown by the large amount of photos that he used from the latter company. It might hence be presumed that his way of taking successful photographs was to some extent influenced by these commercial photographers. It seems that Rousselet was in fact under the influence of Bourne's views, but in a different way that one might expect. A comparison of Bourne's (earlier) views with that of Rousselet reveales that the latter in fact took many of his views from the same point of view as his fellow photographer but successfully managed to capture a different angle and atmosphere. He was careful enough not to re-create Bourne's views.

Rousselet's *L'Inde des Rajahs or India and its Native Princes* was not translated into German but, curiously enough, Rousselet's views are the best known depictions of India in the last quarter of the 19th century in Germany. In 1880 and 1881 Emil Schlagintweit published the first edition of his widely distributed *Indien in Wort und Bild. Eine Schilderung des indischen Kaiserreiches* (India in words and images. A description of the Indian Empire). Schlagintweit's lavishly produced volumes are by a few centimetres larger than the complete French or English editions of Rousselet. Of the few hundred woodcut illustrations that illustrate the two volumes, which in their layout and colour of the covers clearly copy the French editions, only very few are by the famous Schlagintweits themselves. Strangely enough, Emil Schlagintweit in his text mentions exclusively English sources, whereas almost all his illustrations were taken either from Rousselet or two other French travellers, who published their reports in earlier issues of *Le Tour du Monde*: Alfred Grandidier (Voyage dans les Provinces Méridionales de l'Inde) and Guillaume Lejean (Le Pandjab et le Cachemir). Grandidier and Lejean used photographs provided by photographic agencies in India, and not unfrequently, also Rousselet used blocks the prints of which were published in earlier issues of *Le Tour du Monde.*[58] All of Schlagintweit's illustrations of Benares are hence identical with those published earlier by Rousselet.[59]

Variants of Lithographed Photographs

Schlagintweit's plagiat was one of the last large publications on India which was exclusively illustrated by woodcuts. In the 1890s, more and more publications with lithographed photographs came into circulation. In the beginning, publications on the scale of Rousselet's *L'Inde des Rajahs* or Schlagintweit's *Indien in Wort und Bild* were illustrated by both woodcuts as well as lithographed photographs. A landmark with regards to India is Uchtomskij 1894, the illustrations for Benares of which are finely executed woodcuts plus one so-called "Heliogravüre" showing a photograph of the Maharaja of Benares framed by his two eminent Western visitors (Uchtomskij 1894: 459). The French edition of the same, published only a year earlier, has no such lithographed photographs yet (cf. Oukhtomsky 1893).

During the last decade of the 19th century, photographic studios working for the public popped up like mushrooms. A decade earlier almost each ruler of an Indian state employed his own photographer to record the activities of the state *darbar* as well as the architectural monuments or representative animals (mostly elephants or bulls, in arid areas also camels) of the state. It is not surprising that several court-photographers were formerly court-painters, as the taste of the rulers changed in course of time. The court-painter, if he did not want to become unemployed, had hence to become the court-photographer. One example with regards to Benares is Madho Prasad of Ramnagar, of whom painted portraits of the Maharaja of Benares[60] as well as photographic views of the monuments of Benares and Sarnath are known.[61]

It seems that Madho Prasad had a colleague, if not a relative (father or elder brother?) by name of Babu Jageswar Prasad. In 1883 Jageswar Prasad produced a series of albumen photographs on the ghats of the city. His recently published view of the Panchganga ghat with Aurangzeb's mosque is part of that album.[62] For the same view, i.e. another print of the same negative, however, Madho Ram Prasad is credited in an album on Benares in the collection of Maharawal Mahipal Singhji of Dungarpur.[63] For other views as well, two different attributions exist: one to J[ageswar] Prasad and another one to Madho Prasad on the basis of the official printed attribution in the album in the collection of H. H. Dungarpur.[64]

A Benares studio aimed at the needs of the public was operated by "Saeed Bros". This studio started business around the mid 1890s. The activities of that studio were not restricted to Benares alone. The picture postcards which were produced and advertised by that studio—while they were printed in Germany—also include views of Darjeeling (W. B.). The three lavishly illustrated books on Benares that were published between 1905 and 1909[65] all contain numerous photographs supplied by that studio.

At times, a print derived from the same negative is reproduced in two of these three books and one is inclined to assume, that two individual prints from a common negative look more or less identical when lithographically reproduced. Strangely enough, this is not often the case. To give an example: the illustration on p. 81 in Cape 1908 ("The Hindu worships the River and the Sun") is intact, whereas in the book published only a year later, a large part of the lower right hand corner is missing, as probably a piece from the glass plate negative broke away, cf. Greaves 1909, illustration facing p.108: "Bathing Scene". A picture postcard based on a print from the same negative was published by "Saeed Bros., Benares" in the early 20th century. It is titled "No.14—Bathing—Sun worshipping—Ganges, Benares" and shows no defects.

At times, the common negative of two lithographed prints is not recognised, because for the layout of each book a different detail of the photographic print was chosen.[66] Most confusing, however, is a reproduction of a print from a common negative in two different publications with two quite different captions.[67] One print was even reproduced in each of the three books.[68] Finally, two prints from a common negative may look quite differently, if one of them was printed mirror-reversed.[69]

Incorrectly Identified (Lithographed) Photographs

The visit of the German Kronprinz Friedrich Wilhelm von Preußen (1882–1951) to India in 1910–11 was an event, on which several illustrated books were published.[70] In the preparation of this journey, Princess Louise (her full name was Louise Margaret Alexandria Victoria Agnes, 1860–1917), daughter of Prince Frederick Charles of Prussia, who was then, following her marriage to Arthur Saxe-Coburg, Duke of Connaught on March 13th, 1879, the Duchess of Connaught, could have been of great help. The Duchess of Connaught together with her husband represented the Royal house of Britain during the Delhi Coronation Darbar of 1903.[71] Another relative, the Prince of Wales (George Frederick Ernest Albert Wettin [Windsor], 1865–1936), visited India from November 9th, 1905 to March 17th, 1906. The chapter on Benares in the printed programme of that journey contains one of the most comprehensive descriptions of the ghats of that city.[72]

After his return to Berlin, the crown-prince himself published a lavishly produced album of photographs, most of which are by himself. This album exists in at least two edition. The more lavishly produced one is a portfolio which looks like a facsimile of the actual photographs, some of which are even coloured. Each of the 53 sheets of carton contains one or two tipped-in illustrations after photographs, protected by tissue gards in such a way that at a glance the whole looks like a collection of "real (in contrast to: lithographed)" photographs. The caption below each photographic reproduction is heightened in silver. The front-cover of this portfolio is printed in facsimile of the crown-prince's handwriting: "Erinnerungen an meine Reise nach Indien. [...] Kronprinz" with no further references to the author, publisher or date of publication. The less lavishly produced edition is a bound book which has a printed title-page: *Photographische Aufnahmen von der Indien-Reise Seiner Kaiserlichen und Königlichen Hoheit des Kronprinzen des Deutschen Reichs und von Preussen*, and below: "Der Erlös ist für die Invaliden aus Deutsch-Südwestafrika und für die Hinterbliebenen der dort gefallenen Krieger bestimmt". And still below: "Im Auftrage Seiner Kaiserlichen und Königlichen Hoheit des Kronprinzen [by command of His Imperial and Royal Highness, the crown-prince]".

It seems, however, that the German crown-prince did not care for looking into any of the fresh and to him easily available sources on India as mentioned above: a photograph of the gilded roof of the "golden temple" in Benares, of which by then several picture postcards were in circulation,[73] is titled: "Dach des goldenen Tempels in *Delhi* [italics are mine, JKB]".[74] Delhi, for crown-prince Wilhelm or the editor of the album, was probably the dumping ground for anything unaccounted for, as a photograph taken in the fort of Agra is also labelled as being in Delhi.[75] More recently, another photograph of the "golden temple" in Benares, probably dating from the late 19th century, was identified as some temple from Bodh Gaya![76]

When it comes to the incorrect identification of photographs taken at Benares by members of the nobility in Germany, crown-prince Wilhelm of Prussia is not an exception. His contemporary, crown-prince Rupprecht of Bavaria (Generalfeldmarschall Kronprinz Rupprecht von Bayern, 1869–1955) published an often repeated photographic view of the Manikarnika Ghat as being the Sindhia Ghat.[77]

Notes

1 The research on the views of Benares was made possible through a grant ("Sachbeihilfe") of the "Deutsche Forschungsgemeinschaft", to which we want to express our gratitude.

2 We take 1777 as the year of the earliest published view of (a monument in) Benares, cf. Barker 1777 and Volwahsen 2001, p. 86, Fig. 83: "The observatory on the Manmandir in Benares, 18th-century engraving"; p. 86, Fig. 84: "The Great Samrat Yantra in the Benares observatory (top, elevation; bottom, view from sourh-east), 18th century engraving".

3 Tieffenthaler 1786, folded engraving facing p. 228, counted as "VII", titled, in the print "BANARES". Below, in the river: "Ganges". Nr. 061. For the description cf. pp. 228–229. According to p. 5, Tieffenthaler visited Benares in 1765.

4 Bhattacharya 1999, Illustration III.1. (*above*): "Varanasi seen from the eastern bank drawn by an unknown soldier at the end of Islamic rule, original painting at Victoria Memorial Museum, Calcutta". We have consulted Chaudhuri, Gopa & Bhaskar Chunder. *A Comprehensive Catalogue of Water Colours, Pencil Sketches and Pen and Ink Drawings in the Collection of Victoria Memorial.* Calcutta: The Trustees of The Victoria Memorial, (1981), this painting, however, is not listed in this publication. Besides, we have personally enquired about this painting in the Victoria Memorial of Calcutta, also with no result. As Bhattacharya confounded a considerable number of his illustrations we presume that he actually meant "*Anon* (1765) / 1409 The camp of the Mughal emperor, Shah Alam, on the bank of the Ganges opposite Benares (U.P.). 1765", cf. Archer 1969. London: Her Majesty's Stationery Office,. This painting was viewed by us and in all probability is the source of Bhattacharya's illustration, who in fact only offers a rather weak reproduction of a fragment of the painting.

5 Today, it is called "Dharhara Masjid".

6 For a photographic reproduction cf. Bayly et al. 1990, col. illustration p. 202, cat.no.246: "A View of Benares / William Hodges, *c*.1781 / Oil on canvas, 68.5 x 91.4 (27 x 36) / India Office Library and Records /F94); Mahajan 1988, col. illustration p.99 top, text p. 98: "*Benares*, oil on canvas by William Hodges (c.1782). India Office Library, London (By permission of the British Library); Archer 1970, fig. 8, p. 100, centre: "*Aurangzeb's Mosque, Benares* by William Hodges. Oil on canvas, 68 x 91 cm. Formerly in the possession of Warren Hastings and now in the India Office Library"; Archer 1986, Plate XVIII, cat.no.88, titled: "Aurangzeb's Mosque, by William Hodges, 1781". Text p. 68: "*A view of Benares, UP 1781.* ... 27 x 36 ins (68.5 x 91.5 cms) / Purchased October 1904" in the chapter on "Oil paintings by European artists".

7 Hodges 1785–88, Plate 21: "A View of part of the City of Benares, upon the Ganges. (b); *Feby 24th 1787*".

8 Shellim 1979, TD87 BENARES: AURANGZEB'S MOSQUE AND BENIMADHAVA GHAT, U.P. c.1813. / *Oil on canvas, 37 x 50 ins (94 x 127 cms) / National Gallery of Modern Art, New Delhi*, no.1387".

9 Caunter 1834, engraving facing p. 127, titled, lower left: "Drawn by W. Daniell, R.A". Lower right: "Engraved by G. Hollis". Main title: "Mosque at Benares". Below, right: "B. late Dixon". Below, centre: "London. Published Oct' 1, 1833, by Edwd. Bull & Co, 26, Holles Street. Cavendish Square".

10 Archer 1969, coloured frontispiece, described p. vii: "Shivala Ghat, Benares (U.P.). Water-colour. By Thomas Daniell, c. December 1788 (WD 1295; p. 577)". Page 577: "1295 Shivala Ghat, Benares (U.P.). December 1788 or on return journey November or December 1789. Inscribed on front in water-colour: *Shuwallah-Gaut, Bernares*. Water-colour; 15¾ by 24½ ins. Purchased 4 June 1958.

11 Caunter 1834, steel-engraving facing p. 143, artists, as to lower left: "Drawn by W. Daniell, R.A". Lower right: "Engraved by J. Redaway". Title: "Shuwallah Gaut, Benares". Below, right: "B. late Dixon". Description in the text pp. 141–175 ("Insurrection at Benares").

12 Shellim 1979, full-page col. plate p. 105: "WD3 SHIVALA GHAT AND CHEYT SINGH'S HOUSE NEAR BENARES, U.P. 1802. / *oil on canvas*, 31¼ x 49½ *ins* (79 x 126 *cms*) / *Inscribed*: W. Daniell. Shivala Ghaut and Cheyt Singh's House near Benares. / *Private Collection*". Mahajan 1984, col. plate pp. 82–83, XIII, text p. 148: "Shivala Ghat and Raja Chet Singh's House, Varanasi, oil painting by William Daniell, 1802. Private Collection".

13 For an example see: Christie's [sale catalogue of]: *Visions of India.* London, Tuesday 5 October 1999, col. illustration p. 62, left: "{lot} 81 / James Prinsep (1799–1840) / The Mosque at the Panchganga Ghat, Benares / signed, inscribed and dated 'James Prinsep' (lower right), 'Sept 1828' (lower left) and 'Sketch of the Minarets at Benares'. (lower centre) all on the mount / pencil and watercolour, with scratching out / 8¾ x 6⅞ in. (22.3 x 7.5 cm.)".

14 Prinsep 1831ff, "A PREACHER EXPOUNDING THE POORANS In the Temple of Unna-Poorna, Benares". Authorship: "Drawn on Stone by L. Haghe from a Drawing by James Prinsep Esqr".

15 De Jancigny & Raymond 1845, Gravure 76, full-page steel-engraving facing p. 504. Authorship: "Lemaitre direxit". Titled: "Un Pundit expliquant les Saintes Ecritures".

16 Wright 1853, woodcut p. 53, bottom: "No.24. *A Brahmin reading the Shasters*".

17 Wright 1853, full-page woodcut p. 29 [silent pagination]: "No.11. Specimen of Architecture at Benares. *Two of the eight columns which support the vestibule of a Temple, represented by engraving, No.34, page 75*".

18 Hodges 1793, copper-plate opposite p. 63, titled: "A COLUMN taken from the TEMPLE of VIS VISHA at Benares. / Engraved by T. Medland, from a Drawing / made on the Spot by W. Hodges R.A. / London, Published by J. Edwards, Pall Mall, Jany.1.1793".

19 Prinsep 1831ff, lithograph p. 56, titled: "A THAKOOR-DWAREE / Interior of Lala Kashmeree Mul's House, Benares / Dii Penates and Sacra in the Sanctum". Authorship: "Day & Haghe Lthrs to the King 17, Gatest. / Drawn on Stone by L. Haghe from a Sketch by James Prinsep Esqr".

20 Wright 1853, woodcut p. 58, bottom: "No.27. *Interior of a House in the City of Benares*". Signed, lower right: "Howland sc[ulpit]".

21 Prinsep 1831ff, lithograph p. 70, titled: "VIEW OF THE GYAN BAPEE WELL, BENARES". Authorship: "Drawn on Stone by L. Haghe from a Sketch by James Prinsep Esqr".

22 Wright 1853, woodcut p. 75, bottom: "No.34. *A View in the City of Benares*". Signed, lower right: "S.F. Baker [last four letters unreadable]".

23 Prinsep 1831ff, lithograph p. 72, titled: "KUPULDHARA TULAO, BENARES". Authorship: "Drawn on Stone by L. Haghe from a Sketch by James Prinsep Esqr.

24 Wright 1853, woodcut p. 115, top: "No.50. *View in Benares*".

25 Prinsep 1831ff, lithograph p. 64, titled: "ELEVATION OF A HINDOO TEMPLE". Authorship: "Drawn on Sketch [sic] by James Prinsep Esqr".

26 De Jancigny & Raymond 1845, Gravure 36, full-page steel-engraving facing p. 233: "Temple Hindoo". Marked lower left: "Gaucherel del", lower right: "Lemaitre direxit".

27 Prinsep 1831ff, lithograph p. 54, titled: "MUNIKURNIKA GHAT, BENARES / from the river". Authorship: "Drawn on Stone by George Barnard from a Sketch by James Prinsep Esqr. Printed by C. Hullmandel".

28 De Jancigny & Raymond 1845, Gravure 36, full-page steel-engraving facing p. 492: "Munikarnika Ghât à Bénarès". Signed, lower centre: "Lemaitre direxit".

29 Prinsep 1831ff, lithograph p. 54, titled "BENARES / A Brahmin placing a Garland on the holiest Spot in the sacred City". Authorship: "Drawn on Stone by J.D. Harding from a Sketch by James Prinsep Esqr. Printed by C. Hullmandel".

30 Ostertag 1846, frontispiece: "Benares".

31 Cf. e.g. Ostertag 1846, woodcut facing p. 107 titled "Heiligensprung" is based on Solvyns 1808, Dixième Livraison. No.1.: Djhampe. Des Hindoûs se jetant sur des couteaux, sabres".

32 Prinsep 1831ff, "EVE OF AN ECLIPSE OF THE MOON / 25th November 1825". Authorship: "Drawn on Stone by L. Haghe from a Sketch by James Prinsep Esqr. ". In the original edition,

partly visible in the Benares reprint, can be read: "On Stone by L. Haghe" and right: "from a Drawing by Ja[s] Prinsep Esqr".

33 (Société Saint-Augustin): *L'Hindoustan.* Lille: Desclée, de Brouwer et Cie, 1888. Woodcut p. 45, signed lower right "CANEDI", titled: HINDOUSTAN. - Bénarès la veille d'une fête religieuse".

34 Christie's [sale catalogue] London, Tuesday 5 October 1999. Coloured illustration p. 51, top: "{lot} 58 / James Prinsep (1799–1840) / Eve of the eclipse of the Moon, 25th / November 1825, Benares / inscribed indistinctly '25/...and Raj Butubh / Shiwala-(1)/2.Oomraogir Tooshta-/3. Temple / of Lulita Debee' (on the reverse) / pencil, pen and brown ink and watercolour / 7 x 10½ in. (17.8 x 26.7 cm.)".

35 Prinsep 1831ff, lithograph p. 20, titled: "VIEW from UGNESWAR GHAT, BENARES".

36 Christie's [sale catalogue] London, Tuesday 5 October 1999. Coloured illustration p. 51, bottom: "{lot} 59 / James Prinsep (1799–1840) / View of the Ugneswar Ghat, Benares / signed with initials 'J.P. / 15 Sep 1825[?].' (lower / left) and with inscriptions identifying all the buildings (on the reverse) / pencil, pen and brown ink and watercolour / $7^1/_8$ x $10^3/_8$ in. (8,1 x 26,3 cm.)".

37 *Photographic Views* c.1869, p. 24, nos.1165–1184.

38 *Photographic Views* c.1869, p. 61, nos.1185–1198.

39 *Photographic Views* c.1869, p. 24, no.1170.

40 Photographic reproductions include: Reynolds-Ball 1907, Illustration opposite p. 195: "Manikarnika Ghât, Benares". Falconer & Groeneveld 1994, full-page illustration p. 57, cat.no. 30, Text p. 109: "**30** 1170, z.t./untitled [tempels aan de Ganges] / S. Bourne / 1865[ca.] / India, Benares [Varanasi] / albuminedruk/albumen print / 24,1 x 29,1 cm. / inv.nr.F 4436/78"; Worswick / Embree 1976, illustration p. 69, text p. 68: "Samuel Bourne: 'Vishnu Pud' and Other Temples near the Burning Ghat, Banaras, 1865, Item 1170 in the Bourne & Shepherd catalogue". Christie's [sale catalogue] London, Wednesday, 5 June 1996, illustration p. 222, bottom: "{lot} 361 / Samuel Bourne and Bourne & Shepherd / albumen prints, approx. 9 x 11in., one 7 x 12 in. signed *Bourne ... and Bourne & Shepherd* with occasional / numbers and titles in the negatives, two mounted on card, others held in corner mounts, each matted / Including ... "Vishnu Pud" and other temples in Benares"....

41 For references to this lithograph see under 28 supra in addition to Bhattacharya 1999, plate III.17. (*above*): "Maṇikarṇikā, Cakratīrtha and Brahmanālā". (only detail); Gutschow & Michaels 1993, illustration pp. 178–179: "Maṇikarṇikāghāṭ mit den Fußabdrücken Viṣṇus, Lithographie nach einer Skizze von James Prinsep, 1830".

42 For the more often published one see under 39 supra. For the 'second' photograph which is almost identical to the 'first' cf. Pal / Dehejia 1986, illustration p. 199, no.206: "Samuel Bourne. *Benares, Vishnu Pud.* Mid-1860s. Albumen print. Walter Collection".

43 Goblet d'Alviella 1880, woodcut illustration pp. 192–193: "Les Quais du Gange a Bénarès".

44 *Photographic Views* c.1869, p. 24, no.1168. The photograph shows in fact the following Ghats, from left to right: Naya Ghat, Ram Ghat, Balaji Ghat, Panchganga or Madhoray Ghat, Bundi Ghat; photographed from the steps of the Bhonsla Ghat.

45 Cf. e.g. Ollmann 1983, Plate 5: "The Great Mosque of Arungzebe, and Adjoining Ghats, Benares, 1865. (*Bourne #1168.*)". Nr.050. Signed in the negative: "Bourne 1168"; Bourne 2001, illustration pp. 178–179: *"Benares, die Große Moschee des Aurangzeb und angrenzende Ghats*, 1865/66". Bourne cat.no.1168. Collection of Dietmar Siegert. 21,5 x 28,1 cm; Falconer 2001, Illustration p. 121, bottom centre, cat.no.100: Samuel Bourne, 1864 / Albumen print, 23,4 x 29,4 cm / Bourne & Shepherd catalogue no.1168 / Howard and Jane Ricketts Collection"; Christie's [sale catalogue] London, 27 September 2001, illustration p. 53, top, lot 55 "Samuel Bourne, Shepherd & Robertson: Indian Architecture and Scenery, Vol.I".

46 Albumen print, size: 23,0 x 29,5 cm. Private Collection.
47 Cf. note 34 supra.
48 Sala 1875, p. 20 "Benares".
49 Cf. e.g. Khan 2002, full-page plate p. 88, no.53, text p. 89: "The Great Mosque of Auranzebe and adjoining Gates. *Baker & Burke #1030, Varanasi, Uttar Pradesh, 1866–68. Michael Wilson Collection*". Signed: "W Baker 1030" in the negative, lower left; Renié / Vignau 1992, full-page plate p. 90, Planche 158: "le ghât de Madoray / *The ghat of Madoray* -, Bénarès", by Louis Rousselet; Stapp 1994, illustration p. 15, fig.9, top, descriptive text also p. 51, cat.no. 58: "Lala Deen Dayal (Indian, 1844–1910). Ghats Below Aurangabad {sic!} Mosque - Benares, ca.1898. Albumen print. George Eastman House, Rochester, NY, gift of the University of Rochester Library / 82: 2736: 0008"; Tellemann c.1905, p. 116: "Badestellen am Ganges bei Benares. / (Nach einer Photographie.)", just to mention a few.
50 Cf. Friedrich / Leipziger 1888, illustration p. 126. The title of this painting is, following the text on p. 154, identical with the inscription in the painting itself: "K.Auringzeb's Moschee / Benares".
51 Christie's [sale catalogue] London, Thursday, 25 May 1995, col. illustration p. 84, top, titled: "{lot} 133 / Holger H. Jerichau (1861–1900) / A View of Benares with Aurangzeb's Mosque and / the Ghats in the foreground / oil on canvas / 25¾ x 33in. (65,5 x 82,8cm.)". The function of the boat in both the views by Prinsep and Bourne ('first' version) is somewhat taken over by four effectfully introduced palm-trees in this composition. It is, however, unlikely that these trees ever existed at this spot.
52 *Photographic Views* c.1869, p. 24, no.1171.
53 Unmounted albumen print, private collection. Size: 23,0 x 28,4 cm. Signed in the negative: "Bourne 1171".
54 Cf. e.g. Couté & Léger 1989, full-page illustration p. 10, titled: "Palais du raja d'Amber Man / Singh (XVIIe–XVIIIe siècle [sic!]), / photographie vers 1890. / Raja of Amber Man Singh's / Palace (17th–18th century [sic!]) / Photograph circa 1890"., numbered in the negative, mirror reversed: "10019"; Sen 1912, Plate II, facing p. 17, fig.3: "Man Mandil Ghat", a photograph by S. Nasiruddin of Calcutta, just to mention a few.
55 As Rousselet 1882 is produced on a much smaller scale, it lacks almost all full-page illustrations present of the earlier editions.
56 Rousselet 1874, full-page woodcut p. 125: "La fête de Ganésa, à Bénarès. - Dessin de J. Gaildrau, d'après un croquis de M.L. Rousselet".
57 Rousselet 1877, full-page woodcut p. 713, titled: "LA FÊTE DE GANÉSA, A BÉNARÈS". Authorship, following p. 798: "294. Fête de Ganésa, à Bénarès. - Dessin de J. Gaildrau, d'après des peintures indiennes".
58 To give an example: "Mountaineers in action", published in Rousselet 1878, woodcut facing p. 506 was printed from the same block as "Montagnards du Nord-Ouest combattant contre les Anglais" in Lejean 1870, p. 345. This woodcut is based on a now often published photograph by Shepherd & Robertson, cf. Desmond 1982, Plate 55. It reappears, needless to say, in Schlagintweit 1881, woodcut facing p. 220.
59 Schlagintweit 1880, woodcut facing p. 164 = Rousselet 1874, p. 121. Schlagintweit 1881, woodcut facing p. 2 = Rousselet 1874, p. 125. Schlagintweit 1881, woodcut facing p. 54 = Rousselet 1874, p. 119. Schlagintweit 1881, woodcut facing p. 56 = Rousselet 1874, p. 117. Schlagintweit 1881, woodcut p. 58 = Rousselet 1877, p. 716. Schlagintweit 1881, woodcut facing p. 58 = Rousselet 1874, p. 113. Also the view of Benares in the more scientific version of the Schlagintweit travels in India as published by Werner 1884, p. 291: "Benares vom Ganges aus (Benares as seen across the Ganges)", was not taken by any of the Schlagintweit brothers, but

prior to 1872 by an anonymous photographer working for "Frith's / Photo-Pictures / The / Universal Series", no. 3027. For a reproduction of this photograph see Gehring 1908, Vol. 2, p. 181: "Blick auf Benares (rechts die große Moschee)."

60 Cf. Bautze 1998, cat.no.29, col. plate p. 125: "His Highness Ishwari Prasad Narayan Singh, Maharaja of Benares (b.1825, r.1835–1889) seated, with Prabhu Narayan Singh (1855–1931) and Aditya Narayan Singh (1875–1939) standing behind as well as a Portrait of the seated Late Raja of Benares, Udit Narayan Singh (r.1795–1835) on the Wall / by Madho Prasad, Ramnagar, ca.1885. / *Media*: Gouache on paper / *Size*: 14 x 11 in. (35,5 x 27.9cm.) / *Inscribed*: On verso, in red ink: By Madho Prasad / Ramnagar.

61 Cf. Christie's [sale catalogue] London, Wednesday 17 June 1998, illustration p. 183, bottom: "{lot} 242 / Anonymous / 'Views of Benares Presented by the Maharaja of Benares' *circa* 1890s / Album containing forty albumen prints, the majority 8¼ x 11¼ in., typescript title labels on mounts, red cloth (re-lined), titled in gilt on front cover, g.e., oblong 4to. / Titles are given as follows: Fort Ram-nagar, Assisangam Ghat, Bhadaini Water works, Butsrajghat, Shivala-ghat, Kedar-ghat, Amrit-rao Chutter-ghat, Dhunrao Panth-ghat, Rana-mahal Ghat, Munshi-ghat, Ahilya Bai's-ghat, Dasaswamedh-ghat, Observatory of Manmandir, Lalta-ghat and Napalees Temple, Burning-ghat, Munikarnika-ghat (Fig.), Ghosla-ghat, Bajirao-ghat, Minaret and Panch-ganga Ghat, Gai-ghat; views around Rambag and Sumer-mandir; Latif Shah waterfall (Chakia), the Godwalia Temple Gate, Golden Temple and Aurangzeb's mosque, the Town Hall, Prince of Wales Hospital, Ishwari Memorial Hospital, Queen's College, Nadesar House and eight images of Buddhist remains (Sarnath)". The photographs of this album are in all probability by Madho Ram Prasad, as his name appears as the photographer of an almost identical album in the collection of HH Maharawal Mahipal Singhji of Dungarpur. For further published photographic views of Benares by Madho Ram Prasad cf. Havell 1905, p. 126, full-page illustration, titled: "Shivâla Ghât / By permission of H.H. the Maharaja of Benares"; p. 215, full-page illustration, titled "Panchganga (page 150) / By permission of H.H. the Maharaja of Benares". For further individual prints, which are not attributed in to Madho Ram Prasad in the respective publication but are attributed to him in the Benares album in the collection of the Dungarpur royal family cf. Greaves 1909, illustration facing p. 22: "Stone Gateway near Gadaulia". Photograph provided by "His Highness the Maharaja of Benares"; illustration facing p. 82: "Domes of the Golden Temple". Photograph provided by "His Highness the Maharaja of Benares"; illustration facing p. 93: "Ramnagar Fort". Photograph provided by "His Highness the Maharaja of Benares"; illustration opposite p. 95: "Sumeru Temple and Tank at Ramnagar". Photograph provided by "His Highness the Maharaja of Benares".

62 Dehejia 2000, full-page illustration p. 23, text p. 22, Figure 3: "Babu Jageswar Prasad. *The Ghats at Benares*, 1883, albumen print, 21.4 x 27. The British Library, Oriental & India Office Collection photo 177 (29)".

63 Below a photograph of the capital of the Ashokan pillar of Sarnath, now the emblem of the state of India, the album is credited by a printed label to: "Madho Prasad Artist to H.H. the Maharajah of Benares". Since the column and capital of the Ashokan pillar at Sarnath was only excavated during the campaign of 1904-05, the compilation of the album cannot be earlier than that date. A similar if not identical album was offered for sale by Christie's in London, cf. note 64 *supra*.

64 Cf. Couté & Léger 1989, full-page illustration p. 98: "Ramnagar. Le jardin royal de / Ram Bagh, photographie par / J. Prasad (vers 1890). / Ramnagar. The Ram Bagh / Royal Garden. Photograph by / J. Prasad (circa 1890)".

65 These are: Havell 1905, Cape 1908 and Greaves 1909.

66 Cf. e.g. Havell 1905, full-page illustration p. 143 “An Encampment of Sâdhus” which is a detail of: “The Holy Men”, illustration p. 86 in Cape 1908.

67 Cf. e.g. Cape 1908, caption to illustration p. 98 (“The Sādhu’s Wife”) with Havell 1905, caption to illustration p. 157 (“A Vaishnavite Nun Reading The Râmâyana”).

68 Havell 1905, p. 207, full-page illustration: “An idyll of peace and self-satisfaction’ (page 155)”; Cape 1908, illustration p. 93: “The Sādhu’s Calf fears no prodigal’s Return”; Greaves 1909, illustration facing p. 54: “Sadhu and his bed of spikes”. A picture postcard from the same negative was issued by “Saeed Bros., Benares” at about the time these three books were published. It is captioned: “No. 24—A Hindu Fakir on spike’s bed”.

69 For the correctly reproduced view of the Man Mandir see a photograph from the studio of Clifton & Co. in Bombay as reproduced in Noti 1911, plate V (“Die Sternwarte [Man Mandir] des Dschaisingh II in Benares”.) or Reuther 1925, plate 138, top (“Front nach dem Man Mandir-Ghat”.). For the mirror reversed reproduction, the print of which derived from the same negative as the prints reproduced by Noti or Reuther, cf. Garrett 1907, plate I, top (“Das Observatorium in Benares”). There also exists a correctly reproduced picture postcard issued by Clifton & Co. based on the same negative.

70 For books that include photographic reproductions of Benares cf. Bongard 1911, Zache 1911 and Heichen 1911.

71 The personal copy of the detailed darbar programme of the Duchess of Connaught, which contains her diary of the events, shows, how carefully the whole event was planned.

72 Like the detailed programme of the Coronation Darbar of 1903, this programme is bound in red soft leather with marbleized edges. For the chapter on Benares see *Tour* 1905, pp. 168–175.

73 One coloured picture postcard issued by “Clifton & Co., Bombay” titled “Golden Temple, Benares” and one picture postcard issued by “Saeed Bros., Benares”, titled: “No.13—Golden Temple—Benares”. A photograph of the roof of this temple, taken by Madho Ram Prasad, was by then also published in Greaves 1909, illustration opposite p. 82: “Domes of the Golden Temple”. Photograph provided by “His Highness the Maharaja of Benares”. An illustrated Benares guide which the royal party must have come across when in the city, is Heinig 1910, where the “Golden Temple” is reproduced on p. 11.

74 *Photographische Aufnahmen von der Indien-Reise Seiner Kaiserlichen und Königlichen Hoheit des Kronprinzen des Deutschen Reichs und von Preussen* 1911, 40th folio following the title page. The photograph is also present with the same label in the portfolio-edition of the same.

75 *Photographische Aufnahmen von der Indien-Reise Seiner Kaiserlichen und Königlichen Hoheit des Kronprinzen des Deutschen Reichs und von Preussen* 1911, 30th folio following the title page, also present in the portfolio edition.

76 Bonetti & Fusco & Polichetti 1994, full-page plate p. 83, Text p. 262: “Bodhgaya / silver bromide gelatin print*, selenium toning / ING/CI, from original negative / (25.5 x 30.5), inv. cat.6203. Through comparison with other photographs, it has been established that the temple is in Bodhgaya, Bihar (eastern-central India)”.

77 Rupprecht 1920, full-page photographic illustration facing p. 65, titled: “Benares: Sindhia-Ghat / Links in den Bäumen das Nepali Khapra / (Seite 77)”.

References

Archer, Mildred. 1969. *British Drawings in the India Office Library. Volume II: Official and Professional Artists*. London: Her Majesty’s Stationery Office.

— 1986. *The India Office Collection of Paintings and Sculpture*. London: The British Library.

Archer, W[illiam] G[eorge]. 1970. Benares through the Eyes of British Artists. In: *Apollo. The Magazine of the Arts.* August, pp. 96–103.

Barker, R[obert]. [Sir]. 1777. An Account of the Bramin's Observatory at Benares. By Sir Robert Barker. In a letter to Sir John Pringle, *Philosophical Transaction of the Royal Society of London*, Vol. LXVII, part ii, No. xxx: 598f.

Bautze, Joachim K. 1998. *Interaction of Cultures. Indian and Western Painting 1780–1910. The Ehrenfeld Collection.* Alexandria, Virginia: Art Services International.

Bayly, C. U. et al. (eds.). 1990. *The Raj. India and the British 1600–1947.* London: National Portrait Gallery Publications.

Bhattacharya, B[rajamadhava]. 1999. *Varanasi Rediscovered.* New Delhi: Munshiram Manoharlal Publishers.

Bonettti, Maria Francesca, Maria Antonella Fusco & Massimiliano Alessandro Polichetti. 1994. *Federico Peliti (1844–1914). An Italian Photographer in India at the Time of Queen Victoria.* Manchester: Cornerhouse Publications.

Bongard, Oscar. 1911. *Die Reise des Deutschen Kronprinzen durch Ceylon und Indien.* Berlin: Verlag C. U. Schwetschke und Sohn.

Bourne, Samuel. 2001. *Sieben Jahre Indien. Photographien und Reiseberichte 1863–1870. Herausgegeben von Ulrich Pohlmann und Dietmar Siegert* [...]. München: Fotomuseum im Münchener Stadtmuseum und Schirmer/Mosel.

Cape, C[harles] Phillips. n.d. *Benares. The Stronghold of Hinduism.* London: Robert Culley, [probably 1908, cf. preface in Greaves 1909].

Caunter, Hobart. 1834. *The Oriental Annual, or Scenes in India; Comprising Twenty-Five Engravings from Original Drawings by William Daniell, R.A. and a Descriptive Account by ...Hobart Caunter.* London: Edward Bull.

Christie's [sale catalogue of]. *Visions of India, including the Paul F. Walter Collection of Indian Photographs.* London, Wednesday, 5 June 1996.

Christie's [sale catalogue of]. *Visions of India.* London, Wednesday 17 June 1998.

Christie's [sale catalogue of]. *Visions of India.* London, Tuesday 5 October 1999.

Christie's [sale catalogue of]. *Fine 19th Century Indian Photographs from the Collection of Kanwardip Gujral.* London, 27 September 2001.

Couté, Pierre-Daniel & Jean-Michel Léger. 1989. *Bénarès. Un voyage d'architecture / an architectural voyage.* Paris: Éditions Créaphis.

Dehejia, Vidya et al. (eds.). 2000. *India Through the Lens. Photography 1840–1911.* Washington, D. C.: Freer Gallery of Art / Arthur M. Sackler Gallery in association with Mapin Publishing, Ahmedabad and Prestel, Munich, London, New York.

De Jancigny, Dubois & Xavier Raymond. 1845. *Inde.* Paris: Firmin Didot Frères, Éditeurs. (Univers Pittoresque. Histoire et Description de tous les Peuples, des leurs Religions, Mœurs, Coutumes, Industrie.).

Desmond, Ray. 1982. *Victorian India in Focus, a selection of early photographs from the collection in the India Office Library and Records.* London: Her Majesty's Stationery Office.

Falconer, John. 2001. *India: Pioneering Photographers 1840–1911.* London: The British Library and The Howard and Jane Ricketts Collection.

Falconer, John, Steven Wachlin & Anneke Groeneveld. 1994. *Van Bombay tot Shanghai / From Bombay to Shanghai.* Amsterdam: Stichting Fragment Foto (Fotografie uit de collectie van het Museum voor Volkenkunde Rotterdam; dl.7).

Garrett, A. 1907. *Das Jaypur-Observatorium und sein Erbauer. Übersetzt von Professor Dr. Bergholz. Mit zahlreichen Abbildungen im Text und zwei Tafeln.* Berlin: Verlag der Treptow-Sternwarte (Vorträge und Abhandlungen herausgegeben von der Zeitschrift "Das Weltall" unter Leitung von F. S. Archenhold. Heft 19).

Gehring, Hans. n.d. [1907 & 1908]. *Indien. Das alte Wunderland und seine Bewohner. Geschildert von Hans Gehring. Zweite Auflage.* 2 Volumes. Leipzig: Verlag von Otto Spamer.

Goblet d'Alviella, Eugene. 1880. *Inde et Himalaya. Souvenirs de Voyage.* Paris: E. Plon et Cie.

Greaves, Edwin. 1909. *Kashi the City Illustrious or Benares.* Allahabad: The Indian Press.

Gutschow, Niels & Axel Michaels. 1993. *Benares. Tempel und religiöses Leben in der heiligen Stadt der Hindus.* Köln: DuMont Buchverlag.

Havell, E[rnest] B[infield]. 1905. *Benares. The Sacred City. Sketches of Hindu Life and Religion.* London [etc.]: Blackie & Son Limited.

Heichen, Walter. 1911. *Unseres Kronprinzen Fahrt nach Indien. Ein Buch für Volk und Jugend.* Berlin [etc.]: Phönix-Verlag.

Heinig, Edwd. E. 1910. *A Handbook of Benares containing a Description of Excursions in and around Benares City and Suburbs.* Benares Cant.: Printed by A. C. Chakravarti at the M.S.P.S. Press.

Hodges, William. 1785–88. *Select Views in India, drawn on the spot, in the years 1780, 1781, 1782, and executed in aqua tinta.* London: Printed for the author.

— 1793. *Travels in India, during the Years 1780, 1781, 1782, & 1783.* London: Printed for the Author.

Khan, Omar. 2002. *From Kashmir to Kabul. The Photographs of John Burke and William Baker 1860–1900.* Munich/Berlin/London/New York: Prestel.

Lejean, Guillaume. 1870. Le Pandjab et le Cachemir. In: *Le Tour du Monde. Nouveau Journal des Voyages. Publié sous la Direction de M. Édouard Charton, Auteur des Voyageurs anciens et modernes, et illustré par nos plus célèbres Artistes.* Paris: Librairie Hachette et Cie, 11e Année, N° 543, 28 mai 1870, pp. 337–352.

Mahajan, Jagmohan. 1988. *The Raj Landscape. British Views of Indian Cities.* Surrey / New Delhi: Spantech Publishers.

Noti, Severin. 1911. *Land und Volk des königlichen Astronomen Dschaisingh II Maharadscha von Dschaipur.* Berlin: Dietrich Reimer (Ernst Vohsen).

Ollmann, Arthur. 1983. *Samuel Bourne. Images of India.* Carmel, CA: The Friends of Photography.

Ostertag, Albert. 1846. *Leupolt's Erinnerungen an das Missionswerk in Benares, übersetzt und bearbeitet von Albert Ostertag.* Basel: Zu haben im Missionshaus und in Bahnmaier's Buchhandlung.

Oukhtomsky, E[sper] - E[sperovitch] Prince. 1893. *Yoyage en Orient - Grèce, Égypte, Inde, 1890-1891, de Son Altesse Impériale le Césarevitch. Traduction de Louis Leger. Préface de A. Leroy-Beaulieu. Illustré de 178 Compositions de N.-N. Karazine.* Paris: Librairie Charles Delagrave.

Pal, Pratapaditya & Vidya Dehejia. 1986. *From Merchants to Emperors. British Artists and India 1757-1930.* Ithaca and London: Cornell University Press Ltd.

Photographic Views in India by Bourne & Shepherd, Calcutta and Simla [first published about 1869; reprint:] London: Howard Ricketts Limited, n.d.

Photographische Aufnahmen von der Indien-Reise Seiner Kaiserlichen und Königlichen Hoheit des Kronprinzen des Deutschen Reichs und von Preussen [i.e. Kronprinz Wilhelm von Preußen, 1882–1951]. 1911. *Im Auftrage Seiner Kaiserlichen und Königlichen Hoheit des Kronprinzen.* Berlin: verlegt von Georg Stilke, o.J.

Prinsep, James. 1831, 1833. *Benares Illustrated, in A Series of Drawings. Lithographed in England by Eminent Artists.* Calcutta: Printed at the Baptist Mission Press. [Reprint, with an introduction by O. P. Kejariwal: Varanasi: Vishwavidyalaya Prakashan, 1996].

Renié, Pierre-Lin & Sabine du Vignau. 1992. *L'Inde. Photographies de Louis Rousselet 1865~1868.* Bordeaux: Musée Goupil.

Reuther, Oscar. 1925. *Indische Paläste und Wohnhäuser. Mit Beiträgen von Conrad Preusser und Friedrich Wetzel.* Berlin: Leonhard Preiss Verlag.

Reynolds-Ball, Eustace. 1907. *The Tourist's India.* London: Swan Sonnenschein & Co. Limited.

Rousselet, Louis. 1874. L'Inde des Rajahs. Voyage dans l'Inde Centrale et dans la Présidence du Bengale. 1864–1868.—Texte et Dessins inédits. LIII (suite). De Cawnpore a [sic] Bénarès. In: *Le Tour du Monde. Nouveau Journal des Voyages. Publié sous la Direction de M. Édouard Charton, Auteur des Voyageurs anciens et modernes, et illustre [sic] par nos plus celebres [sic] Artistes.* Paris: Librairie Hachette et Cie , 14e Année, N° 685, 21 février 1874, pp. 113–128.

— 1877. *L'Inde des Rajahs. Voyage dans l'Inde Centrale, dans les Présidences de Bombay et du Bengale. Deuxième Édition contenant 317 Gravures sur Bois dessinées par nos plus célèbres Artistes et six Cartes.* Paris: Librairie Hachette et Cie.

— 1878. *India and its Native Princes. Travels in Central India and in the Presidencies of Bombay and Bengal. Carefully revised and edited by Lieut.-Col. Buckle.* London: Bickers and Son.

— 1882. *India and its Native Princes. Travels in Central India and in the Presidencies of Bombay and Bengal. Carefully revised and edited by Lieut.-Col. Buckle.* New Edition. With numerous Illustrations and Maps. London: Bickers and Son.

Rupprecht. 1922. *Reiseerinnerungen aus Indien von Rupprecht, Kronprinz von Bayern.* München: Verlag Josef Kösel & Friedrich Pustet, K.-G.

Sala, George Augustus. 1875. *India and the Prince of Wales.* London: Published at The Illustrated London News Office.

Sen, Rajani Ranjan. 1912. *The Holy City (Benares). [With 58 Illustrations and a Map].* Chittagong: M. R. Sen.

Shellim, Maurice. 1979. *Oil Paintings of India and the East by Thomas Daniell RA 1749–1840 and William Daniell 1769–1837. With a Foreword by Mildred Archer.* London: Inchcape & Co. Limited in conjunction with Spink & Son Ltd.

(Société Saint-Augustin). 1888. *L'Hindoustan.* Lille: Desclée, de Brouwer et Cie.

Solvyns, Baltazard F[rançois]. 1808. *Les Hindoûs, ou Description de leurs Moeurs, Coutumes et Cérémonies.* Tome premier [of 4]. Paris: Chez l'Auteur [...] et chez H. Nicolle.

Stapp, William F. 1994. Souvenirs of Asia. Photography in the Far East, 1840–1920, *Image* 37, 3–4: 2–53.

Tellemann, Cl. von. n.d. Zwei indische Städtebilder. Benares. In: *Reise um die Erde. Heft 5. Herausgegeben von Hauptmann a. D. Karl Tanera und Paul Gisbert* [et alii]. Wien: Verlagsbuchhandlung Leopold Weiß, o.J. [ca.1905], pp. 116–120.

Tieffenthaler, Joseph. 1786. *Description Historique et Géographique de l'Inde, qui présente en trois Volumes, enrichis de 64 Cartes et autres Planches: I. La Géographie de l'Indoustan, écrite en Latin, dans le pays même par le Pere Joseph Tieffenthaler... Tome I.* [of 3] *Contenant la Géographie de l'Indoustan, avec 39. Planches.* Berlin: Imprimerie de Chrétien Sigismond Spener.

Tour of His Royal Highness The Prince of Wales [in] *India. 1905-1906.* 1905. Calcutta: Printed by Thacker, Spink and Co.

Uchtomskij, E[sper Esperovitch]. 1894. *Orientreise Seiner Kaiserlichen Hoheit des Grossfürsten-Thronfolgers Nikolaus Alexandrowitsch von Russland 1890-1891. Im Auftrage Seiner Kaiserlichen Hoheit verfasst von Fürst E. Uchtomskij. Aus dem Russischen übersetzt von Dr. Hermann Brunnhofer.* Erster Band. Leipzig: F.A. Brockhaus.

Volwahsen, Andreas. 2001. *Cosmic Architecture in India. The Astronomical Monuments of Maharaja Jai Singh II.* Munich/London/New York: Prestel-Verlag.

Werner, W. 1884. *Das Kaiserreich Ostindien und die angrenzenden Gebirgsländer. Nach den Reisen der Brüder Schlagintweit und anderer neuerer Forscher dargestellt. Dem Andenken an Hermann von Schlagintweit=Sakünlünski gewidmet.* Jena: Hermann Costenoble.

Worswick, Clark & Ainslee Embree. 1976. *The Last Empire. Photography in British India, 1855-1911.* London: Gordon Fraser Gallery, Ltd.

Wright, Caleb A. M. 1853. *India and Its Inhabitants. The Author visited India and traveled extensively there, for the express Purpose of Collecting the Information contained in this Volume. Seventh Edition* [first edition: 1848]. *Illustrated by numerous Engravings.* Cinicinnati, Ohio: Published by J.A. Brainerd, 1853.

Zache, Hans. 1911. *Mit dem Kronprinzen durch Indien. 2. Auflage.* Berlin: Süd-West-Verlag.

Sandria B. Freitag

Visualizing Cities by Modern Citizens: Banaras Compared to Jaipur and Lucknow

Introduction

This essay focuses especially on the consumption of visualizations of urban space as an aspect of the emergence of modernity. As such, it examines cities not as primordial entities, but as historically contingent places shaped by the active "production" of perceived purposes and products relevant to particular times and contexts. In the acts of consumption (of these purposes and products) lies revealing evidence about the processes by which consumers become "modern" citizens of a larger world.

Other essays in this collection explore the consumption of perceptions of Banaras by its inhabitants, such as *dhobis* (see Schütte, this volume). I would like to complement those pictures by examining how urban perceptions of spaces are projected for external as well as internal consumption. These perceptions, when examined for just one urban site, often seem to be specific to that place and relatively limited in their focus. Yet, when we combine or layer this viewpoint by applying it to more than one city, we begin to discern larger patterns, relating to the linkages between everyday life and state activities, civil surveillance over the state, socio-political movements, and the like. These larger patterns often are best discerned when we examine comparatively the distinctive cultural productions designed to project particular perceptions of place.

Thus we will examine three urban places here—Jaipur, Lucknow and Banaras—for the differing images or visualizations of space/place that they have projected from the early modern period on. I chose these three because each proffers rich visual evidence produced and consumed by Indians that historians can analyze productively. To encompass three sites in such a short essay, we must necessarily use analytical shorthand for discussions of the roles of urban centres, cultural production, and linkages between identity-formation, consumption, and modern urban cultural production: I apologize for these abbreviated treatments and refer readers to other work that discusses them in more detail (see references in each section, below).

The Urban Site

This essay suggests how Banaras fits within a larger pattern by which the visualizations of cities become "consumed". This consumption, I have argued elsewhere (Freitag 2001), simultaneously expresses individual interests and collective concerns. Consequently, such consumption shapes participation by modern individuals in a larger civil society, which is both specific to a place and capable of linking participants to larger polities.

Cities are valuable sites for tracing historical developments that change the socio-political formations in a nation-state and beyond. Serving as intersections for a wide range of new patterns, cities provide revealing evidence for understanding a civilization, especially as it modernizes. Such places often encourage ideological formulations more abstract and encompassing than those prevalent in the small face-to-face communities of the countryside. Although they generally use the same conceptual building blocks—e.g., kinship—they provide the opportunities to extrapolate those concepts into larger and more inclusive frames of belonging. In turn, "belonging" enables forms of mobilization that are more elaborated, with one set of identities and loyalties overlapping but differentiated when necessary from another.

The images produced in these circumstances contribute to public activities that shape the emergence of a modern civil society whose motive forces become a significant measure of the organization of polity and locality.[1] Perhaps the most vital source of image- and identity-formation emerges from what we might call "cultural production"—forms of cultural activities and products that are produced not just once but repeatedly, over time. The genealogies of such forms by which the culture of a place is reproduced and consumed can be traced in ways that provide historians and consumers alike with material evidence of collective identities and values, and how these are perpetuated as well as re-formed to meet changing circumstances. We may use cultural production as the lens through which to refract these meanings. Local understandings of what it means to belong to Banaras are constructed through images of the city exported and held by others—yet this process of cultural production is deliberately shaped by the residents of that city. To make this process visible, we will compare the cultural production of particular regional identities that can be traced to the urban sites of Banaras, Jaipur, and Lucknow.

Images of the cities exported and held by others are shaped in many ways. For most of the influential South Asian cities, there are early-modern royal roots for these perceptions: royal houses worked in large part through cultural patronage, both broad and specific ("specific" in terms of particular forms of production, from poetry to painting to pottery). Cultural production thus made the mark of a dynasty through both the support of the producers—artists and artisans alike—and the specific arts and crafts products that came to be seen as characteristic of that place. Simultaneously, rulers searched for public activities and other collective cultural expressions to underscore shared values and identification of interests with their local subjects. In these more abstract forms of cultural production, consumption through participation in public reinforced the ties between rulers and ruled, in a dialogic process shaped as much by participants as by royal patrons. In addition, the actions of local residents contributed in the aggregate to these images. Sometimes these actions related to the professions/work of resident specialists—from photographers to priests to artisans creating goods distinctive to the region—whose cultural productions contributed to the perceived visualizations of that urban site by those outside the city. Sometimes they related to local expectations, as services and goods were created to satisfy local demands and the internal economies of a particular place.

Taken together, however, these activities conveyed certain understandings about an urban place: what functions it performed in a larger world of inter-related places that formed a cognitive whole; what meanings were attributed to its spatial arrangements; who was welcome within it and for what purposes; and so on. Under these twin impulses of production and consumption, most cities took on a special sense of themselves—residents understood as distinctive and significant the way people in their city did things (e.g., kiteflying in Jaipur; devotional "shauk" around public performances of Ram Lila in Banaras; poetry in Lucknow, and so on). Thus cultural production in this context may be seen as resulting from the contributions by myriad actors in an urban place. These contributions shape the image of a place that then becomes available for visualizing and consumption by visitors and residents alike.

Perhaps this contribution of urban sites has functioned for centuries (some of the ancient texts produced for a place like Banaras and cited in this volume certainly suggest that this is the case); what is new in the modern period is the expansion and concentration of urban places, with their undisputed role in creating responses that can accommodate notions of the modern both through cultural production and through the consumption of these products. We will examine here the similarities and differences in how such cities are being visualized. Let us begin by examining the way Jaipur and Lucknow have emerged from the early modern ferment of the 17th and 18th centuries.

Cultural Production and City Site: Jaipur

From its inception, the capital city of Jaipur has stood as the expression of a singular understanding of a capital city held by a line of long-lived, creative, even visionary Maharajas. Characterized recently by one of the few scholars of Jaipur as the first "secular city" of India (Asher forthcoming),[2] the initial design of Jaipur reflected the desire of Maharaja Jai Singh II (ruled 1699–1743) to shape an urban space as an integrated, inclusive world, to which he invited cultural specialists and Jain merchants to join his courtly elite.[3] It seems likely that the city provided a series of differing templates or mental maps when viewed by marble sculptors, hand-block printers of fabric, jewelers and gem merchants, inhabitants of the Hawa Mahal, or devotees of Krishna (this latter occupied a template (Masselos 1991) anchored especially to the Govinda Devi temple). Yet the Maharaja aimed especially at creating a city capable of sustaining itself and its nobility in certain style and living standards. The resulting emphasis on those involved in the production of elite consumer goods was especially marked.

This development grew out of the ability of Jai Singh II and other influential denizens to simultaneously pursue two cultural strands—Mughal and Rajput—in several realms, ranging from individual buildings to the decorative arts. For example, Jaipur's famous "blue pottery" became established there after two potters' families were sent to the Mughal court, where they resided until they had mastered this distinctive medium as well as the chemistry needed to produce the colours particular to the art form. (In turn, of course, blue pottery had reached the Mughal court through a complicated pathway shaped by Persian participation in Silk Road trading competition with China.) In a similar example, we can see the long-term influence of wall-painting through examples on the temple complex in Galta and the Sisodia Rani ka Bagh (palace built by Madhu Singh I in 1774), to the Shekwati *havelis* built by Calcutta-based Marwari merchants.[4]

Indeed, by the mid-19th century, the strong commitment to crafts of the region by Jaipur's Maharajas led Ram Singh II (ruled 1851–1880) to establish, in the modern way, a School for the Industrial Arts. That is, the Maharaja wished to augment traditional modes of family training with a "modern" approach, while recognizing and solidifying the production of goods that traditionally represented Jaipur (and the surrounding Rajasthani region). School curriculum encompassed miniature painting; hand-block printing of fabric; blue pottery, etc. Noteworthy in implementing this vision was the addition of the modern art of photography in the next decade.

Fig. 1: Postcard: Street View in Jeypore, IOL&R Photo 514: Postcards. Many similar shots were produced, including "Souvenir of Jaipur" Photo album, dated 1905, with photographs by Gobindram and Oodeyram, photo no. 8.

It seems clear that two goals prompted this royal policy: (1) to ensure good economic conditions for the artisans of the city; but also (2) to project an image of Jaipur as a cultured place distinctive for its visual ambience (both two- and three-dimensional). Deliberate fostering of an image emerges, too, in the "face" put on the city through the initial design and layout (including the juxtaposition of the Govinda Devi temple and the court; the temple story linking Krishna to the Maharaja's dynasty to reflect the population's deep devotion to Krishna, etc.).

By the mid-19th century, this concern with the cityscape was directed to external audiences as well as local ones. Not only did the Maharaja have the entire city centre (the city palace and all of its environs along the main roads) painted a distinctive "pink" (a combination of red sandstone and whitewash), but he invited most of the prominent photography studios across the subcontinent to come and photograph it. That is, he deliberately created and projected out a singular vision of his city that has endured—many of the same, iconic photographic angles (Fig. 1) used to capture the essence of the city on that first occasion are still perpetuated today.[5]

In a related example that was set in motion by this Maharaja and carried out by his successor, Madho Singh II (r. 1880–1930), Jaipur built the Albert Hall Museum to house the most impressive artifacts of cultural production from the region. Layout, design and even documentation of the Museum and its first exhibition were shaped by close connection to the British, yet this was clearly a project of the Maharajas, reflecting their embrace of the modern in India as an aspect of Jaipuri identity.[6]

The sense of Jaipur as unique, in large part because of its central role in cultural production, continues down to today. Travelers are lured to Jaipur through images of the "pink city", the painted *havelis*, and the special products of the region, from jewelry to hand-block printed fabric, hand-made paper, blue pottery and the like. Yet this legacy is often not a continuous tradition. Extraordinary efforts have been made in industries such as blue pottery and hand-made paper to entirely re-invent these "Jaipuri" crafts.[7] Restorers of the tradition of blue pottery, for instance, had to scientifically replicate the colours used after the earlier formulas died with the original families; the craft has now been taken over largely by social uplift NGOs. Hand-made paper had been a declining industry until it was reconfigured entirely as an "eco-friendly" enterprise aimed at a global market.

Looking more broadly, however, the ability of these enterprises to reinvent themselves in modern guises and for new markets, is an extension of what was created in the 18th and 19th centuries. Especially important in understanding this continuity is situating these modern crafts in the context of independence, for in 1947 the arts curriculum in all of the schools for arts in various regions was standardized as a "national" expression of India's artistic heritage. The only way regional crafts could be successfully preserved, perpetuated and marketed as visualizations of a regional space, was through the localized efforts of Non-Governmental Organizations. The success of this utterly new form of cultural patronage and resulting production, nevertheless, had its roots established in the reconfiguration of Jaipur as a publically-consumable city. Jaipur's markings of place rested on an ambience of craft-production perpetuated in modern ways (first in a school, then augmented by social uplift work in villages), and consumed through visual practices that had a public as well as an economic impact (through a modern museum, craft marketing on a global scale, etc.). Although Jaipur is not alone in utilizing these modern practices, it can stand for us as a singularly successful example of this model—the visualization of a city through production and consumption of its regional arts and crafts.

Cultural Production and City Site: Lucknow

An active node on an important trade route during the Mughal period, Lucknow came into its own as the capital city of a successor state. The cultural base of this state, formed by the Nawab of Awadh, was designed to fit into north Indian cultural practices in order to connect, at least indirectly, with popular culture. At the same time, Awadh stood, clearly distinct, from both the Hinduized culture of the region (focused especially on the story of Ram) and the Indo-Persian culture fostered by the Mughals. To some extent the *nawabs* elaborated from a Shi'i cultural base, but to an equal extent, they created a unique set of cultural expressions that came to be seen as distinctively Lakhnawi.

Fig. 2: Abbas Ali, *The Lucknow Album*, 1874, "containing a series of 50 Photographic Views of Lucknow and Its Environs together with a large-sized plan of the city, executed by Darogah Ubbas Alli, Asst. Municipal Engineer...". Calcutta: printed by G. M. Rouse, Baptists Mission Press, 1874. Photo no. 46, Principle building of the Husainabad *imambara*. (Courtesy of Alkazi Collection of Photography).

And yet, it is significant that success in this endeavor may be measured not by local consumption of Lucknow's image, but by the profound influence Lakhnawi cultural practices had on the nationalist statements of Muslims during the struggle against British imperialism in the 1920s and 1930s, and the influential legacy still shaping the understanding of what it means to be a Muslim in contemporary India. That this nexus of cultural symbolism and praxis should have been shaped by a minority (Shi'as) within the minority community (i.e., Muslims, predominantly Sunni) in modern India tells us much about the capacity of cultural formulations to serve community and/or nationalist interests far beyond and long after the existence of their royal founders. Once again, the example of Lucknow also shows us how cultural production distinguishing a place comes to be consumed by those both within and outside the urban centre. In the modernizing circumstances of the 19th and early 20th centuries, this consumption demonstrates an explicitly concrete process of linking consumption, identity and what become larger ideological movements.

The example of Lucknow also helps to suggest why the royal ability to deploy a consumable vision of a particular urban centre, especially when it functioned as a state capital, should prove centrally important for us. The cultural system developed by the nawabs became a form in which symbolic support for the regime and its ideology, expressed through public exercises and in public forums, underscored a statement of shared community that transcended the regime and that led to the emergence of modern civil society. The cultural sleight of hand that finessed state and community proved remarkably successful in Lucknow at a historical moment when courtly culture could provide the "glue" for this cultural system. Among the early actions of the nawabs, for instance, was the linking of both the regime and the elaborations of Muharram[8] to the shrine of Hazrat Abbas[9] (one of the most prominent sites among the networks of saints' tombs in the subcontinent). The initial observance in the civic Muharram events became a procession of *tazias*[10] and *alams*[11] to the shrine to be blessed (Ali 1917).

Similarly, the recitation of poetry in the *imambaras* built to house the *tazias* between Muharram observances, and the central role of the *imambara* buildings in civic life, signified other underpinnings of this shared civic culture.[12] These physical sites (e.g., Fig. 2), in a city dramatically imprinted by the construction of "political architecture" to satisfy the *nawab's* "building mania" (Llewellyn-Jones 1985), played an obvious role in forging a shared Lakhnawi culture. Significantly, the symbolic importance remained even when the nawwabi court was exiled to Calcutta and the British took over in the wake of 1857. Yet new forces worked to reshape the cityscape itself and even affected the perceptions of place associated with Lucknow under these new circumstances. Let us look briefly at the resulting tensions between the new and the old understandings of this urban place.

The "Mutiny" of 1857 provided an entirely different image of Lucknow, that as one of the most significant locales for the British narrative of Indian history. Murray's *Handbook for Travellers*[13] stands as representative example of the British-propelled perception of Lucknow. Murray's includes two maps—one of the city itself, and one of the "intrenched positions of the British garrison". It devotes one page to "history" (i.e., the *nawabs*) followed by eight pages describing the "Mutiny" events in the city. In the five pages describing city sites, Murray's concentrates on the sites that loom large in the British "Mutiny" narrative. Of the buildings constructed by the *nawabs*, Murrays' comments generally follow the line (as for the Kaiser Bagh), "the palace was built by Wajid 'Ali Shah with the utmost extravagance and in the vilest of taste" (p. 294).

In contrast, we see a modern Indian perception of Lucknow through a guidebook published in 1874, the first to self-consciously connect *photographs* of sites with text

Fig. 3: Abbas Ali, The *The Lucknow Album*, 1874 (see Fig. 2), Karbala Talkatora, Photo no. 50. (Courtesy of Alkazi Collection of Photography).

descriptions (in English) for the visitor. The guidebook stands out for its clear sense of two audiences, the "educated Native" and the "traveling public of all nations", both motivated by "the spirit of inquiry" (Ali 1874). The author, Darogah Ubbas Ali, an engineer, was uniquely suited to speak to these two audiences.[14] He retains much of the "Mutiny" narrative, but also resists the British construction in part by recalling the cultural capital created by the *nawabs*, especially through the inclusion of sites important to Muslims that were not usually included in guidebooks.[15] Similarly, in implicit argument with British aesthetic condemnations, he underscores that there was a point to extravagant cultural patronage. The true extent of his effort to counter the British re-presentation of Lucknow as simply a "Mutiny" site is reflected especially, however, in his focus on Muharram sites (e.g. Fig. 3)[16] for it is particularly through the elaboration of Muharram that the *nawabs* had defined a cultural identity for their regime and their capitol city in a way that could include a large and diverse population.[17] Not least important in this resistance of the British images of Lucknow is the nostalgia and hence sense of loss associated with the evocation of the pre-"Mutiny" culture of the city. Indeed, as the nationalist movement developed and Muslims explored their way into an independent India where numbers counted more than cultural impact, the sense of loss itself became an important identifier for Muslims (Freitag 1989a).

These competing images continue to have salience for more than the next century, as the Lakhnawi's view of the city struggles to remain visible under the onslaught of British (re)constructions (both literal and figural). Even the photographic record underscores this contestation: unlike Jaipur, photographic representations of Lucknow present no integrated vision of a city, but only a congeries of particular sites, almost always glossed quite differently, depending on whether the emphasis is Successor State cultural system or "Mutiny" site. Although the *nawabs* pursued strategies often paralleled to those of Jaipur's Maharaja—for example, citywide consistency in style of decoration, in colour of the buildings, and so on—these were never captured or perpetuated as symbolic of the city itself. That is, no "views of the city" became standard photographic fare; it was not possible to show a cluster of buildings constructed in the nawwabi style to convey a consistent and continuing vision of place that could be invoked, independent of British re-orientation.

Fig. 4: Husainabad Imambara under flood (note Duldul, sinking). IOL&R Photo 10/8, part of Marquess of Reading Collection. M. M. Beg, Art Photographer, Lucknow. Photo no. 4.

Lucknow today, even considering the obvious changes introduced over the last half century, nevertheless continues to be visually characterized by an extension of the particular cultural style created under the *nawabs*. It is telling that the aspects of cultural production with lasting impact have also been the ones in which the nawwabi succeeded in linking up to popular culture. The abstracted understandings of Muharram, as that has shaped the larger Indo-Persian cultural style, continue to influence the non-material cultural production for Muslims in India. Poetry and music, although most often performed privately in individual elite homes, remain alive and influential (not least in their impact on early film-making). There is, as Ram Advani, the eminent bookseller in Lucknow, commented recently, "a place for Muslims here" that is unique and constitutes a crucial element in Lucknow's vision of place.

Thanks in large part to Darogah Ubbas Ali, particular places do invoke this sense of Lucknow: the Husainabad Imambara continues to be emblematic from this early period, and the construction of Hazratgunj in the early 20th century added to the stock of non-"Mutiny" images important in evoking that sense of place that can become a form of identity.[18] Photographs documenting a dramatic flood in 1923, for instance, present haunting images of the Husainabad Imambara with Husain's horse sinking under the floodwaters (Fig. 4). No other imagery could so well have conveyed the city's plight at that moment of devastation. For our purposes, then, the model represented by the vision of Lucknow created and consumed in modern India helps us to understand the reservoir of visual invocations associated with cultured Muslims—and with loss. Perhaps the *imambaras* are most effective in this role, since their associations range from Muharram to the public recitation of Urdu poetry, both key markers of being Muslim in modern India.

Cultural Production and City Site: Banaras

Several articles in this volume suggest the sources for Banaras' image as a primordial city of ancient, sacred significance. Nevertheless, for purposes of comparison in the modern period, we need to recognize that the Banaras with which we are familiar is almost entirely a construction (both literally and figuratively) of the 18th century. The rebuilding of Banaras at that time reflects a number of historically contingent developments, though the notion of a timeless Banaras fulfilled in crucial ways the purposes of those who rebuilt Banaras at that point in time.

To place modern Banaras in its historical context, then, we must recognize several developments of the period, including the patronage by Maratha leaders in pursuit of both territorial expansion and legitimation through sacred sponsorship. The proximity to Awadh and its capital, Lucknow, also provided a crucial model for forms of patronage that incorporated public displays linking an increasingly independent dynasty to popular culture in its area of cultural production characterizing modern Banaras.[19] Crucially important to the Bhumihar success was the alliance it forged with merchant-bankers and Gosains. We will return to these points, below.

Indeed, the very similarity of developments in Lucknow and Banaras suggests that much larger processes were at work. These processes brought together: (1) the devolution of local power and political authority from the Mughals to a number of successor states operating regionally; (2) the effects of increasing urbanism in the 18th and 19th centuries, resulting from shifting trade routes and economic actors as well as expanded commercialization;[20] and (3) the function of urban sites in fostering ideological movements that helped define "community" through certain cultural constructions and practices (Freitag 1989b). The specifics for Banaras illustrate how configurations unique to that city shaped the distinctive identities and communities there, in a local expression of the larger patterns and developments characteristic of modernizing urban India.

Although it shares important expressions with Jaipur and Lucknow, emerging from the influence of royal roots, modern Banaras also stands out in certain key respects. As noted, the image of a sacred city we now recognize was constructed in the 18th century, before the Maharaja of Banaras had achieved his official recognition and broad influence. Consequently, a great deal of the (re)creation of Banaras was shaped as much by Maratha rulers and other patrons from far-flung locales as by a resident royal house.[21] The stamp put on the city at this time emerged in large part from Maratha construction of dharmashalas for pilgrims, temples and feasting to support Brahman priests, and palaces for themselves and their local kin and agents. Such patronage, because it addressed claims for legitimacy by Maratha sponsors, made much of Banaras' ancient role as a sacred site (Fig. 5).

Fig. 5: *View of Banaras*, presentation album by Maharaja of Banaras, photographer Raghnath (or possibly Madho) Prasad, active 1890s–1910. Photo number 15, Burning Ghat. (Courtesy of Alkazi Collection of Photography).

Similar in influence is the presence of mendicant-traders, usually referred to in the literature as "Gosains" (e.g., Cohn 1974: 175–82; Bayly 1983: 181–84), who were the largest owners of urban property in Banaras in the late 18th century. As a "religio-commercial sect, militarized to some degree, and organized according to the guru-chela principle" (Kolff 1971: 213–17), they doubtless proved especially important in the visualizing of a modern city that could combine the commercial and the sacred, including the organization of communal life. Similarly, myriad interactions between pilgrims and local religious specialists contributed to the same predominant perception of Banaras as sacred. In the process, modern Banaras became shaped to fit age-old expectations.

Vibrant commercial life as a perceived aspect of Banaras should not be underemphasized. Complementary to the religious aspects of envisioning Banaras are the other consumable products associated with the city, not least Banarsi silk (produced by communities of Muslim weavers who have survived there in ways unparalleled among other weaver communities throughout India). This latter image of the city relates to the position of Banaras on the main west-to-east trade route, which guaranteed the flow of Banaras-produced goods (and, hence, images of an urban centre) to distant places.

Visions of the city, consumed by locals, far-flung patrons, and pilgrims alike, then, were shaped by deliberate actions on the parts of many participants, most of whom had something to gain by an emphasis on the city's sacred functions. From the *panda*s recording the family histories of pilgrims from far south, to sculptors who crafted small images of the gods to be taken home as souvenirs of pilgrimage, to the huge casts involved in the staging of Ram Lila—each of these actors contributed consciously to the envisioning of Banaras within the sacred roles it played. It is hardly surprising that these long-lived contributions amalgamated so convincingly to the modern tourist industry and its public relations campaigns, which similarly packaged Banaras and sold it to potential consumers across the subcontinent and beyond.

We should note that this fashioning of a coherent image of Banaras to be consumed was accomplished through a huge range of activities that did not need to be coherent to be effective. Indeed, many of the activities created mini-visions of Banaras, compelling only to a subset of the larger audience of consumers—as attested by Gaenszle and Sharma's discussion of Nepali places, and Wilke's analysis of the Durga cycle. That there was no need for a single, coherent image has enabled the city to create images and niches to make many feel welcome, from Muslim weavers to Nepali short- and long-term residents, to South Indian pilgrims at Annapurna.

Generally speaking, then, the predominant characteristic of modern Banaras we should keep in mind, in its historicized context, is of a multi-layered, even syncretic and internally contradictory urban culture. Only at certain moments have the anomalies built into this complex set of images and expectations led to tension: it is noteworthy that a site supposedly important for a particular religious significance has, in fact, witnessed very few large-scale conflicts on religious grounds (Freitag 1989b). I would argue that this fact owes much to the complexity of visions of Banaras. Even among religious specialists, particular castes and the organizations regulating their work of cultural production aim at particular audiences, fine-tuning this production to fit the needs and expectations of certain niches of clients (Parry 1980; van der Veer, 1989). As illustrated by Chalier-Visuvalingam's article in this volume, the very significance attributed to particular spaces may have differing meanings and contexts for different groups. Similarly, the weavers of Banaras—strong in their identity as stalwarts of Islam (Pandey 1989)—use their practices of work and leisure to support their activism (Kumar

1989) in a distinctly Banarsi-pan way. Indeed, Rana P. B. Singh noted that, in a city presumed to be archetypically Hindu, Muslim shrines and mosques number some 1400 to the 3000 Hindu shrines and temples.

The capacity, nevertheless, to project a unified vision to be consumed, drawn from the complexity behind Banaras' cultural production, stems from two complementary sources. First is the shared pool of values which finesses the myriad differences of meaning behind them. How this works is nicely characterized in Steve Derne's ethnography of merchant-banker families in Banaras, which examines how family members negotiate between larger normative values and individual "common sense" (Derne 1995). Second is the influence of royal patronage, which has welded these disparate pieces or visions together, particularly around the story of Ram, to which the Bhumihar dynasty connected itself. In a dialogic process, the Maharajas have collaborated with Banarsi specialists over the last century to elevate public performance (Fig. 6) of a particular story in a way that now provides a larger context for the long-standing tradition of publicly-enacted spectacle so characteristic of Banaras.[22]

Fig. 6: Drawing of a Ram Lila observance, 1831. James Prinsep, *Banaras Illustrated*, Calcutta 1833. (Courtesy of Ames Library, University of Minnesota).

Indeed, we may draw interesting parallels between many of the developments in Banaras and in near-by Lucknow. The form of cultural identity forged for the capital city reflected in both places the efforts by successor states[23] to draw on the kingly style established by the Mughals.[24] For instance, the elaboration of a civic expression of an important sacred story—in this case, the Ram Lila—reflected a successful effort to link a ruling house with local popular culture. The emphasis on the connection between the royal house and Ram—in a region and city known for its identification with Siva—makes this elaboration of Ram Lila especially telling: like the Nawabs of Awadh, the Raja of Banaras deliberately established an occasion for popular devotion that distinguished itself from the culture of the area[25] while simultaneously connecting to popular concerns. For our purposes, it is especially important to note that this elaboration reinforced the growing ties between the triumvirate of power and influence in the city—the Bhumihar dynasty, the merchant-bankers, and those trader-mendicants known as the Gosains (Freitag 1989b).

Parallels with Jaipur also can be seen in the fostering of certain (elite) consumer goods and broader processes of cultural production, the most obvious of which might be Banarsi silk, and the patronage of many artists producing forms of Ram's story (from illustrated manuscripts to the permanent construction and work related to Ramnagar as a site for Ram Lila). Two points about these concrete forms of cultural production should be noted. To begin with, the processes that produced the actual goods also represented important alliances within the city. Through silk, for instance, we can trace the tight interrelations between merchant-banker families and Muslim weavers (who made up more than a quarter of the city's population). Furthermore, for our purposes, we want to emphasize again that many of these artifacts of cultural production depend on consumption/ participation in public, and that people come to the city for these purposes.

Success in solidifying the position of the Bhumihar dynasty, then, grew at least in part from the alliances forged by the Maharaja (as he was known after 1857[26]) with merchant-bankers and trader-mendicants ("Gosains"). In its turn, this alliance resulted in not only a "Hindu renaissance" for which the Bhumihar house became famous, but a particular style of cultural production that marked products, performances, participation by large numbers of locals and pilgrims, and the place itself as distinctively "sacred". The missing presence in Banaras of an Indo-Persian courtier elite intensified this process, and also created the necessity for making room for lower-class Muslims within the single, inclusive cultural system that emerged.

The legacy in the 20th century of these successes can only be noted briefly in passing. Banaras Hindu University, as the centre where much of the intellectual and cultural underpinnings of the renaissance have been based, has contributed greatly to redefinitions of the Indian cultural system in the circumstances of nationalism and independence. The model of a complex city, where overlapping understandings of meaning and signification co-exist, must have been an important contribution during the Nehruvian era, when an emphasis on (so-called) "secularism" demanded more room for other, modernizing values and developments in the new nation.

More recently, however, other significant implications have emerged from these characteristics of the city's cultural system. The first implication is that the vision of the city that has been consumed over the last century or more provides the potential for combining a desire to be modern with Sanskritizing elements, such as patronage of religious personnel, buildings (especially the construction, maintenance, and elaboration of temples), and ceremonials. The appeal to primordial images simultaneously offers a legitimacy born of "tradition" with a reservoir of values to be drawn on when negotiating change under modern circumstances. As a consequence, a second point is the importance of reformist activities that frequently exhorted the lower castes to adopt Sanskritizing values and modes of behavior. Taken together in the modern (20th and 21st centuries) context, they direct our attention to the third, and perhaps most significant, point: it is precisely these characteristics that most often become linked to political movements.

Thus the cultural production of a vision for the city carries with it important ramifications in contemporary India, and make of Banaras a third and distinctive model in our analysis. For these ramifications stand in compelling contradistinction to the implications of consumable visions for Jaipur and Lucknow. As Derne's work illuminates, much of this Hindu renaissance—especially as it made room for modernizing individuals to search for new personal as well as collective meanings—becomes available for harnessing to the Hindutva movement. Given the complexity behind the image of the sacred in Banaras, this is neither a straight-line development nor an inevitable outcome. Yet the ability to tailor consumable images of sacredness around the city of Banaras illustrates the capacity of those involved in producing such images to respond to new needs and desires.

Conclusion

This potential seems ironic, in some respects, in a city where one of the most important elements of Banarsi culture is provided by the large population of Muslim craftsmen. Given the challenge to Indian Muslims to define a safe and acceptable vision of themselves in the face of aggressive Hindu fundamentalism, the potential also suggests at least one place we should watch carefully for changes—these are changes that not only delineate the local impact of Hindutva, but that could have profound influence on a dramatically altered vision of the city of Banaras itself. For if we watch from the perspective of the last two centuries, we cannot help but notice that visions of modern Banaras that have come down to us are of a society marked in important ways by both "syncretic" and "overlapping" understandings of space and practice. If this changes under contemporary pressures to redefine what it means to be Indian, then the visualizing of Banaras will mark a very new and disturbing development in the use of cities to link the personal and the political, to situate individual actions within a larger civil society.

One of our important tasks in understanding how a city's image is visualized and consumed, then, includes recognizing the complexities that enable certain kinds of visualizations to be produced internally. Yet we need also to understand the extent to which the consumption or pursuit of these images is changed and shaped through everyday uses by a broad spectrum of both local and external society. Perhaps most importantly, we ought to remind ourselves that no single set of actions, however carefully intentioned by its actors, can contain the ultimate consumable vision of a city as complex as Banaras.

Notes

1 For a more detailed definition of "public" see Freitag 1989a.

2 This argument takes on directly the traditional understanding that Jaipur was constructed to a Sanskritic sacred city model: sources using this assumption are cited by Catherine Asher as she argues against this understanding.

3 See, for instance, the Muslim synods documented by Asher. I am immensely grateful to the extraordinary research into the founding and layout of the city accomplished in recent years by Catherine Asher: much of my understanding of this urban place, and the larger context in which it emerges, rests on her labors and interpretations. See especially Asher 1992 (for context) and Asher 2000 (for Jaipur).

4 Cf. Hardgrove 2002. Interestingly, the Marwaris chose this *haveli* form of cultural patronage and expression as status statements of belonging to a region far removed from their actual domiciles, and seldom actually used by them.

5 The story of façade and mosque is analyzed in C. Asher (forthc.).

6 The first Exhibition in the space, held in 1881, also functioned as a way to bring the rest of the world to Jaipur, as exemplified by the display of photographs of recent archeological discoveries in Egypt.

7 In the case of blue pottery, this is because the formulas died out with the demise of last descendants of the original artisans trained in the Mughal Court. The efforts to re-create Blue Pottery was spearheaded by a small, private Arts College established after independence in Jaipur.

8 Muharram is month and therefore the name of a mourning observance that takes place at that time, that retells the story of the martyrdom of Hussain. The observances take place in most major cities, and involve not only Shi'as (for whom the story is most relevant) but Sunnis as well; in South Asia it is seen as a primary statement of general Muslim identity. Hindus of those towns also participate (more in the past than at present). In most locations the observances include evening private majalis or gatherings in which the story is told sequentially over the several days (these private observances make it possible to offer very different interpretations for each group), as well as daily processions (in which *tazia*s are carried, and groups often are led by those holding *alams* [see definitions, below], while the marching mourners beat their breasts in a syncopated rhythm to the chant of "Husain, Hassan". Onlookers provide refreshments. Ultimately, at the end of the ten days, the *tazia*s are borne to Karbala grounds for concluding ceremonies.

9 This process was strikingly similar between Jaipur and Lucknow, in that the founding stories describe the ruler in each place finding a relic or figure of the reverenced subject and bringing it to the capital city.

10 *Tazias* are replicas of the tombs of those martyred at Karbala. In Lucknow these are very large floats that are paraded annually and stored in the imambaras for the rest of the year. In other locations—for instance, Banaras—most are smaller, shoulder-borne compositions that often are built for use in one year's activities and then "buried" at the end of the mourning period in the "karbala grounds" of that city.

11 An *alam* is a flag of Husain, or a brass version of a "flag" held aloft on a pole during a procession. The *alam* symbolizes and proclaims the identity of the group that marches with it; often the *alam* is decorated with the hand, indicating the family of the Prophet.

12 For nobles, for instance, *imambaras* functioned as places of worship and ritual mourning, literary salons, personal monuments, family cemetaries and the like. For the general populace, they functioned as the storage buildings for *tazias*, and were elaborately decorated for the public. See Freitag 1989a.

13 First published c. 1903; quotations here are from the 1911 edition, representative of a long tradition of "Mutiny" narratives.

14 Other productions of his include, for instance, an album of the taluqdars of Awadh, suggesting that his audience generally included both the British (such as local administrators who relied on the album for guidance as a "Who's Who") and elite Indian audiences. A carte de visite, functioning as a modern business card, identifies him as "[Pensioned Municipal Engineer] Photographer under the Patronage of His Honor the Lieutenant Governor of the N.W. Provinces and Oudh" (London: Alkazi collection 2001.14.0229).

15 His very modern commentaries on the engineering of the bridges is also a form of resistence, as it insists on his own expertise as an engineer.

16 Each city with a Muharram observance created grounds called "Karbala" at which many of the processions concluded. This site is Karbala Talkatora, the main such site in Lucknow. See Freitag (1989a) for disputes over this site and the establishment of a new one in Lucknow.

17 There is no space here, for instance, to indicate the extent to which Muharram was one of the few glues that united the taluqdars (large landlords) with the ruling house. In any case, given the composition of the population of the area, the observance of Muharram historically involved more Hindus than Muslims, more Sunnis than Shi'as. The private *majlis* portions of the observance enabled groups with different motives and understandings of the events to participate; where they differed in these understandings could be pursued in the more private gatherings where the story of Husain could be told in more coloured terms. See Freitag (1989a) for more detail.

18 It is arguable that this general characteristic of Lucknow changed with the construction of Hazratganj in the early 20th century. Although this fancy shopping area was off-limits to Indians in its early days, certainly by World War II "ganjing" was the most popular form of leisure activity, and the distinctly modern architecture of the central street (including its movie theatre) could be considered emblematic of the city. As Imtiaz Ahmed (1997) comments, Lucknow is perhaps best known for its ability to enable its residents at all levels of society to enjoy leisure. Another mark in this landscape of special importance has been Ram Advani's bookstore, where intellectuals of all income levels gained access to a larger world.

19 This increased prominence was finally confirmed by the British in the early 20th century, when the Raja was officially recognized—through an increase in number of guns fired in salute, and other markers—as holding the rank of Maharaja. This official elevation was a deliberate policy move by the British to balance out the increasing influence of the nationalists.

20 Population grew rapidly during the last half of the 18th century. Banaras' 200,000 residents placed it among the largest north Indian centres during the decline of Mughal imperial centres.

21 In return for an alliance with the Nawabs of Awadh against the Rohillas, the Marathas were awarded influential bases in the old north Indian centres on the major pilgrimage routes. The influence and control over Banaras lasted longer than elsewhere.

22 This volume includes several examples of public spectacle organized for benefit of particular niches of believers. The same tradition, I would argue, has shaped much of Banaras' public life even when it is unconnected to overtly sacred purposes; here I would point to the house tax protest of the early 19th century, as well as the protest against technology (and its implied patron, the British empire) in the 1890s (Freitag 1989b).

23 Literally, of course, Banaras began the Successor State period as part of the Nawab of Awadh's polity. But the elevation of the Raja of Banaras, first in practical terms and then formally as a result of British efforts to buttress their power there by posing the Raja against the nationalists, reflected a pulling away from the Nawwabi and an effort at expressing a unique sense of Banaras

that had been developing since the 18th century. See Freitag (ed.) 1989b for details about the three levels of rule operating at this time in north India.

24 Bayly has commented on the "mughalizing" processes affecting north Indian urban centres at this time. Of all the cities that grew up in the wake of Mughal decline, only Banaras lacked a significant Indo-Persian courtier element in its "corporate" life. See Freitag (1989b) for more detailed discussion.

25 As noted elsewhere, this had to do with the establishment of the Bhumihars, and this family in particular, as integral parts of a Bhojpuri regional culture rooted in landownership and collaboration with merchant-bankers and Gosains. Freitag (1989b).

26 The British left the Raja in place as largest landholder and de facto local ruler—responsible only for paying tribute to the Nawabs of Awadh when they had replaced the Mughals. For his loyalty during 1857, the Raja was rewarded with the title of "Maharaja". Much of his formal power was removed during the 19th century but, in the 1910s, the British installed him with much greater powers and the status of other princes, in large part as a counterweight to the growing nationalist movement centred in north Indian cities.

References

Ahmed, Imtiaz. 1997. Through the Eye of a Street. In: V. Graff (ed.), *Lucknow: Memories of a City*. New Delhi: Oxford University Press.

Ali, Darogah Ubbas. 1874. *The Lucknow Album. "containing a series of 50 Photographic Views of Lucknow and Its Environs together with a large-sized plan of the city, executed by Darogah Ubbas Alli, Asst. Municipal Engineer...".* Calcutta: printed by G. M. Rouse, Baptists Mission Press.

Ali, Mrs. Meer Hassan [Mir Hasan]. 1917. *Observations on the Mussalmanns of India* (original ed: 1832). Revised and edited by William Crooke in 1917. New Delhi: Deep Publications, 1975.

Asher, Catherine. 1992. Mughal Sub-Imperial Patronage: The Architecture of Raja Man Singh. In: B. S. Miller (ed.), *The Powers of Art: Patronage in Indian Culture*. New Delhi: Oxford University Press, pp. 183–201.

— 2000. Mapping Hindu-Muslim Identities Through the Architecture of Shahjahanabad and Jaipur. In: David Gilmartin & Bruce B. Lawrence (eds.), *Beyond Turk and Hindu: Rethinking Religious Identities in Islamicate South Asia*. Gainesville: University Press of Florida, pp. 121–148.

— (forthcoming). Making the Old New: The Jaipur Jami Mosque and the Creation of Muslim Identity. In: D. Gilmartin, B. Lawrence, & T. Stewart (eds.), *Visions of Community: The South Asian Muslim Imaginaire*.

Bayly, C. A. 1983. *Rulers, Townsmen and Bazaars*. Cambridge: Cambridge University Press.

Cohn, Bernard. 1964. The Role of the Gosains in the Economy of Eighteenth and Nineteenth Century Upper India, *Indian Economic and Social History Review* 1, 4: 175–82.

Derné, Steve. 1995. *Culture in Action: Family Life, Emotion, and Male Dominance in Banaras, India*. Albany: State University of New York Press.

Freitag, Sandria. 1989a. *Collective Action and Community: Public Arenas and the Emergence of Communalism in North India*. Berkeley: University of California Press.

— 1989b. (ed.). *Culture and Power in Banaras: Community, Performance, and Environment, 1800–1980*. Berkeley: University of California Press.

— 2001. Visions of the Nation: Theorizing the Nexus between Creation, Consumption and Participation in the Public Sphere. In: R. Dwyer & C. Pinney (eds.), *Pleasure and the Nation: The History, Politics and Consumption of Public Culture in India*. New Delhi: Oxford University Press, pp. 35–75.

Hardgrove, Anne. 2002. Instituting Origin in a Retrospective Move. In: Sumathi Ramaswamy (ed.), *Beyond Appearances? Visual Practices & Ideologies in Modern India*. New Delhi: Sage, pp. 323–364.

Kolff, D. A. 1971. Sannyasi Trader-Soldiers, *Indian Economic and Social History Review* 8: 213–18.

Kumar, Nita. 1989. Work and Leisure in the Formation of Identity: Muslim Weavers in a Hindu City. In: Sandria Freitag (ed.) 1989b: pp. 147–170.

Llewellyn-Jones, Rosie. 1985. *A Fatal Friendship: The Nawabs, the British, and the City of Lucknow*. Delhi: Oxford University Press.

Masselos, Jim. 1991. Appropriating Urban Space. In: Sandria Freitag (ed.), *South Asia* (special issue on the public sphere in South Asia), pp. 33–64.

Murray, John. 1911. *A Handbook for Travellers in India, Burma and Ceylon*. London: John Murray.

Pandey, Gyanendra. 1989. *The Construction of Communalism in Colonial North India*. New Delhi: Oxford University Press.

Parry, Jonathan. 1980. Ghosts, Greed and Sin: The Occupational Identity of the Banaras Funeral Priests, *Man (N.S.)* 15: 88–111.

Van der Veer, Peter. 1989. *Gods on Earth: The Management of Religious Experience and Identity in a North Indian Pilgrimage Centre*. New Delhi: Oxford University Press.

Social Practice and Everyday Life

Nita Kumar

The Space of the Child: The Nation, the Neighbourhood, and the Home

In this preliminary anthropological study of the structure and imagining of space in the lives of children, I go through the following stages of discussion. First, I consider how the construction of an environment is used to discipline the body and the self. The child, who finds herself subject to a certain space in the home, is being taught to be a certain kind of being. This is the politics of age domination. The child can only maneuver within it, and it is only as adult that she can manipulate or change it, and then in turn make it an instrument of teaching for her children. Given this positioning vis-à-vis space, I feel confident in talking of a specific experience called "childhood", and in regarding it as a specific aspect of education that can be retrieved.

Two, I discuss the location of the child in the nation, and how an initial difference appears in the imagining of the nation between the schooled and unschooled child. But the difference produced by schooling is offset by the learning that is in common, for instance from the media; and partly offset by the sheer poverty of schooling across the whole range of schools. This pedagogic poverty equals the failure of the nation-state. The nation is far more interestingly imagined in films and songs.

Three, other spaces that produce a common experience for all children are the neighbourhood and the city. My city is Banaras (Varanasi), the provincial city *par excellence*, characterized by garbage. Whereas the metropolitan one might have public garbage too, this is not one of its structural characteristics, as it is of the provincial one. There are commonalities in the technology of space, as well as in metaphor and imagery, that cut across class. The neighbourhood, at least of the provincial city, might be one of the most powerful sites for the learning of a common habitus.

Four, given all these commonalities, and turning back to the home, I find that the most crucial lines of division within a class called children are between girls and boys: girls privilege the home, boys the neighbourhood; girls know they have more than one home, boys that they will forever wander. Their persona, and their lives, as male and female, are shaped by the spaces they find themselves in, the spaces as they read and live them, and the spaces they anticipate.

Age and Control: The Fact of Childhood

In seeking to talk about children, I am proposing a belief in some form of "age culture", and that "children" or "childhood" are somehow a reality. As a discursive proposition, I have argued elsewhere, this reality needs questioning for South Asia (Kumar 2000). However, I will postulate here, as an anthropologist, that a demonstratable "children's culture" and "world of the child" exists, in that there is observable action performed by children, games and activities, participation in social life and indeed, lifestyles (for Germany, see Fuchs 1994). Whether these are accepted and sanctioned by adults is relevant, but not central to the proposition.

What is central is the creation of much of the child's world by adults, or the politics of age. Space, in the home, the neighbourhood, the city, and the nation, is constructed and provided by adults for children. Even when involuntarily done, this involves control and domination of the child, and the effort at disciplining. Taking it at its most benign, we may call it "teaching". The uses of space by adults are all aspects of a teaching strategy by which effective socialization is sought to be carried on, and the body is adapted to the constructed environment.

The Case of Zeenat, a Weaver's Daughter

In class 3 of the *Madrasa* Hamidia Rizvia for girls, I watched how the teaching consisted of writing answers on the blackboard for children to copy—some explanations having been thrown in for my benefit—and then the teacher going into a kind of doze while children wrote at their own speed, or not at all. Those at the back played, gossiped, exchanged objects of value. Those in front must have been the better, or best, students, because their entertainment consisted of jerking their pens to form pools of ink, scrambling to share the ink, asking each other for ink, discussing the shortage of ink, and then with a pout bending down to write.

Zeenat was right under my nose so I picked on her. Or, rather the teacher did first, asking for her book and her copy to show me. She would not have chosen a bad student, so the choice of Zeenat was partly because of her location and partly her scholarship. She had indeed a fine hand and also an aesthetic sense, writing all the questions in green and the answers in blue. She looked very serious and would not respond when I tried to make her smile or laugh. Through her teacher I got her to invite me home.

Her house was just across the road, an excellent arrangement by all considerations. She seemed so "good" that I took for granted that she was from a "good" family, by which I meant quite unreflectingly a comfortable, middle-class, upwardly mobile family in a clean, spacious set of rooms that I could look at approvingly. We climbed up four flights of steep stairs of a house with an opening in the middle like a well and rooms all around. It was the all-too familiar design of all old homes in Banaras, seemingly more of Muslim homes but that only because more Muslim homes were old-fashioned. We climbed up and up and on each floor women and children gathered and stared at us without reservation. (I mean, stared at me.)

The room she took me to was equally familiar. There was a cooking space on the landing outside. The room had no furniture and was barely seven or eight feet wide on each side. The father sat on his haunches against a wall. The mother sat against another. A mat was spread against the third wall. The fourth wall was taken up by the door and a window through which peeped family members constantly. In other words, it was cramped and poor (see Fig. 1).

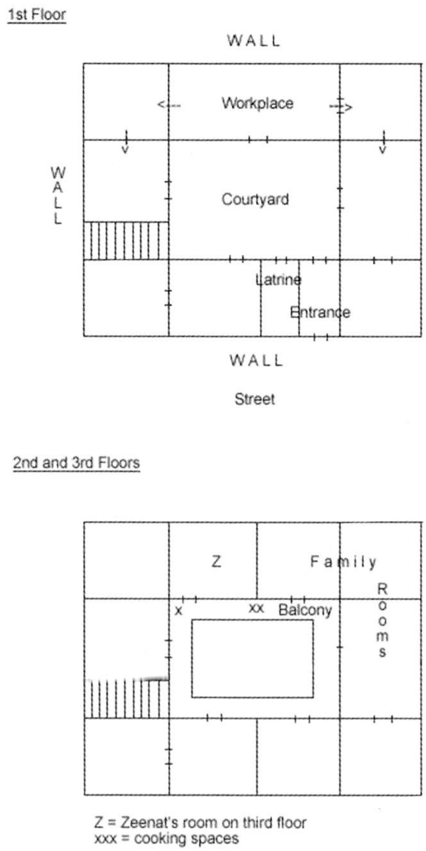

Fig. 1: House plan of Zeenat, a weaver's daughter.

They had five children, Zeenat being the middle one. The oldest boy, Wasim Riaz, and the girl, Shahina Parveen, had both dropped out of school early because they did not have the inclination for studies. They were bright in other things? I prompted. Neither mother nor father responded, although we had excellent rapport otherwise. Wasim wove and Shahina did housework. Zeenat and her immediately younger brother Mukhtar Naim went to school, she in class 3 in Hamidia Rizvia and he in class 3 in *madrasa* Jamia Islamiya. Rizvia is a Barelwi *madrasa* and Islamiya a Deobandi one (cf. Kumar 2000, Metcalf 1982), and regardless of the importance of sectarian differences as presented by *maulanas* (intellectual and religious leaders), this family like many, did not seem to regard them as significant.

The parents were very articulate on the question of some children studying and some not. By and large school education is seen as useless and a burden. While the child studies she can do nothing else, such as sewing or cooking. Because of the burden of studies many families keep tutors also. Anyway, some children "can" study—they have the *zehen* or inclination—and some "cannot". Zeenat and Mukhtar can, and Wasim and Shahina cannot. All four are equally accepted, however. The youngest, I was told, a toddler of two, Shagufta Yasmin, was going to be the best scholar of all—she was already so interested in books. When I asked about the whole place of studies in their lives, Waheeda Bibi explained candidly that nowadays when a marriage was being fixed, the groom's side asked, how far has she studied? Did she complete the Quran? Indeed, in order to make Shahina properly marriageable, she was teaching her at home. The two others had completed their Quran courses.

Quran courses were held in the house of her *bari Amma*, or senior mother, grandmother, or aunt, with at least twenty girls gathered under a maulani. Classes took two hours a day, from 2.30 to 4.30 or 5. There were no charges, but there was a gift given on graduation, and sweets at other occasions, and when the student would get married, a set of clothes. The *ustad* was to be honoured next only to the parents.

Zeenat was sitting close to her mother and looking more serious by the minute, so I tried to speak to her at which she seemed to draw into herself all the more. She must have been so hungry, having eaten whatever she did before 7 a.m. and then carried a mere tiffin to school. But there was no sign of lunch; indeed there were chunks of raw meat showing no input of preparation. Perhaps that was for dinner. So, lunch? I asked carefully. Oh, it's early yet, I was told. We'll eat at 1.30 or 2. They served me Crack-jack biscuits and tea. I hurriedly gave the two school children a biscuit each, under protest. The little baby wanted tea and got some.

I turned to Zeenat because I was her guest after all and there because of her. What were her games, her toys? She didn't budge—too hungry?—but her brother brought down a lovely doll's chair and bed, and then a doll. Then he brought down a suitcase full of little clothes stitched by the children, every bit as good as those sold world-wide by companies like Mattel. The baby played with everything as I exclaimed over it. I suddenly realized that Zeenat was too old; by a rough calculation, the youngest was two, the next 10, Zeenat 12, the next children 18 and maybe 19. None looked their age. All were small built, thin, and tired looking, except the mother who was plump and glowing. Wahida Bibi was educated, since she had spoken of teaching her daughter. But she had not gone to school or *madrasa*, rather learnt at home from female family members, as they had learnt before them.

She stitched everything at home: her clothes, the children's clothes, the linens, everything. Only the gent's clothes had to be bought. How had she learnt? How would her daughters? She had learnt just by doing. That was how everyone ever learnt. Zeenat was too busy with her studies to sew, but when the *madrasa* was closed she did whatever she wanted. By such doing, one learnt. Her sister had lots of time once she left school and was an expert seamstress.

The room in which the family lived presented a huge problem. What was the little Zeenat learning? In school, watching her, I had thought that she was destined to study far and was perhaps nurturing ambitions inside her serious little brain and that she would break her family tradition in some way. After all, I had only just met several women teachers, such as Safia Sultana, who had completed her *maulvi* (equivalent of High School), *alim* (Intermediate) and *fazil* (B.A.) degrees while living in a hostel in Maunath Bhanjan, sent there by her family in north Banaras, and who had been teaching since, even with two sons one and two years old.

But upon seeing Zeenat's family, I realized that the little girl's seriousness at school was destined to pass like a happy dream. Whatever she ever did, she would do very well. What she did now was schoolwork; what she would do later was housework and child rearing. All the time she was learning as conscientiously at home as in school. In school, she was learning the reading and writing of three languages, and some other subjects. At home, she was learning cooking, sewing, keeping baby, serving, tidying up, running for errands, all the details of how a home works and what in their "culture" and "community" a homemaker should do. She had finished with the Quran and although that is learnt only by rote, there was other intangible teaching that goes along with it that she had imbibed.

Zeenat was learning how to deploy herself in the one room her family lives in, to share the bathroom on the landing with the extended family, to cook with her mother, and gradually by herself in the space outside their door, to clean up each time and leave the space unmarked, and otherwise practice a total non-specialization of domestic spaces. The room and its outside were swept morning and evening or when littered, whichever came first. All materials were put away after use. No space was called after its function, as the place "for" some particular activity, including sleeping. The room became, in turn, a bedroom, a dining room, a sitting room, a den, a study, a workplace, a dining room again, and again a bedroom. Zeenat was mastering the discourse perfectly without a word being spoken on the subject. She was learning further where and how to spread out the rugs for those who would work or sleep, and those who visit. She was learning how every niche and cranny of the room has its uses.

Zeenat is, as a child, and will grow up into an adult as, a being perfectly comfortable and "natural". When she is transported into another, foreign space is when she might, miraculously, become awkward and unnatural, as described by Abu-Lughod of the Bedouin family in a glamorous upper class home full of artifacts (Abu-Lughod 1993: 120).

But she will also go to another space that will be interpreted as "natural", a place that will become "home" to the extent of her replacing this home with that one. Her married home will be a home configured initially by others, such as her mother-in-law, and then gradually (and dialectically, we hope) both be constructed by her as it is constructing her too. As I note about a different household and its mistress elsewhere (Shree 2000: 183–86; see Fig. 2), it is the very design of the house that can impose a discipline on the daughter-in-law. In the novel *Mai* the house has a central courtyard with verandahs and rooms on three sides, that ensures that the mistress of the house is in confinement because she can always be observed and criticized by the bed-ridden mother-in-law, sitting strategically on one side of the courtyard. If *Mai*, our protagonist in this story, had initially had a closed kitchen with modern gadgets and a counter—all of which she got, but too late—she would not only not have had the bent spine she got, but would have had the privacy to order her activities her own way. But there are other, more mixed, results of the house design. In the absence of a master bedroom and the companionate marriage that this implies, *Mai* has the freedom to spend time with her children, in their room, hers and theirs, and build an inter-generational solidarity. Such a solidarity is the stuff of the socio-cultural training which constitutes "tradition" in families, equaling a lack of freedoms in a certain perspective but other freedoms in a different perspective. For the daughter in this story, Sunaina, and maybe for Zeenat in the future in ours, there comes to be an identification of the house with childhood, with life itself, with something that haunts and keeps coming back. The house bears the children, it binds them, it makes them grow, it suffocates them. They dream of escaping from it. When they escape, they dream of coming back to it. Homes, even more than parents, control lives, and become for the children the locus of childhood and the past.

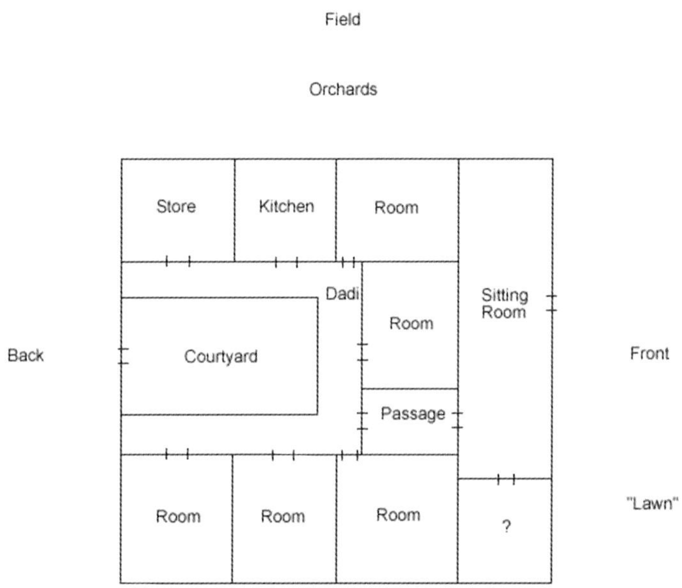

Fig. 2: House plan in *Mai*.

Much of the culture that anthropologists locate, including lifestyles, values such as of interpersonal relations, personal identity, honour, and shame, are all aspects of the functioning of the environment and its bio-politics. We have insufficient discussion in Anthropology yet of how spaces create and control lives, even as Henrietta Moore pointed out in her interesting study of the Marakwet of Kenya, "It is now axiomatic that spatial relations represent and reproduce social relations, and it is the view that relations of likeness exist between social distinctions and spatial boundaries that links the study of gender to the study of space" (Moore 1986: xi). The thrust of the research on space, or rather, buildings, as a political technology for the exercise of power by organizing and canalizing bodies in space derives its inspiration from Foucault (1975, 1984; Rabinow 1984). This is taken forth to the case of colonies like India to show how architecture, city planning, and aesthetic forms were used for culturally and epistemologically oriented colonial domination (Rabinow 1982; Oldenburg 1984). When brought forward to national periods, the exercise of power by national states is seen in the continuation of the same techniques of architecture and planning of the colonial periods, leading to even more elaborate control of the domains of daily life (Holston 1989). I am proposing a discussion of a more subtle, everyday politics: the colonization of children by adults that also translates into an extraordinarily strong socialization, resulting in inter-generational continuity in values. The question that remains to be asked alongside is about how children reappropriate space, and what their tactics are, overt and clandestine, to escape the nets of disciplining. How do they themselves participate in their socialization? Because... "the pedestrian's walking, like the flaneur of Walter Benjamin (...) is the spatial acting out of place, creating and representing public space rather than being subjected to it" (Low 1999: 113–14).

Useful for my approach is also the work of Bourdieu (1977) and the concept of *habitus*. "In his examples, the Kabyle house becomes the setting in which body space and cosmic space are integrated through metaphor and symbolic homologous structures. Through the experience of living in the spatial symbolism of the home, social structure becomes embodied and naturalized in everyday practice (...). Bourdieu also examines resistances and the effect of feedback on the social system. Since the concept of habitus spatially links social structure to the human body and bodily practices, the possibility of resistance to these practices becomes more apparent (Giroux 1983)" (Low 1999: 114).

Education: Flaws in Teaching the Nation

Because, specifically, of the colonial history of India in the last two centuries and the ways that modernity was defined and used, schooling came to play the part of the archetypical signifier of the nation-state, progress, and upward mobility. The space of the school came to be a fetishised one. It became a ritual space where a wonder was supposedly enacted. Because it was, and continues to be, closed and hallow, no one may look inside and almost no one has a precise idea of what is enacted within, especially not the parents. But a magic was/is supposed to occur, indeed, the biggest magic of modernity: education can apparently change a person from one kind of being to another.

One would expect, then, that the child's understanding of space also varies according to whether the child is schooled or unschooled. This interesting proposition of the school as marking the social division in India between two parts of the population can be investigated with regard to the nation.

The nation for the schoolchild is visualized as a geographical space: a map of a certain space that can be brought to life with images of mountains and rivers, plains and countrysides. This map and this collage of images could be encountered routinely by the child in the school, and in a spectrum running from strong to weak to nonexistent, its presence gets diluted from the most modern (a missionary school) to the least modern school (a municipal school). The nation, as I have discussed elsewhere (Kumar 2001, 2002) does not exist as a physical, material concept for the unschooled child.

Further, a child going to school can answer questions about the nation, that is, he can *imagine* the nation. An ironic characteristic of schooling in India is that the more powerful or effective the schooling is, the more it has to ignore the local. Schooling does not so much battle with the local as render it invisible, which is of course one of the strongest symbolic weapons to wield. "Progress" and "modernity" are measured with regard to the distance from the local. This distance could carry one well beyond the nation into the terrains beyond (called the global by us, but not my informants), and this is exactly what happens (Kumar 2001).

Let us look very briefly at the play of national and spatial symbols in three schools (primary and secondary in each case): St Johns, Gopi Radha, and Jamia Islamiya. St Johns has "foreign" pictures of romantic landscapes, from the same repertoire as its others of cuddly animals, blonde infants, and bouquets of flowers. It also has a large concrete map of India on the wall, discoloured by being outside where the assemblies are held, and a flag of India that is unfurled on ritual occasions. The "houses" students are divided up into for competitive sports and other activities are named after national heroes: Gandhi, Nehru, and Tagore. Elsewhere, in Loreto, they are Carmel, Fatima, and Lourdes, and in the Ramakrishna Mission School there is Vivekanand and Netaji Bose. When questioned, the students had no idea what role "houses" were supposed to play in their lives, and the map of India was consensually relegated to the unwanted margins, but both do exist.

In Gopi Radha and provincial schools which are either Hindi medium or are English medium only "in name" (not in fact), there are pictures of Krishna in various contexts, the choice falling, ironically, more on Krishna as adult than as child. There are pictures of Saraswati, the patron of learning and the arts, blessing without discrimination. There are idols closer in history, such as Gandhi, Nehru, and Tagore, always straight-faced and forbidding. There are landscapes of mountains and fountains, the idealized scenery never typically seen by the children. The principle seems to be to maximize the distance from the mundane and the real, and immerse them in an adult world of the desirable. But "immerse" is hardly the right term because the effort is technologically so flawed. The posters and pictures are not at children's eye levels but at adult height. Anything remotely prone to damage, such as of glass, clay, or paper, is safely removed from within touching distance. There are no colours, patterns, or images in tune with children's mental worlds, and certainly nothing produced by children themselves.

Jamia Islamiya, like all *madrasas* and Sanskrit schools, eschews deliberate symbolism, and effortlessly creates a continuity of sorts with the child's home environment (Fig. 3). This includes floor sitting, a demonstration of poverty in unwashed walls and mats, a noise level and disorder that is more benign than a modernist disciplining, and the absence of any modern artifacts or images, including a common sense of the very impossibility of them. In the novel *Mai*, we are told: "Dada's sitting room had been converted into a drawing room. His two portraits and sword in a leather sheath were still on the wall. But around them were English landscapes framed by babu in golden frames. Apart from this drawing room there were pictures on the wall only in the puja room. *We had not developed a tradition of hanging things on walls with a view to decorating the place*" (Shree 2000: 99; italics mine).

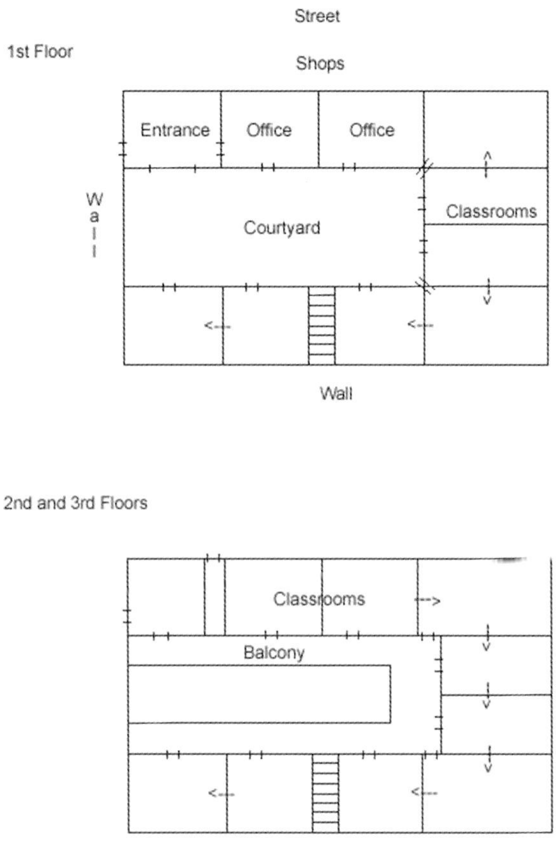

Fig. 3: Madrasa Jamia Islamia.

This lack of wall decoration is a closely held ideal in the *madrasa*, and something that matches exactly the lifestyle of every student and teacher. I must not be misunderstood as suggesting that the *madrasa* is not a "modern" institution. In all those aspects which characterize a modern education: age-graded classes, pre-determined and centralized curricula, examinations and report cards, the dominance of the textbook and the irrelevance of any particular teacher, it is as modern as all the other educational institutions which proclaim themselves as that. It is also modern in the historical Indian way that I have mentioned: in spite of a cultural-spatial continuity with the home, the *madrasa* school, like all schools, has a strained relationship with guardians. Each party in the relationship, the school and the home, blames the other for the inadequate socialization of the child into the self-conscious, confident, participating citizen of the modern nation-state.

When we go beyond the symbolism of the spaces and look at the curricula and rituals, all the schools, from *madrasa* to missionary present a set of patterns. The schooled child learns early on that the name of her country is *India*, that it is *her* country, and that it is *great*. The child sings songs to that effect, maybe marches or parades or salutes the flag, and takes part in celebrations on Independence day and Republic Day. By class 3 or 4 the child studies about the thirty-two different states of the Indian union, the boundaries, physical and economic features of each, and a bit about the costumes and festivals. Apart from the two all-India Boards that are popular in all towns, there are separate State Boards which give more attention to the geography and history of the particular state the school is located in.

My argument, however, is that the pedagogy of the nation is flawed. The efforts of all those who seek to create a new Indian citizen in the school have historically been lacking in technology, both its execution and its very conception. Their politics, for nationalists, have been correct—we must build up consciousness of and loyalty towards the nation—but their pedagogy has been insufficiently developed. Consequently, what has actually been offered to the child in the classroom has not been of the standard to produce the results desired. The schooled child in India has not developed that consciousness of the nation or that loyalty which was supposed to be produced. Although this argument could be demonstrated by citing the assessments of educators themselves, I have tried to look into explanations for it as well. Through ethnography in the classrooms, I show elsewhere that children in *madrasa*, private school, and public school, all experience a distancing from their history and geography that is a product of poor pedagogy. They cannot envision their country, cannot draw it, cannot refer visually to its physical features, cannot relate to its characteristics, or conceptualize any continuities—even ironic or subversive between it, the nation, and themselves in their experienced, lived-in lives. Through interaction and interviews with children I found that even those children who had mastered the arts of answering adult questions and spinning narratives within test procedures ("Name the five largest rivers of India"; "Where does it rain the most?") showed no ability to construct connections between their own biographies and their nation's history (Kumar 2001).

If adults in India have not thought of pedagogic strategies that could successfully teach the nation, how do they react when they are directly *taught* the strategies, as in a workshop for the interactive teaching of nationalist history, such as I witnessed?

There were several reactions. The first was of agreement, a confirmation of knowing the same approaches, perhaps of having attended several workshops, of having taken training courses and got degrees. There was an agreement which seemed duplicitous, because there was a slight masking in the eyes and lack of volubility about precise work. There was an emphasized separation into "theory" and "practice" and the response was saying, "These are great ideas in theory *but not in practice*. I know the theory quite well, thank you, but I am clever enough to also know that practice is different". If we translate the discourse, it runs as follows. *Of course India is a nation, and a great, ancient one, the best in the world. But we Indians are too clever to fuss around with that. The real challenge to us lies in our confrontation with the problems of everyday life. Let us go on with resolving those. That is the practical. All this rhetoric about the nation etcetera is fine, but is only theory. (Not that the nation is theory, but to talk about it in this solemn, quasi-religious way will never happen with us, and is thus theoretical).*

Education Outside the School: Here Comes the Nation

When the child does not go to school, there are two main ways in which she encounters the nation, which are shared by school-going children too. One is in movies and the other is in songs, typically heard over radios, television, or over loudspeakers amplifying their sounds generously in streets and neighbourhoods. In contrast to the movies of the fifties and sixties, there are fewer nationalist plots today that elaborate on citizens' duties towards national development: the building of roads, dams, and high structures. That was perhaps not "realism", but only idealism. Setting up a city in the jungle is a long-standing cultural ideal and was transferred from the king to the government. Movies of the nineties onwards are continuous with the older ones in their evocation of the natural beauties of the land: mountains, deserts, and waterfalls. But they are new in their direction of responsibility, not towards development, but to the protection of the virtue of the land and of its borders. Enemies are all those located outside: Pakistan which would seek to plant terrorism and attack the borders; and Europe, America, Australia and New Zealand, which are enticingly beautiful just like India but fail in virtue when compared with India. In movies such as *Pardes, Dilwale Dulhaniya Le Jayenge, Jeans, Ham Dil De Chuke Sanam, Pyar to Hona hi Tha, Kabhi Khushi Kabhi Gham*, India is not celebrated through its nature, but with a new confidence, through its value.

In this, the media, movies, television, and taped music, are all more advanced in nationalist rhetoric than school textbooks, which still continue to equate nation to nature. Textbook poetry asks: "What does 'nature' teach us?" and then answers with reference to the different parts of the national imaginary: "The Himalayas teach us to stand firm. The rivers teach us to nourish. The trees teach us to protect. The clouds teach us to share (...)". Which is so much less resonating than the dialogue from one of the above films: "Are you a German?" "No". "Are you an American?" "No". "Are you a Russian?" "No". "Then?" "I am *an Indian*". The latter speaker has decided to fight for the girl's virtue, in contrast to the other nationalities evoked who presumably would have failed to. There are numerous variations of this theme.

The Neighbourhood: Garbage and Beauty

A trip through the *mohallas* or neighbourhoods of the city, along its main roads as well as its several lanes, inside the campuses and compounds of its main institutions and then a very close investigation of the official, commercial, residential, and recreational areas shows us that Banaras is a modern city with a particularly unified character. That is, not that it is not recognizably divided into inner and outer, upper and lower class, ethnic or caste, and educationally and occupationally separated areas. It is that the divisions, insofar as they exist, are secondary to other unifying characteristics, which, in the interest of brevity, are the following, illustrated by ethnographic glimpses.

When I watched from my balcony at seven a.m., I saw five or six women winding their way to and from between the houses of my colony. They were maids who worked in different houses, typically as cleaners, to do the sweeping, mopping, washing of dishes, and occasionally clothes. I focused on one woman who was just leaving a house. She must have been one of the five a.m. workers. She was carrying a packet of garbage in her hands. I knew exactly what would happen. She walked to the empty plot one house away in front of our house and aimed unenthusiastically towards the ground within. She missed, as do most women, and men. The garbage landed on the roadside, to join the growing piles on both sides, most of which had spilled out from its containers and blown and spread around. The twenty or thirty yards of the road at that part was a garbage dump itself.

Similarly, I watched and talked, with this woman and that, including our landlady, our neighbour, our maid, miscellaneous workers and loiterers, and confirmed that *all* of them took it as a law of nature, god, and society that garbage should be thrown wherever no one personally stood as occupant of the land. It is the first unifying fact about the neighbourhoods of the city. The difference that could arise, say out of regionalism is slight, and subverted by other processes.

Vishvajit Chatterji, in class 9 of Anglo-Bengali Inter College, lives with his family in an ancient, crumbling house in a tiny *gali*. The house was in the corner, so they had some open space around their first floor rooms, but there were other equally tall or taller and older, houses a few yards away, so the room upstairs was dark and closed, in spite of its windows and balconies. More than half the room was taken up by a double bed, on which most of the social and educational activities were carried on. On the other side there was a fridge, a steel almirah and a dressing table. Vishvajit and his brother Surajit cannot help noticing that their home looks different to their non-Bengali friends' homes, and similar to their Bengali friends'. Other points that might strike them as part of their Bengali identity are that their mother chops and cuts with a *boti*, not a knife. She cooks rice all the time, and fish regularly, as well as several other dishes in a Bengali way. They speak Bengali at home, and the children seem as fluent in Bengali as in Hindi. They celebrate Durga Puja in a big way. They go to Calcutta once a year and have relatives there. The mother's friends are almost all Bengali, although the father, being also a BJP member, moves across all sectarian lines.

Where they are no different to all their neighbours is in their respect for the city's ethical codes about garbage, and its relationship to freedom. This is not articulated by them (as much of the above is not) and can only be observed. Father and son share the city's ethos. The provincial city does not teach a negative proposition such as "We do not respect public places, but only our private ones". Rather, in a myriad different ways it teaches the positive proposition that to litter anywhere desired is a mark of freedom. Every informant boasts of the special qualities of Banaras as a city that permits the freedom to act thus, wilfully, unlike the metropolitan one.

Fig. 4: The mazar and the mohalla's space.

Second, for the most part, there is no discernible separation into public and private spheres of activities. That which may be considered a private activity at one place, for instance, bathing or massage, is explicitly a public one at another, and vice versa. That is to say, the code for public and private does not exist, and the city's ethos is unified in that it holds *one model* of the public and private: the lack of interest in a fixed division into public and private.

There is therefore no specialization of spaces by activities. Inside the house, cooking, eating, resting, and sleeping, gives place to working or meeting, without prejudice to any principle. Outside, every space from side of street to centre of street to park or empty public land may be used interchangeably for work, celebration, meeting, storage, and so on. There is perfect unity of social code and social practice.

Third, some arguably modern values, or more precisely, metropolitan modern values, are recognizably absent, including: the observance of rules of precedence in traffic, recognition of symbols for discipline, obedience to appointed lawmakers, and a sense of that internalized control that would preclude, for instance, throwing garbage on any public space.

It has been my argument (Kumar 2001a) that the nature of schooling, across the spectrum of all kinds of schools, is such that the discipline often boasted about or alluded to positively by schools and educators is not the discipline of Foucault's argument, that permeates the self and transcends conflict and consciousness, that empowers even as it controls, and makes the disciplined person a true subject of modernity. Rather the child in a provincial school learns only to be subjected to authority, and this authority is arbitrary both in composition and actual action, and arbitrary in its deployment of symbols. When free of supervision, the lessons of the school such as punctuality, straight lines, silence, cleanliness, or maintenance of codes of discipline, are totally unobserved by him. They are a kind of game that must be played only if those in authority are watching. This is not the phenomenon described by Paul Willis of working class kids reproducing their class through resisting a certain socialization (Willis 1977). Rather, the children in a provincial city collaborate with the adult authorities to put on a show of learning discipline, and both collaborate in a kind of humanity that precludes learning the discipline, and in terms of the child's experience actually empowers him—within the provincial world.

But there is more to the teachings of the neighbourhood, and my fourth point is about the aesthetics of everyday life. A religious institution called The Gayatri Shaktipeeth had a grand celebration lasting for three days. It is one of hundreds of such sectarian public organizations, and they all have their periodic celebrations. The Shaktipeeth had dozens of loudspeakers spread over four neighbourhoods. They had two full days of chanting, lectures, yoga, meditation, puja, sermons, prasad, and singing. The theme was "*Mahila jagriti*" or the arousal of women. The speakers explained how important it was for women to rise and take up their full share of building up the society and the nation. The nation was taken to be homogenous, with no difference of caste, class, or creed, as the cliché goes. The event was crowded. The arrangements were expensive and efficient. The sermons were articulate and powerful. The singing was tiresome.

In the field of a local school, a tent was erected where the public sat, women, or the sisters, on one side, and the men, or the brothers, on another. There were stalls of power-building paste, *chyavanprash*, of incense and *yantras* and rings, of pictures of the founding teachers, and of books on the topics of the day. The stage on one side had six young women in saffron, playing, singing, and preaching. The audience was absorbed and often clapped or chanted alongside. At the edges children ran around playing their games.

What did these children learn from this event? Regardless of its specialized nature, it is one with large scale feasts and rituals like weddings. Children could certainly catch odd things from the mikes on *mahakal* (cyclic time), on *shakti* (power), on the problems of contemporary society and their solutions. But what else? I would like to interpolate here an observation based on myself as an informant.

When I was small I often hung around till late at night at such events, sometimes falling asleep on the sofa or *dari* (mat), running around with cousins if at home, with utter strangers if elsewhere. I always caught something of what was going on. I saw Indian dances first at such occasions, heard non-film Indian music first on such stages through such loudspeakers, and built up involuntarily my intellectual world with knowledge such as: what was so distinct about each of the Pandava brothers? Why did Mira hate her husband's family? How did Yashoda react to her son's naughtiness?

I learnt these things because I saw them repeatedly and not just once. But even more, I learnt how such events should "feel", and the correct, normal ambience in them. If as an adult I came to arrange any large-scale ritual, it did not feel successful if it did not feel like that. Like what? With a tent (or marquee, as Salman Rushdie calls it) in the night, food served outdoors, people coming and going, amplified lights and sounds, and a sense of liminality. A oneness with nature, a continuity of space, an endlessness of time, the outdoors transformed into the indoors, but so much more exciting than the everyday indoors. These were things I learnt that were as concrete as the other.

The Home: Class

The sensual, symbolic, and metaphoric world of the provincial child learnt by her in the neighbourhood is shared across gender and class. By sensual I mean experiences such as those of the monsoons, of eating mangoes, of swimming in the river, and of playing in the streets. The symbolic is a continuation of the same, in that it constitutes and does not merely denote: the food, colour, sound, and liminality of festive occasions as necessary for both ritual correctness and the good life. By metaphoric, I mean the almost classical-by-now metaphors of the *ghazal* and the cinematic world, such as the cluster around love: stars, night, thirsty eyes, sleeplessness, tossing in bed, crying, calling out a name, responding. This world of the senses and metaphors is learnt also in the home, simply because the home is continuous with the neighbourhood.

The home of Gulistan and Gulfesa (Fig. 5), ten and six years, was surrounded by a sea of water one day in August. The only reference I heard to the water was that the two girls would not be going for their tuition that day because of the water. Perhaps that was why they were happily playing cards instead. Gulistan and Gulfesa do not experience shortage of space or difficulty in any job in home because they have grown up in this home and they are totally used to it. This is how space and time are, as normal as sun is sun and water is water.

When I was small and visited my ancestral house in Kutchery Road, Lucknow, I never thought of its surroundings as a crowded street with open gutters on either side, and food shops which were made on wooden benches on top of these, the refuse from these shops going directly into the gutters underneath. There was a Thakur who brought around his cart every evening, ringing his bell as a special signal, and we would rush, if we could afford it, to slurp up his shaved ice and cream and syrup, never imagining that it was less than sanitary. The *only* consideration, ever, was, can I afford it? Only when I was grown up did I hear this beloved Kutchery Road described as "dirty" and children being warned not to eat the "*nali ki chat*" (snacks of the gutters) or hang around by the roadsides. Moreover, all my cousins in this Kutchery Road house slept in smaller, darker bedrooms than did I in my cantonment bungalow, but I did not think their situation pitiable. Rather, it seemed wonderful and enviable. The eyes I have now to discern clean and dirty, spacious and crowded, well-lit and dingy, are not the eyes that children ever have, no matter who they are (see Fig. 6).

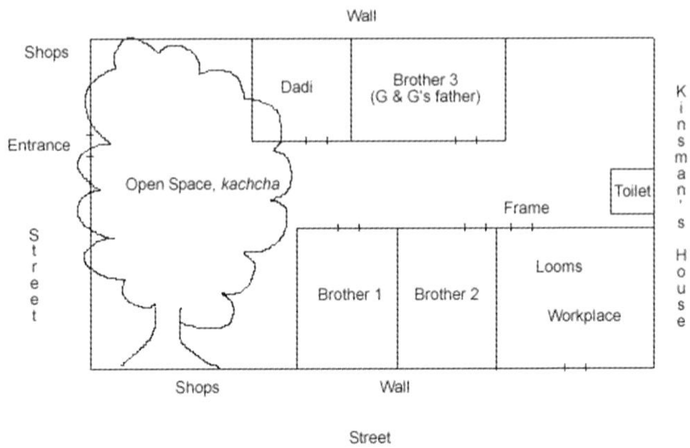

Fig. 5: House plan of Gulistan and Gulfesa, a weaver's daughters.

When I was small, the rainy season was a time of magic. Clouded skies, breezes, drizzles and sprinkles, a kind of magic quality in the air. There was a chance of "rainy days" and school being unexpectedly closed. It was possible to sit in the verandah and work all afternoon, without being bothered by heat or cold. A child, after all, does not have to be conscious of adult monsoon problems like mildew and mould. I remember many play kitchens, with little clay or metal cooking utensils, leaf and clay foods, picnics, real and make-believe. I remember swinging on the charpoy swing, or being safe under a roof, soft greens outside.

Maybe Gulistan and Gulfesa are enjoying their childhood monsoons equally. The delight on their faces certainly seems to bear witness to it. This, despite the fact that their house is broken down, very cramped, surrounded by filthy water, and they are themselves resource-less, poorly fed and nightmarishly educated.

Of course class is important. The toilets of poor people are a problem—for the outsider, but maybe for them too. Some are totally outside the home, in the worse cases being the side of the nearby pond. Little children in poor localities are encouraged to defecate on the wayside, and almost all children to urinate anywhere. When slightly better, homes have toilets in the courtyards, entrance ways, and staircases. They are dark, with non-flushable commodes in the ground, of the squatting "Indian style", with stored water to be dispensed with a mug. The child learns to maneuver among all this matter-of-factly, and never actively suffers.

Fig. 6: A child's rendition of Banaras.

Bathing places are not specialized either. The tap or well is not in a closed space. Men and women bathe routinely in courtyards, keeping some clothing on. In other cases, there is no private well or tap and buckets are brought in from outside. There is a preferred place for keeping them, but obviously a bucket could be kept anywhere. Thus a child could one day be bathed on the roof and another day in the corner of a room. The Ganga is a vast reservoir for one's personal use, including brushing of teeth, washing of clothes, and soaping and scrubbing oneself clean.

The night spaces in the cramped rooms get messy and are slowly transformed. A thorough *jharu* (sweeping) is done on all the floors, by the mother or an elder sister, never by the father. Sheets and blankets are folded up and stacked away. Mattresses and *chatais* (mats) are rolled up and put away. Some beds are stood up; others are left flat for daytime rest. Where there are permanently placed beds they serve as sofas, couches, diwans, and are just like the floor in bed-less places. The floor everywhere is made clean enough to sit and eat at. Nowhere do people wear shoes into the house or living areas.

But class socialization is much weaker than gender socialization. Unlike with gender where both the parents are simultaneously present for comparison, there is no other model before the child except that of the home she is born into. Those whom she visits are other members of the family or close friends, and are typically of the same class. Or she assimilates, as in my childhood visits to the ancestral house, two different lifestyles as equally "natural". Only the adult suffers pragmatically from shortcomings, and only the outsider interprets spaces, processes etc. to be limited or cramped. For the child they are normalcy itself and she finds them not remarkable in the least.

The House of Bachhe Lal, a Painter, and the Process of Gendering

What I am suggesting is that all children do *not* share the same experience of space, nor express themselves similarly with regard to it, but that the lines of division are not those that are the strong, evident ones of education and class. Neither the difference between the schooled and unschooled child, or the child in elite public school versus the child in the *madrasa*; nor the difference between the elite and lower class child, if both products of the provincial city, is the crucial axis of division. Children are, on the national plane, divided according to whether they belong to metropolitan or provincial cities, towns, and villages, because the nature of neighbourhood and community life and teaching differs in each case. Within the provincial city, the experience of space and the expression of space varies according to the sex of the child.

In the house of Bachhe Lal (Fig. 7), a potter, there live his mother and father, and his five brothers, of whom two are married and have seven children among them. The house is an open space with three separate rooms or huts, all *kachcha*, of mud. The open space is like the water in a map of the earth with some land interspersed in between. On one side is usually spread a curtain on which Bachche Lal paints typical studio scenes of buildings and roads with artificial flowers sprouting in the foreground. It is very iconical of the much discussed modernist tropes of the road leading to the new

India (see Srivastava 1998). On the other side lie some masks drying, in symmetrical rows of deer, bears, monkeys, lions, and so on.

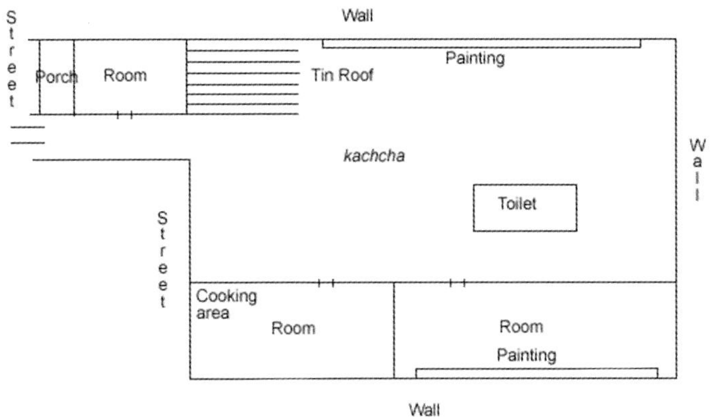

Fig. 7: Bacche Lal's house.

The treatment of children in this family was a puzzle. Again and again I was assured that nothing of the family craft was taught to them until they were much older, and that even then they just picked it up from watching. The children seemed to have little to do, with no toys (Bachche Lal chose to interpret my question about toys as the masks and dolls made in the family and shrugged about how satiated the children were with those) and nothing child-size in the house that I could see. They were not allowed to touch any of the materials. There were of course no separate spaces for children at all. It seemed self-defeating to not provide them with anything of their own and then scold them incessantly for trespassing into the adults' worlds. They would stand around, feel the pressure of bottled-up energy and surge forward to touch something or act upon something, and be shooed or struck out of the invaded space again. Adults were forever creating a centre from which children were barred and of which they were supposed to stay in the margins.

Bachhe Lal had actually made a children's landscape which was ordered, purchased, and now hung up in the neighbourhood school. He took me to see it. *It was exactly like the landscapes he made for his photo studio clients.* There were mountains in the background, a stream in front with boats on it, a bridge on one side, trees and a settlement with huts on the other. There were bright colours, a sunset, flowers, shadows. There were no people. It was abstract and distance and frozen in time.

On Diwali day, it felt that the activities going on were no different to other days. The house was not cleaned up or decorated. The family behaved like the producers of the objects being celebrated—images, clay lamps and toys—with attention to others' consumption in the market rather than on their own consumption. In the middle of my mid-morning visit, they packed up two crates with these things, ready to go. One was picked up by the youngest brother, a boy of about eight, one by the father. They headed for Khojwa market, where two other sons had set up Diwali stalls. I followed. The little boy walked fastest, and led the procession. Khojwa, even after hundreds of visits, can remain a mystery with its lanes and byways, all suffused with garbage. The last time I had to find a place, I was also led by a boy of maybe ten, who led me all along back lanes in the interest of saving time and showed me another world than I had known existed, this one with even more garbage.

Then I remembered other incidents from my fieldwork. I remembered all the scores of times I had been sent to wherever I needed to go under the guidance of little boys. Or, when I had asked directions at a shop or on the road and a little boy had stepped forward with alacrity to guide me, and adults had taken it as a given that he should. Then other consistent patterns: little boys of even four or five sent out for pan and tea to the corner shops. Women's voices echoing throughout the day inside homes, calling out to sons, nephews and little brothers to fetch them this or that from the market, from outside.

Fig. 8: Free little citizens of the city.

The outside. Before beginning an analysis of the inside-outside division one must repeat that in India there is no exclusive space for children, of whatever age. When they are small they are not taken anywhere special by their mothers/parents, as, for instance, to a park but to places where everyone goes, such as markets, or others' homes, or the *ghats*. When they are independent, from just three or four onwards, they wander into each others' homes, congregating wherever convenient, say on rooftops, or in the lanes themselves. At slightly older, they seem to take over vacant lots, neighbourhood squares, and burial grounds, but only temporarily, because they can be and are chased out regularly from these places. The same spaces are used for work, such as drying yarn, selling or demonstrating wares, and for adult recreation, ranging from the most informal to the most formal. Then, when they are adolescents, the children have absolutely nowhere to go. Embarrassed to play in open places, they congregate in tea or pan shops, on ghats and streets.

At the early age of six or seven, the category of "children" gets divided into "boys" and "girls". The Hindi word for children, *bachhe*, tends to be used thereafter only for boys, and thus have I used "children" above. The separation is on many counts: clothes, speech, body language, activities, and even food. But its most dramatic dimension is in its location of children: in the "inside" for the girl and the "outside" for the boy.

At the times of the day when school begins and ends, both girls and boys will be visible on the streets, coming from or going to school. Girls walk openly, or cycle, or ride rickshaws, scooters, or buses, apparently in equal numbers to boys. Sometimes they walk in groups and seem to take over parts of the street like boys do. The difference lies in where this takes them, if I might be forgiven this pun of mobility. Apart from travelling to and from school, mobility is not a structural characteristic for girls. They do not go out for pleasure, for hanging out, or to pass the time. There would be nowhere to go. A street side, pan shop, tea shop, or any open urban space is not the provenance of the female of any age, unless they are professionals or otherwise have work. From early morning to late night, females are visible on streets as sweepers, garbage collectors, vegetable sellers, and hawkers, and of course, shoppers in transit. But they do not stand around without work, as men and boys do.

The little boy in Banaras gets the training to be a flaneur early on in his life. The city is his oyster, starting from his neighbourhood. One could write extensively on the subject of men's activities in the city (Kumar 1988); what would remain to be added is that the socialization into this occurs early on for children. There is scanty literature on this, even as for girls there is the proposition that in colonial times they came to belong to the inside, to stand for the home, purity, and the non-colonizable nation. At this point I can only give tiny ethnographic insights regarding the gendered ways of the working of space.

All children, for example, wake up to the sound of their parents. Gender socialization begins at that point, because the mother is typically filling water, doing the dishes, or sweeping the floor, and soon lighting the fire for the first meal. The father is either still asleep, or also busy. In some cases his work entails going before dawn to buy the raw materials for his business, such as pan, vegetables, or milk. In some cases, he has to start setting up his shop by carrying things out. In other cases, he might spend the first few hours of the day in grooming himself and preparing for the day, whether with a newspaper and cup of tea, or a leisurely toilet, or preparing what he needs for the day at home. *In no case* is the father the person who is responsible for the morning meal and the children getting fed and sent off with their tiffin boxes.

The only exception is when the mother is ill, or dead, or absent because of an emergency somewhere. Then the father does take over, ready to pass on the job to the oldest female. An occasional informant, such as Lajjavati, might narrate how her father brought her up completely after her mother's death and how he remains the best cook she ever knew. All the women of Assi, my research neighbourhood, get a welcome break from cooking when they are ill or having their periods. As their daughters grow up, however, such as Lajjavati's Chunni, which means when she is as little as nine or ten, the daughter takes over the substitution work from the father.

The mechanisms of all this are clear enough to the child, and the gender socialization is stern and unalterable. The gender socialization, as with class, seems natural and is not remarked upon. Almost all children take for granted that the mother should provide all the services at home, as well as be around the whole day like an anchor for the various vessels to come back to and tie themselves up temporarily. They do not consciously think that the woman's situation could be compared to the man's or any constructive ideas generated from the comparison.

But there may be more commonalities than we imagine between the sexes too. I will end with one woman's and one man's song. The first is a *vidai git* or the lament of parting from her family by the bride as she leaves for her new home. The second is a *ghazal*, about the man lost in the world, ready to wander. Uprooted as they both are, could it be that the pain is so great only because they would both *like* to be fixed in a place?

> The woman sings:
> *... sakhi sakhiyan bachpan ka ye angna guriya jhule koyi bhi to hoga sabnam*
> *chhupaungi aansu kaise bhigege kangna*
> *sakhi sakhiyan re...*
>
> ...my friends, my girlfriends, the courtyard of my childhood, the dolls, the swings....
> how will I hide my tears [as I depart]?—my wedding bracelets will get soaked...
> my friends, my girlfriends.... (Akhtar 2002)
>
> The man sings:
> *ai dil mujhe aisi jagah le chal jahan koyi na ho*
> *apna paraya meherban ya meherban koyi na ho...*
> *duniya mujhe dhunde magar mera nishan koyi na ho*
>
> ah, my heart, take me to a place where there is no one
> there is no one of mine, no one other, no friend or supporter at all...
> where the world would keep searching for me but there is no sign of me... (Sultanpuri 1950)

References

Abu-Lughod, Lila. 1993. *Writing Women's Worlds: Bedouin Stories*. Berkeley: University of California Press.

Akhtar, Javed. 2002. Mere yaar ki shaadi hai. In: *Yaar ki shaadi hai*. Music by Jeet-Pritam, singer: Alka Yagnik.

Bourdieu, Pierre. 1977. *Outline of a Theory of Practice*. Cambridge: Cambridge University Press.

— 1990 [1970]. The Kabyle House or the World Reversed. In: *The Logic of Practice*. Oxford: Polity.

De Certeau, Michel. 1984. *The Practice of Everyday Life*. Berkeley: University of California Press.

Donald, James. 1996. The Citizen and the Man about Town. In: Stuart Hall & Paul du Gay (eds.), *Questions of Cultural Identity*. London: Sage.

Fabian, S.M. 1992. *Space-Time of the Bororo of Brazil*. Gainesville: University of Florida Press.

Foucault, Michel. 1975. *Discipline and Punish: The Birth of the Prison*. New York: Vintage Books.

Fuchs, Dieter. 1995. Training and Continued Education—Children's Culture. *European Education Summer* 26, 2: 5–17.

Giroux, Henry A. 1983. Theories of Reproduction and Resistance in the New Sociology of Education: Critical Analysis. *Harvard Educational Review* 33: 257–93.

Holston, James. 1989. *The Modernist City: An Anthropological Critique of Brasilia*. Chicago: University of Chicago Press.

Kumar, Nita. 2000. *Lessons from Schools: The History of Education in Banaras*. Delhi and Thousand Oaks, Calif.: Sage.

— 2001. Children and the Partition: History for Citizenship. In: Suvir Kaul (ed.), *The Partitions of Memory*. Delhi: Permanent Black.

— 2001a. Learning Modernity? The Technologies of Education in India. In: *Les jeunes: hantise de l'espace public dans les societies du Sud? Autrepart* 18: 85–100.

— 2002. Why do Hindus and Muslims Fight? Children and History in India. In: Diana P. Mines & Sarah Lamb (ed.), *Everyday Life in South Asia*. Bloomington and Indianopolis: Indiana University Press.

Littlejohn, J. 1967 [1960]. The Temne House. In: J. Middleton (ed.), *Myth and Cosmos: Readings in Mythology and Symbolism*. NY: American Museum of Natural History.

Low, Setha M. 1999. Spatializing Culture: The Social Production and Social Construction of Public Space in Costa Rica. In: Setha M. Low (ed.), *Theorizing the City: The New Urban Anthropology Reader*. New Brunswick, NJ: Rutgers University Press, pp 111–37.

Metcalf, Barbara. 1982. *Islamic Revival in British India: Deoband 1860–1900*. Princeton: Princeton University Press.

Moore, Henrietta L. 1986. *Space, Text and Gender: An Anthropological Study of the Marakwet of Kenya*. Cambridge: Cambridge University Press.

Rabinow, Paul. 1982. Ordinance, Discipline, Regulation: Some Reflections on Urbanism. *Humanities in Society* 5: 267–78.

— (ed.). 1984. *The Foucault Reader*. New York: Pantheon Books.

Redfield, Robert & Milton Singer. 1954. The Cultural Role of Cities. *Economic Development and Cultural Change*. 3: 53–73.

Shree, Geetanjali. 2000. *Mai: A Novel*. Translated from Hindi by Nita Kumar with an afterword. Delhi: Kali for Women.

Srivastava, Sanjay. 1998. *Constructing Post-colonial India: National Character and the Doon School*. New York: Routledge.

Sultanpuri, Majrooh. 1950. Ai dil mujhe kahin aur le chal. In: *Arzoo*. Music by Anil Biswas, singer: Talat Mahmood.

Willis, Paul. 1977. *Learning to Labour: How Working Class Kids get Working Class Jobs*. Aldershot, U.K.: Grover.

Stefan Schütte

The Social Landscape of the Washermen in Banaras

We are illiterate people and that is why we have to organise ourselves. In that we find the strength and unity to survive.
Ramu Kannaujiā, 19.12.2001

Introduction

This essay deals with the washermen ("Dhobī") of Banaras and the ways they collectively order their living space and community life. The study is based on one year's empirical research[1] and seeks to discover how traditional forms of social organization in urban India continue into the 21st century, and to what extent these still shape everyday life, social interactions, and spatial perceptions. It is argued that the example of the washermen in Banaras demonstrates the actual significance of traditional orientation systems which help to build social security, dignity, create opportunity and facilitate collective action, and go hand in hand with a distinctive spatial framework. It thus gives a secular perspective on Banaras and the various ways in which the city is ordered and perceived according to the needs and aspirations of a subordinate caste group. What are the mechanisms that enable the washermen to carry on with their timeworn social practice? How have they been able to cope with the numerous social, economic and political changes and influences that have certainly impinged on their system? Answers to these questions will not only give important insights into processes that allow certain modes of social networking in urban India to adapt to changing conditions, but furthermore hint at the overall significance of social organization for leading a life in well-being and security. The latter point is central to the far-reaching contemporary discussions in various disciplines about the concept of social capital—a term that, very generally speaking, points to the potential virtues of social relations for enhancing security and economic performance, and by putting it in a contextual framework, also assesses the discursive and political economic background in which collective action occurs (see Bebbington & Perreault 1999). These considerations are elaborated more fully elsewhere (see Schütte 2003). The main intention here is to lay down the spatial ordering of an encompassing social network and to assess its practice and functions in terms of indigenous concepts of space and social relations.

Indigenous Connotations

The washermen's arrangement of space and social relations emphasizes the importance of collective organization to cope with the mundane reality of everyday life. It represents collective efforts to contend with and manage their place and position in urban society and is an essential asset for achieving a life of security and content. Even so the Dhobīs' method and its related knowledge of Banaras remains enclosed in caste-boundaries, which are socially and historically constructed and denote power relations amongst the community and beyond it. Any attempt to elucidate the social and spatial topography of the Dhobī must therefore be anchored in indigenous connotations of space, spatial relations and social organization. These are expressed mainly via a complex concept which the washermen call Ṭāṭ—a notion that carries a variety of meanings and which outlines the basic principle of what I would call the washermen's "social landscape".

However, what is expressed through this practical example has a theoretical basis in the various discourses of human geography regarding how space is treated in relation to social theory (e.g. Läpple 1993; Gregory 1994a, 1994b; Simonsen 1996; Soja 1999). A prominent view is to conceptualize space as "social spatiality"—that it is socially produced in a variety of ways and consequently has to be treated as a social category. Apparently the abstract theoretical notion of social spatiality resembles very much the actual social practice of the washermen, since here spaces of social relations are institutionalized and constructed in the shape of social localities. These localities do not have any immediate physical form but are directly associated with a collective practice that creates a distinctive social form of space.

The Dhobī in Banaras

Banaras municipality accommodates about 15,000 washermen[2], a number that has been roughly estimated during fieldwork. In Uttar Pradesh they are recognized as a so-called "scheduled caste". This term basically serves as an official identifier of what in the dominant orthodox Hindu worldview are the so-called "untouchable" ("polluted") castes. Like a legal compensation for their subordinate position in the social structure of Indian society, the status entitles them to certain reservation quotas for public employment and access to education (see Mendelsohn & Vicziany 2000). However, in the urban environment of present-day Banaras the practice of untouchability has become less obvious and Dhobī today are not subjected to any open social discrimination.[3]

The traditional occupation of a Dhobī is to wash and iron clothes, and although there have been many attempts to provide higher education for the younger generation and thus allow them to move into different strata of work, most washermen have not succeeded in doing so. A current statistic published in a local newspaper (*Dainik Jagran*, October 8th 2001) claims that no more than 8 percent of the total Dhobī population in Uttar Pradesh[4] has found an occupation outside the washing business. As this figure refers to the entire state, the situation in Banaras itself might be somehow different, but it gives a good indication of the extent to which occupation and caste are still congruent.

Fig. 1: An interior washing-*ghāṭ*, the Bhavānī Tālāb. The place is located near the waterworks, which supplies excess water from the River Ganges to the small pond. It is always very busy and attracts Dhobī from all over Banaras. Beside its use as a washing-*ghāṭ* the place serves as one of the communities' meeting places and has a guarding deity (Daitrābir Bābā) protecting the place. (Photo: Schütte, Dec. 2001).

Physical Dimensions of the Dhobīs' Spatial Practice

Washing is a process consisting of many successive steps, and it is best to consider the case with the example of one bundle of clothes in need of washing. Most likely the Dhobī who is assigned to do the job will first visit the customer and collect the bundle. Either he will walk, if the distance is close, but often he will use a bicycle as the customer may not live in a nearby neighbourhood. The bundle has then to be prepared for washing: it will be soaked in soap and soda water or, more traditionally, in a mixture of sheep-dung and the fine earth-crumb taken from fallow land. These steps usually take place in front of the washermen's houses. From there the bundle is taken to wash at the *ghāṭs* of the Ganges or Varana River or else at one of the very few interior Dhobī-*ghāṭs* in the city.[5] Each Dhobī has his regular washing place, often since many generations and near his residence, but in numerous cases a considerable distance away. And since it regularly poses a problem to dry the load at the washing place itself, many of them after washing go to a separate drying place. Having waited there for the clothes to dry, they are then carried back home or, possibly, to a small clothes pressing shop somewhere in the narrow streets of Banaras. Finally the bundle makes its way back to the customer's house. It is a time-consuming process, with the tasks shared among family members. Except for the washing itself, which is considered men's work, both sexes participate in the various steps.

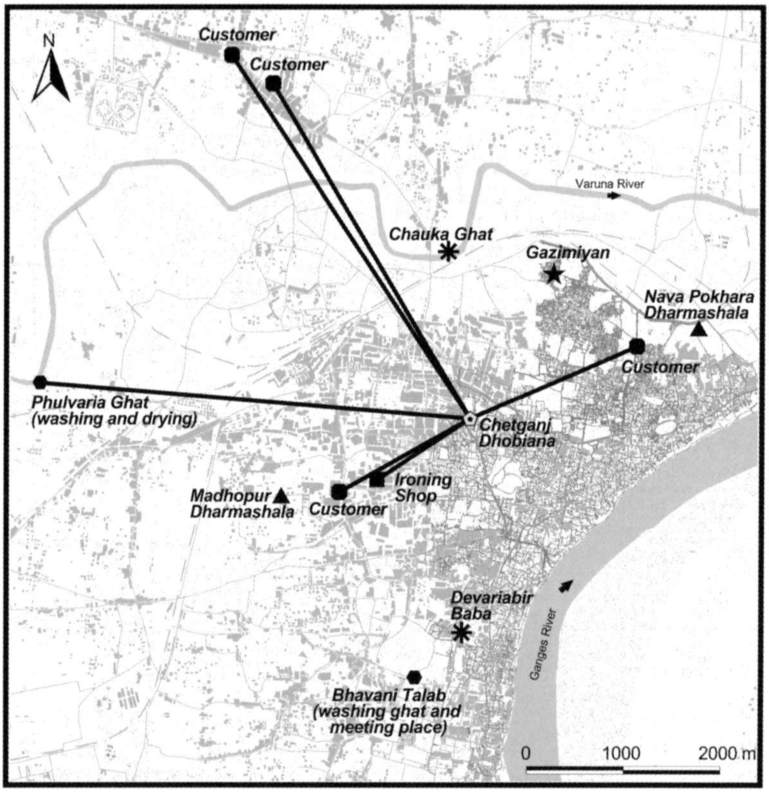

Fig. 2: Action spaces and central locales—the example of one Dhobī-household living in Cetgañj. Huge distances have to be covered regularly for conducting the traditional occupation and the example-case is no exception. (Design: Schütte 2002, based on Benares Guide Map, 3rd Edition, Surveyed 1928–1930).

This short description of a typical washing sequence[6] demonstrates that washing is not a stationary process at all—it requires spatial mobility, sometimes on a comparatively large scale. Washermen carry their bundles through the city to work at different places, and depending on the location of their customers' residences frequently may cross the whole city for delivery and collection of clothing. This coherency is represented in Fig. 2. The map shows representatively the regular mobility required to fulfil the professional duties of one Dhobī-family living in Cetgañj, right in the centre of Banaras. The representation allows visualization of a spatial practice constituting specific action spaces. These may differ individually in scale and width, but they are generally determined by the needs of the traditional profession. Further activity places depicted in the map show central localities used for regular religious and social activities of that particular family, and for Dhobī in general.

It is interesting to note that many Hindu washermen worship at the Muslim shrine of Ghāzī Miyā̃ located in Bari Bazar. Habitually they would visit the shrine every Sunday to make offerings and leave a wish for the saint. Similarly, the yearly festival in the Hindu month Jeth at that particular shrine is celebrated by Dhobī, other Hindu low caste groups and Muslims together.[7] But the Dhobī also celebrate their very own famous festival called Pyalā, which takes place at Caukā Ghāṭ (Varana River) and one particular temple in southern Banaras (Devariābir Baba). Muslim musicians (Dafāli by caste) assist during the minute ritual, and the festival day is used for extensive picnicking at Caukā Ghāṭ. It is a socializing event celebrated annually.

As social focal points the washermen maintain two Dharamśālās.[8] Both of them serve to accommodate caste fellows from outside and especially as central locales for frequent community assemblies. The "Bhavānī Tālāb" in southern Banaras is another important meeting place besides being the busiest interior washing *ghāṭ* (see Fig. 1 and 10).

Approaching an Extensive Indigenous System

Having detailed the physical dimensions of a Dhobī's spatial practice in Banaras, the next step is to delineate its social dimensions as well. As already indicated, notions of space and sociality among the washermen are signified through their interactive concept of Ṭāṭ.[9] This concept generally refers to patterns of internal caste government including a strong element of decentralization, as has been practiced already over long periods of time. The British surveyor Blunt in his general assessment of the Indian caste system actually identifies the term (among others) as expressing a clear demarcation of local boundaries of jurisdiction over certain areas, presided over by caste councils (Blunt 1969: 105).[10] In his account, Blunt provides numerous examples of local caste government among predominantly low caste groups in North India, thus asserting that the practice must have been very widespread. In today's Banaras, however, permanent caste councils are hard to find, and the elaborate practice of the Dhobī in the form of the Ṭāṭ-system clearly represents a peculiarity. Various low caste groups have been approached during field research (Dom-Sweeper, Camār, Kumhār, Mallāh, Yādav), and though in most cases the respective elders remember that there has been a practice of caste jurisdiction sometime in the past there is no sign of its continuity. Or as people ascertain: "*is paramparā khatm ho gayā*"—"this tradition has disappeared". The reason why these councils are still able to exercise power and jurisdiction among the Dhobī in Banaras may be located in their adaptability to changing circumstances and the effectiveness in taking up new directions while at the same time they are conserving traditional functions. Furthermore, these councils are composed not only of a few powerful decision-makers but from members of each household in a certain area of jurisdiction and thus truly represent a collective assembly where all members are allowed to raise their voice and express their opinion.

As a first approximation Ṭāṭ may thus be defined as a hierarchically layered system of caste councils and as such has above all a very practical dimension—it generally denotes the ways in which the Dhobī organize themselves socially and spatially by means of regular social interaction. The collective aspect is also expressed in a metaphorical daily usage of the term. It may be translated as "carpet" or "mat", and people in that sense refer to Ṭāṭ as "the thread that keeps our community together", an analogy to the thread from which a carpet is woven. A similar usage is expressed when people say "Ṭāṭ is the place where we all sit together"—on a carpet which is reserved for washermen only. This metaphorical usage can be stressed even further when acknowledging the various practical connotations of the concept which resemble a patchwork pattern of meanings in continuous interplay. These can be expressed visually as a carpet made out of various segments representing its main characteristics.

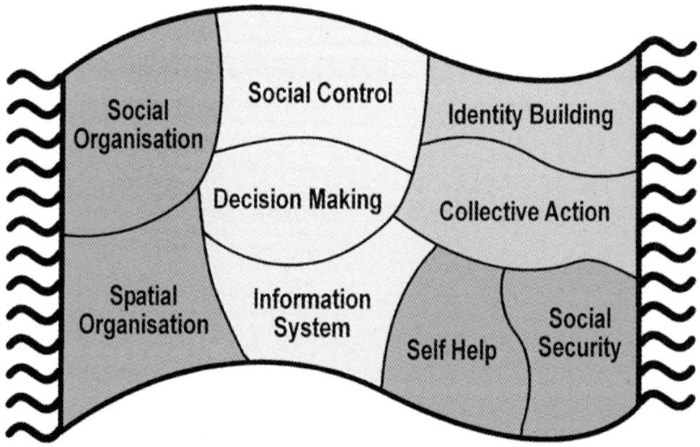

Fig. 3: The various aspects of the Ṭāṭ system represented as a "carpet", which is a literal translation of the term. As can be noticed, the concept takes over many different aspects which govern the local caste-society of the washermen. These aspects are not to be separated in their practical functioning, but point towards the complexity of the system.

A short look at the different aspects encompassed by the Ṭāṭ structure reveals its extensive nature as an adaptive regulatory system—a system that allows for a self-determined regulation of all important community matters. However, it is not the objective of this essay to elaborate in detail the various practices of caste governance and their interplay, as it is indicated in the patchwork of the carpet. Instead I will concentrate on its spatial practice since the whole concept enfolds itself in a socio-spatial framework as it is laid out below. Before doing so a brief summary of the other aspects of Ṭāṭ is in order; the comprehensive examination of these aspects takes place elsewhere (Schütte 2003).

Ṭāṭ as an Institutional Arrangement

The Ṭāṭ-system works as an extensive social institution—its various councils define rules, obligations, codes of conduct and measures of social control for the community members in the spheres of social and professional life. These are collectively agreed upon in regular meetings and any misbehaviour is sanctioned with monetary fines, or social boycott as the most severe form of punishment. Jurisdiction is carried out, decisions and judgements are made, and a fast flow of information is channelled through the community by a lucid messaging system.

Ṭāṭ as a Social Security System

The Ṭāṭ-system provides means of self-help carried out through institutionalized mutual support. Collective funding, rotating money saving schemes, credit schemes, and measures to support individual emergencies help to create a sense of security among the community members.

Ṭāṭ as a Forum of Collective Identity and Collective Action

The Ṭāṭ system as a collective device with its various layers builds up identity, dignity and self-confidence and it serves as a vehicle for collective action. People are conscious of the value of their unity, which allows them to articulate and enforce interests as well as it creates a feeling of strength among them.[11] Or in the words of an interview partner: "Ṭāṭ is above the individual but together we are working as one" (Bhargav Kannaujiā, January 4, 2002).

Making Geography Work: Spaces of Social Relations among the Washermen

Ṭāṭ defies any singular definition. Its complexity is administered through an organizing principle that pins down its essentials to a socio-spatial shape, enabling people to make use of their system. This principle is implemented by means of a spatial practice that actively creates social spaces. Dhobī construct their own multileveled localities that are, as Arjun Appadurai has put it generally, "... primarily relational and contextual rather than ... scalar or spatial" (Appadurai 1995: 204). Moreover, the localities of the washermen reflect hierarchically ordered and institutionalized spaces of social networking, in which the above mentioned aspects of the system are operated at different layers. Dhobī have developed a socio-spatial system of localities that work at the heart of their social organization and to whose layers they assign particular names. The different levels of the overall structure ranging from lower to higher strata, with each one encompassing the preceding, are summarized in the following illustration (Fig. 4).

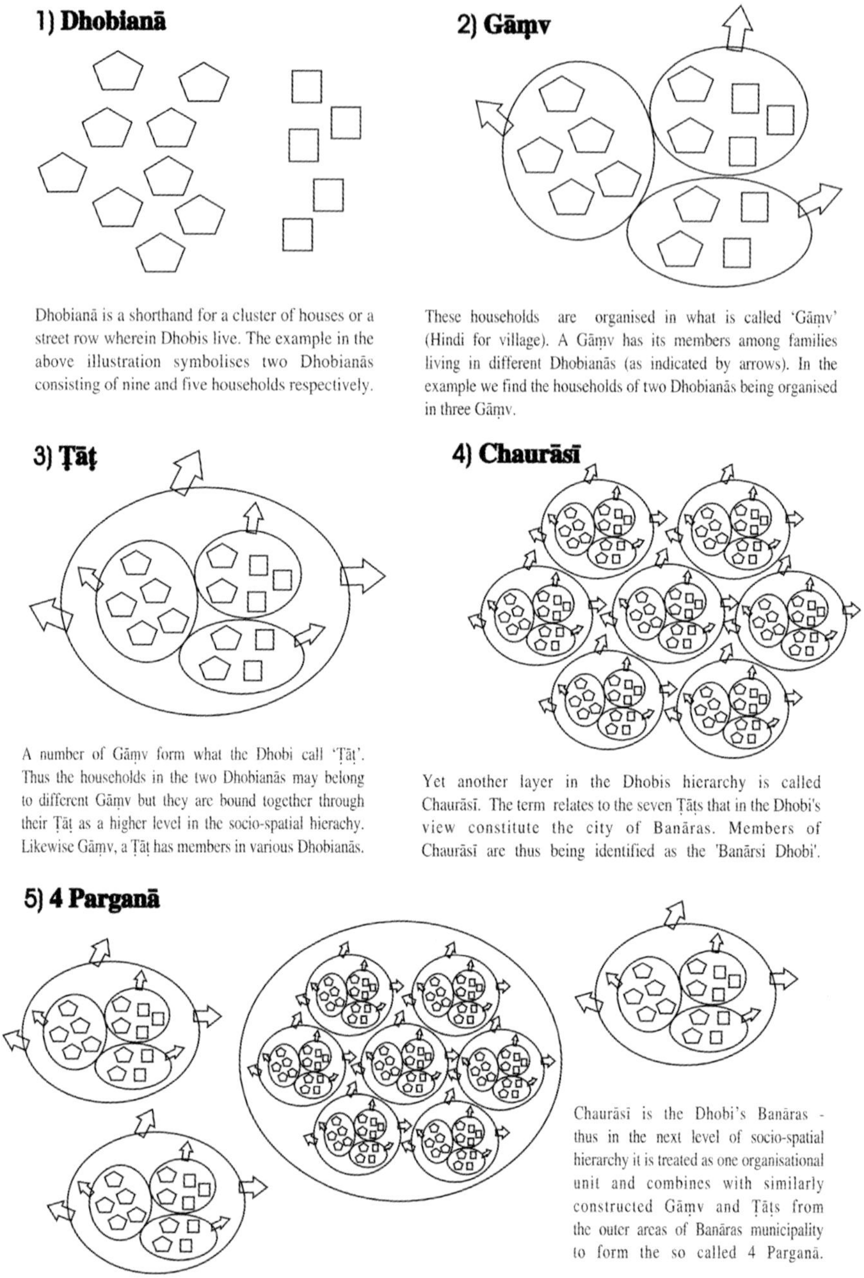

Fig. 4: A schematized model of the different layers constituting the socio-spatial system of the Dhobī in Banaras. This system is hierarchical: the lower entities add up to form higher, more encompassing forms of socio-spatial organization, though every unit also retains a sense of autonomy. (Source: own draft).

However, it has to be kept in mind that the notion of Ṭāṭ denominates the whole hierarchy of localities as well as a specific layer of the overall system.

Apart from the first level of the Dhobianā, which has a material structure and thus is easy to locate in the city's space (see Fig. 6), these spatial layers are purely social. They signify spaces of social relations, jurisdiction and economic-political contexts.

Dhobianā

Urban spaces occupied by washermen are called Dhobianā. The term refers to a cluster of houses, a street row, or sometimes a whole quarter wherein Dhobī reside. These settlements range in size between 5 and 50 households, and spread all over Banaras.

Fig. 5: Typical scene in a Dhobianā. Dhobī from Teliyānālā in northern Banaras prepare the clothes for washing. (Photo: Schütte, Dec. 2001).

A Dhobianā borrows its name from the Mohallā, the city quarter, in which it is located. Here Dhobī usually live in close proximity to different caste groups—apart from a few exceptions their settlements are socially and spatially not segregated, as it is the case with, for example, most sweeper colonies in Banaras. However, apart from being recognized by its location in physical space the Dhobianās also have their position in social space. The various locality constructions can be characterised as multiple layers of social networks that connect a certain number of households residing in different Dhobianās and thus build up well-defined "spaces of social relations".

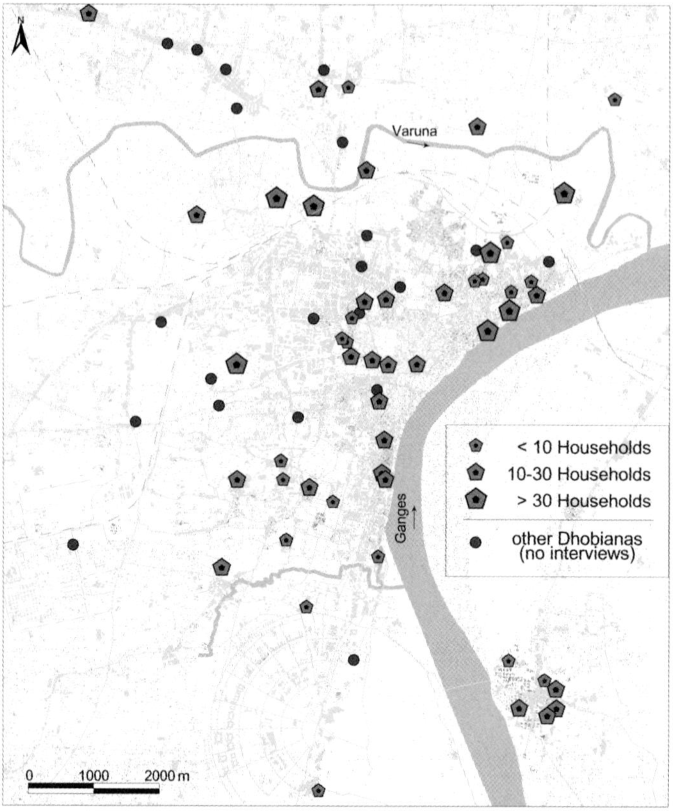

Fig. 6: Dhobianās, the residential areas of washermen, in Banaras. They take their name from the Mohallā wherein they are located and refer to a street row or a cluster of houses occupied by Dhobī. (Design: Schütte 2002, based on Benares Guide Map, 3rd Edition, Surveyed 1928–1930).

Gāṃv

Gāṃv (the Hindi word for village) is the smallest layer of social networking among Dhobī. It is a network linking groups of 1–10 households located in different Dhobianās, totalling up to 25 households which together constitute one Gāṃv. In Fig. 7 these localities are represented in a series of six Gāṃv (all located in southern Banaras) that form one Ṭāṭ as the subsequent layer in the Dhobīs' hierarchical system.[12]

Neighbouring households of one Dhobianā may belong to different Gāṃv. Thus social relations, not residential proximity, define the shape of a "village". This structure reproduces itself at higher levels of organization. Every Gāṃv is led by one chief (Caudharī) whose name serves as the identifier of the village as well, e.g. Kishun's Gāṃv, Lullan's Gāṃv or Mahangiram's Gāṃv. Attached to the hereditary position of a

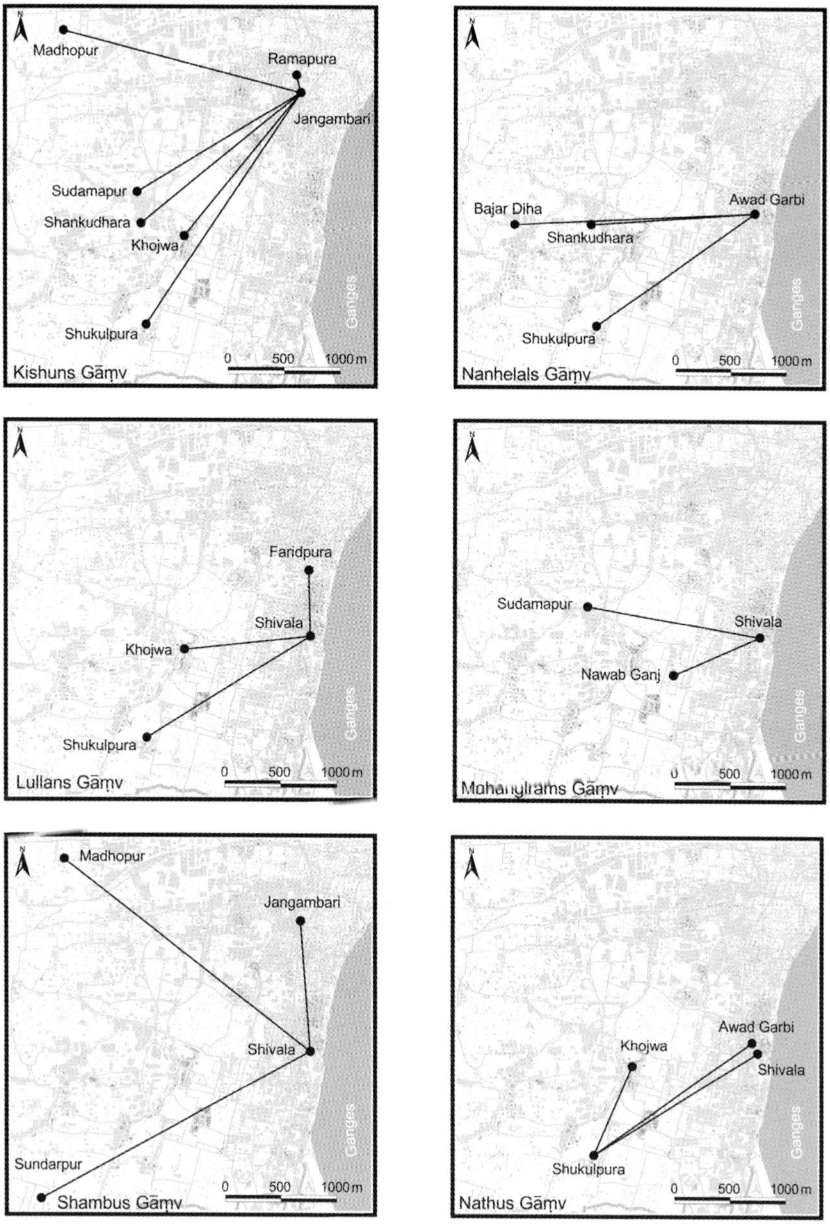

Fig. 7: The series of illustrations shows six Gāṃv with its respective members living in various Dhobianās. Every Gāṃv is named after its respective Chief, whose place of residence takes the centre of each representation. In their associated interplay these six Gāṃv form the Bhadaini-Ṭāṭ in southern Banaras (design: Schütte 2002, based on Benares Guide Map, 3rd Edition, Surveyed 1928–1930).

Caudharī are certain responsibilities: to inform the village members of important matters, to act as a representative at marriage or death ceremonies, and to preside over regular and collective funding meetings. The position of a Caudharī is ascriptive and assigned with a great deal of prestige. Their families are cultivating the role with dignity and are quite often the most elaborate spokesmen in the frequent community assemblies, though many are illiterate.

Fig. 8: Mahangiram Caudharī in front of his house at Śivālā (photo: Schütte, Jan. 2002).

The head of each household is required to attend the monthly Gāṃv meetings. This practice of coming together keeps the whole system at work, as it is during these meetings that decisions are made, specific rules discussed and enacted, penalties imposed, and money from each member collected. The meetings are regularly and strictly scheduled; absentees as well as latecomers are fined a small amount, the money is deposited in a collective cash box. But the system is composed only of men; women attend the meetings only when a respective household is headed by a female, and then only if a pressing issue needs to be discussed.

As it turns out, a Gāṃv is a strongly institutionalized network of households—there are rules and regulations to be followed that may differ in every such unit. But above all this social construction, just like the other layers of the system, serves to maintain a sense of unity and security. Its traditional roles of supporting marriage festivities, and to maintaining, e.g., a centralized storehouse of kitchen utensils for any such solemnity is still acknowledged. But today people are also attempting to extend the traditional functions of a Gāṃv: money saving for special events, information exchange (e.g. about reservation quotas, overall political situation, marriage, gossip) and regular activities like picnicking and going outside[13] are efforts to keep the system going under modern conditions. Further, the matter of unity ("Dhobī-Ektā") in the community is an attribute very often referred to by people. Solidarity and collectivity is understood as the base of the system, as the informant Lullan Caudharī stated in January 2002 during a meeting: "Only together can we bring our community forward. In our Gāṃv and our Ṭāṭ we discuss and decide any matter together, we go together on the street and demonstrate, we celebrate and we suffer together, we help each other and sometimes we quarrel—but our unity is our strength". Any threat to unity is rigorously punished. Members who do not follow the rules or misbehave are fined or, as the most severe means of sanctioning, are temporarily (or permanently) excommunicated.[14]

Ṭāṭ

The term Ṭāṭ denotes a specific and well-defined layer as well as the overall system of the washermen. In forming the council of various Gāṃv, the organizational pattern of a particular Ṭāṭ reproduces the structures as they have been described in respect to Gāṃv on a higher level. But it is the Ṭāṭ level at which the Dhobī traditionally find their place in the world. For many it is still the main reference point, rather than the physical city quarter in which they live, the "Mohallā", and it thus carries a high identificatory potential. People literally belong to a Ṭāṭ, they are subject to its rule and by means of regular social interaction continuously reproduce its structures and inform its regulations. But Ṭāṭ as locality is not really a mark in material space; it denotes a network of Gāṃv, led by its respective chiefs and combined of different households that may live in several localities.

Fig. 9 shows the Dhobianās of Banaras as they are distributed according to their prominent Ṭāṭ affiliation. The names of these social entities are in some cases derived from city quarters (like Cetgañj, Bhadaini or Teliyānālā, among others), which indicates that the Ṭāṭs have their origin in certain areas of Banaras. The rise in population and residential mobility have subsequently widened the extension of a specific Ṭāṭ-rule, as movements in physical space do not affect the membership in a certain Ṭāṭ. Other Ṭāṭ-names like Tirvani, Kampu or Bispaṭṭā do not directly refer to a Mohallā and their origin is difficult to trace, though they correspond to a certain area or a specific place.[15] The relation between Ṭāṭ and Mohallā is probably best explained by Mahangiram Caudharī (Dec. 2001): "Ṭāṭ is an alliance of Mohallās, and the Mohallā is the place where my house stands".

However, the representation in Fig. 9 does not take into account that in some Dhobianās members of various Ṭāṭs may live. Every household is exposed to the institutional arrangement of its particular Ṭāṭ. A specific social order is thus conducted which demonstrates that Ṭāṭs with their inherent clusters of Gāṃv are well-defined localities as perceived by the washermen. They are distributed quite evenly over the city's space and are reproduced again and again through meetings, since the members of each Ṭāṭ assemble at least monthly or even more frequently (see Fig. 10).

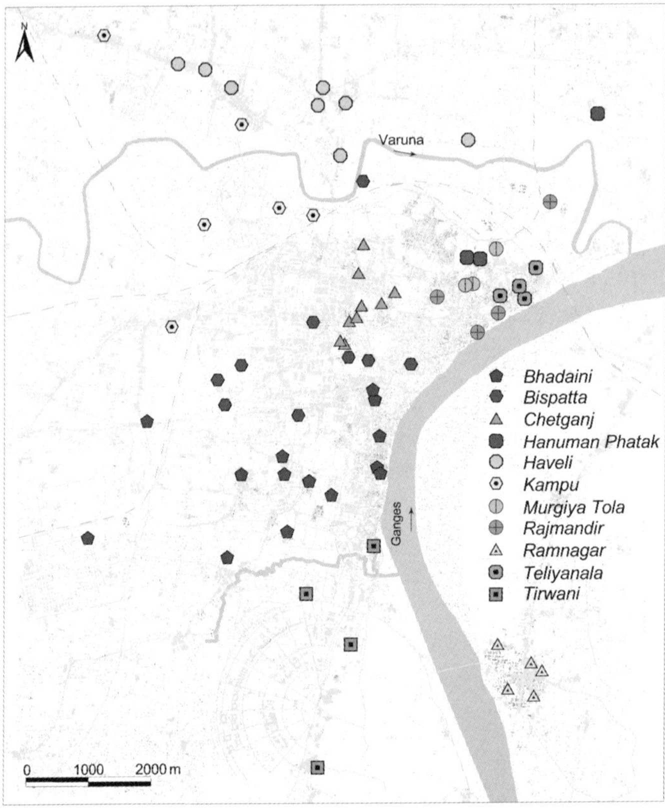

Fig. 9: The representation shows the Dhobianās of Banaras municipality marked according to their most prominent Ṭāṭ affiliation. Thus, Ṭāṭ relates a group of Dhobianās socially and spatially, e.g. Bhadaini is in the south and Bispaṭṭā in the centre of Banaras (design: Schütte 2002, based on Benares Guide Map, 3rd Edition, Surveyed 1928–1930).

The assemblies gather not only on a regular basis to discuss actual and general issues (such as professional difficulties, the availability of new government schemes or the organization of rallies and festivities). They may also be approached in case of personal conflicts between families to find a solution which is collectively agreed upon. Most negotiated cases are about marriage problems, divorce, violence and property disputes, but also about accusations of inflicting ghosts or supernatural beings ("*bhūt-pret*") on somebody, which is treated as a serious matter. In that sense the Ṭāṭ assemblies serve as an internal jurisdiction. Rather than approaching the courts people try to solve their problems in the community meetings.

These meetings are enacted with considerable emotional intensity and are guided by the so-called "Barikh"—a sort of "chairman" who calms down the discussion if it becomes too heated, and who designates who will speak at any given time. The general procedure of conflict-solving follows a certain pattern: the opponents both stake their claims and describe the problem from their viewpoint. During the discussion that follows, each member is allowed to voice his opinion. After the debate, the Barikh suggests a solution; one party may be fined, both may be found guilty for the problem and be penalized equally, or the decision may be postponed to be discussed again. However, it is necessary that the opposing parties both agree on the suggested solution. If that is not the case the matter will be handed over to a higher level of decision making, i.e. to the assembly of all Ṭāṭs in Banaras commonly known as Caurāsī. Finally at the end of each meeting the date and time for the next gathering will be discussed and announced by the Barikh.

Caurāsī

The unit of Caurāsī (Hindi for "84") comprises the seven Ṭāṭs which collectively identify the "Banarsi Dhobī".[16] Caurāsī is a regularly used platform of information exchange and higher jurisdiction. Here rules are enforced that cover the whole town. A famous example is the introduction of a holiday on the 11th in every month, where no Dhobī is allowed to work. Instead this day is used for conducting meetings, to keep money saving schemes going[17] or to go out for picnicking. Somebody caught working on an 11th is fined, and there used to be even a kind of "patrol" controlling the washing *ghāṭs* to ensure that the "holiday-rule" was observed.

Caurāsī assemblies also gather to coordinate collective action such as protest marches and strikes. One such strike occurred last in 1998 when a number of washermen were arrested by the police for washing in the Ganges River. Caurāsī assemblies are used to inform members of proposed activities and to discuss the rules of festivities. As indicated above, they can be approached if lower level meetings could not solve the problem. In principle, Caurāsī meetings where juridical affairs are debated follow the same pattern as has been described for the Ṭāṭ-level. And again, if there is no agreement, another hierarchical level has to be approached, the 4 Parganā.

4 Parganā

Caurāsī resembles what in the view of washermen belongs to Banaras. Because of this evidence it is treated as one organizational unit in the next level of the Dhobī's organizational hierarchy. The 4 Parganā thus include Caurāsī and three more urban-based units as a decision-making unit (see Fig. 9, where the Dhobianās belonging to Kampu,

Haveli and Ramnagar are depicted). The term Parganā derives from the times of Mughal rule over North India (see Cohn 1987: 489f.). During the Mughal period it referred to revenue subdivisions; the British took over this terminology from their predecessors, and today Parganā survives as part of the Dhobī system. If approached, this assembly decides about unresolved juridical issues. It does so in the usual way, but because it reaches across Banaras proper, the institution of 4 Parganā is also used to distribute relevant information (about government schemes, proposed actions, festivities and the like) and to promote the process of exchange and discussion within the entire urban community.

Fig. 10: People from Bhadaini Ṭāṭ hold a meeting at an interior washing-*ghāṭ* (Bhavānī Tālāb) near the waterworks. The meeting took place on February 15th 2002, just a couple of days before the state elections in Uttar Pradesh. That specific meeting was used to discuss the possibilities of a collective vote. The Barikh collects a five Rupees fine from each late-coming member and the name of each absentee is recorded to penalize him afterwards (photo: Schütte, Feb. 2002).

In that way the organizational hierarchy reflects a distinct spatial order. Starting from a group of households (Gāṃv) to a number of these groups (Ṭāṭ) in certain parts of town and finally stretching over the entire city of Banaras (Caurāsī) and its outer areas (4 Parganā), the various levels of the Dhobīs' system successively reach the whole urban community. Through Ṭāṭ washermen are collectively organized in a multi-layered institutional framework—but yet another level binds also the rural areas of Banaras District in the decision-making process.

Bāvanī

Bāvanī (Hindi for "52"), the Dhobīs' "High-Court", assembles for the superior decision-making in any unresolved matter. It reflects the rural-urban linkage connecting washermen from the countryside of Banaras district, the latter independently organized in another 14 Parganā, to their counterparts in the urban-based 4 Parganā. Each of these 14 rural units consists of a number of villages in the district but also certain parts of Banaras municipality.[18] During a Bāvanī meeting on February 3, 2002, around 800 washermen from the entire Banaras district assembled near Ramnagar to find a solution to a marriage problem concerning two families. A girl refused to live in her husband's house, and her father accused the bridegroom's family of having "thrown a ghost" on her. An "Ojhā" (a person capable of controlling ghosts and exorcizing them from bodies and houses) had been employed and neighbouring families were assigned to keep an eye on the further course of marriage life and to report to their assembly in case of any mishaps.

After handling that particular matter, the meeting was used to discuss political strategies. The effects of a recent protest march to the state capital of Lucknow were evaluated and further collective activities proposed, which points to the fact that the platforms of traditional organization have become political arenas as well.[19] This socio-spatial layering in an all-embracing network, forming hierarchical and at the same time autonomous entities in the shape of Ṭāṭ and ultimately uniting the whole Dhobī community of Banaras district finds no comparison among other low caste groups in Banaras.[20] Somehow Dhobī have been able to preserve a mode of internal organization, whose roots reach back at least to the 19th century but probably even to earlier times.[21] It is obvious that the traditional form of Ṭāṭ today still has its value and that it is able to adapt to changing circumstances. Though sometimes referred to as being a "system for illiterates" (Ramu Kannaujiā, 19th Dec. 2001) that compensates for a lack of education through collectivity and local self-organization, Ṭāṭ is increasingly being used by educated members to agitate and unite the caste also in political issues.

Ṭāṭ as the Property of Social Life

After having assessed the components of the Dhobi system it is easily comprehended that Ṭāṭ is a social institution which is constricted by caste boundaries. Ṭāṭ serves caste interests, and organizes daily life in a subordinate community. It handles problems arising from the Dhobī profession, passes judgements in conflicts among community members, decides what appropriate public behaviour is, and determines how to arrange the basic stages of life, such as marriage and death rituals.[22] Similarly, Ṭāṭ promotes forms of resistance such as protest against perceived injustice and increasingly directs political action. It brings people together to speak with a common voice. This institution of social life has a partial autonomy. Every Ṭāṭ has its slightly different set of rules and thus independently conducts its respective institutional regime. In relation to Banaras urban society as a whole the Ṭāṭ system appears as a local sub-society that works by self-governance inside the overall framework of the urban social structure.

The social organization thus resembles a twofold geography of power: Internal power relations may shape the explicit layout of rules, create leadership and an obligatory regime that binds the community into certain patterns. In relation to urban society beyond their caste the Dhobī exercise the power to handle their lives rather self-determinedly and are able to oppose any threats emerging from outside—be it eviction from the riverbank, police violence or injustice to individual community members.

Sure enough, the picture drawn here may appear somehow too perfect, and quite naturally there are frequent internal disagreements and conflicts that may take a long time to solve, both between individual members and entire Ṭāṭs. More substantially, the process of social change will possibly modify the system altogether. Translated into the vocabulary used by Rudolph & Rudolph (1969), Ṭāṭ can be regarded as the sophisticated form of an "ascriptive association", but it is already on its way to turn into a so-called "paracommunity" (*ibid.*: 33). The activities of the "U.P. Dhobī Kalyāṇ Mahāsabhā" in Banaras, voluntarily organized, politically oriented and trying to form a Dhobī-Union covering the whole of Uttar Pradesh, have increasingly encroached on the traditional system in the last couple of years and thus put forward a contested politicization of their caste (see Schütte 2003).[23] But still, Ṭāṭ as the traditional way of organization up to now resists disappearing. It serves as a marker of identity because Dhobī strongly associate it with positive measures of unity, strength, community life, prestige, security and collectivism. It therefore contributes to a feeling of self-confidence: due to their strong organization Dhobī feel no need to downplay their disrespected profession and they are able to enforce own interests. Nevertheless, washermen increasingly switch over to other occupations because the future of washing seems at least precarious, and naturally the hard working life is by no means appreciated but felt as a given necessity. However, many working Dhobī accept their working conditions and struggle for improvements while their system provides a type of social and economic security.

But what are then immediate reasons for the prevalence of Ṭāṭ? As indicated, the subjective value attached to it by community members is essential for its survival. Likewise the regularity and permanency of the meetings contributes to its continuance. But even so, I would argue that the general openness for political questions and its modification to a network increasingly directing collective political action is even more important. For here a change from sheer local self-organization of caste defined issues to the invocation of a wider public has taken place. Dhobī raise their voice not only in their caste councils, but they enter the streets to make themselves heard to a wider audience; they are approaching a different stage that reaches beyond their local necessities. This process definitely has been accelerated by the Mahāsabhā, whose arguments are willingly accepted and translated into collective action, so far without making obsolete the traditional functions of the system. As such the concept of Ṭāṭ may well be conceived of as a form of social capital that is deployed to stake claims, to regulate social life, to establish a safety net and in its modified shape of the Mahāsabhā to advance the overall position of the washermen in the social, economic and political environment of urban Banaras and northern India.[24]

Conclusion: Ṭāṭ as Social Spatiality

In the beginning it was briefly outlined that the theoretical notion of social spatiality is a possible way to integrate space in social theory. It is not the place here to dwell deeply upon this discussion but instead to draw on the theoretical importance of the washermen's socio-spatial practice in this respect. A shorthand definition will clarify a few points:

Ṭāṭs are social and spatial entities, which are perceived as such only by Dhobī. They correspond to different areas, wherein Dhobī established a self-determined institutional regime that only has importance for them. Spatial perception is thus shaped and defined in relation to social organization and social context.

There are two important clues in this definition:

— Space refers more to a network of social relations and less to a bounded physical area;

— Space is defined and conceptualized in accordance to social structure; that means boundaries are social, not spatial, and perceptions are constructed via processes of social positioning and affiliation and have thus particular social forms.

That again implies that space and spatiality are not any given precondition determining everyday life and human agency. Rather they refer to a model of perceiving and arranging the world, a model which is actively and collectively constructed and therefore is based on communicative processes. In the washermen's system these processes are constantly created and re-created through social interactions—on the everyday level, where set rules are lived by and members of Gāṃv and Ṭāṭs regularly come together in the course of their daily chores as well as at various meetings. Social space, thus, is formed out of "stretched out" social relations, as Doreen Massey (1995: 54) has argued. Naturally this is, as any model of the world, subject to change and has constantly to be communicated and protected by measures of sanctioning in order to reproduce it. But Ṭāṭ is also a hybrid construction, thus virtually anticipating what currently constitutes the postmodern discourse in geography and other social sciences. It does so by representing a bounded locality of institutional rule, thus referring to the well-defined spaces of jurisdiction, but it also constructs the virtual spaces of social networking that are maintained to link the community over sometimes large areas and connecting people living far away from each other.[25] Thus Ṭāṭ has a solid material base in assigning specific areas of interaction, but at the same time, and even more so, it demarcates the social spaces of the washermen, defined on the basis of social relations spreading over the whole Banaras district.

Another general finding which can be derived from the Ṭāṭ system pertains to the question of boundaries: a socio-spatial order requires borders to delimit a difference. These are usually fluid and porous, but they are always constituted socially. This is not altogether a new insight and holds true for any kind of boundary one may think of (see Massey 1995: 67f.); yet in the case of urban India the actual drawing of boundaries that separate social and physical spaces points directly to the institution of caste as the central element of structuring society on the subcontinent. The Ṭāṭ system belongs exclusively to the Dhobī from Banaras—its boundary is a caste line, which is continuously reproduced by it. But this line denotes spaces of social subordination as well: washermen are strongly organized in that specific form also in response to a low educational and social status and economic hardship; it enables them to lead a life in which people collectively protect themselves from disrespect, exploitation and insecurity. Seeing it this way, the social transactions directed by Ṭāṭ reflect the needs, fears and aspirations of a community that is continuously struggling to maintain its place in both geographical and social space. The Dhobī's place in the city of Banaras needs to be fought for, and this is made possible via Ṭāṭ as a social system and a social locality, where power and prestige are expressed and key issues are debated.

Notes

1 I thank the German Research Foundation for providing substantial funding to the project and my field assistant Mr. Hemant Sarna, without whom the daily fieldwork would have been difficult.

2 Dhobī or washerman is an aggregated term that points solely to the profession. Internally the group is divided into numerous endogamous sub-castes. In Banaras almost all Hindu Dhobīs carry the name "Kannuajiyā" as their caste title. In addition to them approximately a hundred families of Muslim Dhobī are also living in Banaras.

3 This is generally the case, although in a few neighbourhoods of Banaras certain temples and tea-shops are still "forbidden spaces" for Dhobī.

4 The Dhobī population of the federal state of Uttar Pradesh is estimated at around 2.2 million (*Dainik Jagran*, October 8th 2001).

5 Washermen mostly take a bicycle to move the package of clothes. They previously kept donkeys for that purpose, but today only a few families in Banaras retain that tradition.

6 Naturally there are exceptions and variations of that routine—people sometimes concentrate only on ironing or have been able to establish a dry cleaning shop. There are also seasonal varieties, as during the monsoon washing in a flowing river is almost impossible. That leads to overcrowded interior washing *ghāṭs*, and many families have decided to stop washing altogether in the rainy season.

7 Hindu deities worshipped during specific festivals are Ghāṭvārin Devī, the goddess residing in the washing stone at the *ghāṭs*, whose celebration with sheep sacrifice takes place yearly in the Hindu month of Śrāvaṇ on many Dhobī *ghāṭs* in town, and Phulmatti Devī, celebrated during Navarātra and also associated with animal offerings.

8 A Dharamśālā is a pilgrim hostel—there are hundreds in Banaras, and two of them are reserved for the Dhobī community in Banaras.

9 The term is generally used as a signifier for the whole concept, though it also refers to a specific layer of the system.

10 These caste councils are often referred to as "*pañcāyat*". Blunt, however, does list the Dhobī as being one of the numerous groups having such a council, but he is not relating the term Ṭāṭ directly to them. See also Dumont (1998: 172f.) describing generally the practice of internal caste government in his "Homo Hierarchicus" and Lynch's elaborations (1969: 182f.) on the fading urban system of councils among the Jatav in Agra called "Thok".

11 A recent incident was a huge protest march to the city's magistrate house in 1998. People were claiming their rights to wash at the Ganges Ghāṭs which had been banned by the municipality shortly before. A ten-day strike followed and finally Dhobī were able to retain their claim to wash

in the Ganges as long as the government is not supplying alternatives in the form of washing *ghāṭs* inside the city.

12 Some Ṭāṭs are very small and do therefore not sub-organize themselves into Gāṃv.

13 Every couple of years or so members of Gāṃv or small Tāṭs celebrate a three day holiday with picnicking events for the whole family, the expenses for it being paid out of the collective cash box. Apart from this activity in the circle of a Gāṃv the culture of "going outside" (see Kumar 1988) generally is very popular among Dhobī.

14 This practice is called "Hukkā Pāni Band", meaning that the affected person is deprived of using pipe (*hukkā*) and drinking water (*pāni*) with his caste fellows. He will not be allowed to have any kind of social interaction with anyone from his community until the imposed penalty has been requited.

15 Kampu for example is congruent with the British Cantonment and Bispaṭṭā is said to have been the name of a hospital long before. The latter especially seems to be known only by Dhobī—no other person could locate the term, and even the most experienced Riksha-drivers never had heard of it.

16 The question remains how the number 84 came to symbolize a specific caste-defined perception of Banaras, a perception that excludes large parts of the present municipality. No Dhobī could explain, as the name has been passed down over many generations and its origin has been forgotten. Its meaning, though, is common knowledge among the washermen. An entry in William Crooke's "Rural and Agricultural Glossary" (1888: 62) helps to clarify the term. It refers to "Caurāsī" as "the whole brotherhood in castes that have a pañcāyat"—so that the term must have been used widely by many traditionally organized caste groups in the 19th century. This is congruent with the account from Lynch (1969: 190) about Jatav in Agra, whose city assembly has also been called Caurāsī.

17 Washermen operate what they call their "community bank": Representatives from a number of families (ranging from 50 to 200) gather monthly on the 11th to deposit a fixed amount of money (between 200 and 500 Rs) at one place, which then will be distributed among two or three participants of the scheme. This is common practice among washermen, leading to a considerable amount of cash at hand. More than twenty such self-organized banks cover almost every Dhobī household in Banaras.

18 See on Fig. 9 the Dhobianās of Tirvani, which is one of the 14 Parganā.

19 The growing political awareness of the Dhobī community in recent times is reflected through the formation of the "U.P. Dhobī Kalyāṇ Mahāsabhā (Union for the welfare of the Dhobī community in U.P.) in Banaras (founded 1999), which is led and directed by educated members of the caste (see Schütte 2003).

20 Recently attempts are being made by educated members of the Dom-Sweeper caste (also listed as Scheduled Caste) to organize their community in Banaras. In an invitation letter they explicitly refer to the social progress of other "untouchable" groups due to their united appearance in public. Up to March 2002 two preliminary meetings with representatives from all Dom-Sweeper colonies in town have been conducted to discuss how the community may find compulsory ways to get closer together and improve their living standards.

21 See Heitler (1972) for considerations with regard to the early existence of sophisticated organizational forms among many caste groups in Banaras. The effects of it turned out publicly during the so-called "Varanasi House Tax Strike" in 1810–11. Though Heitler does not mention the Dhobī in his article, it seems very much likely that they have participated in the event as well.

22 There are different customs in different Ṭāṭs: e.g. the Dhobī from Kampu bury and not cremate their dead.

23 At the time of field research the Bhadaini-Ṭāṭ had already split with the Caurāsī for more than one year. Its members are accusing the Mahāsabhā, which is strongly supported by the other six Ṭāṭs of Caurāsī, of involving the community too much in party politics. A deeper reason, however, can be located in leadership conflict.

24 Mohan and Mohan (2002: 197) raise the question if there can be a geography of social capital. One may be tempted to interpret the washermen's system as an example of such a geography, which localizes varying degrees of associational life in the social spaces of their Ṭāṭs and Gāṃv.

25 In drawing on Srinivas' work in South India, Louis Dumont (1998: 175) mentions the "immense area of jurisdiction" covered by the washermen in Mysore, who, by being represented only in small numbers in each village, inevitably had to spread their jurisprudence over larger areas than other groups in order to maintain the practice.

References

Appadurai, Arjun. 1995. The Production of Locality. In: R. Fardon (ed.), *Counterworks—Managing the Diversity of Knowledge*. London: Routledge, pp. 204–225.

Bebbington, Anthony & Thomas Perreault. 1999. Social Capital, Development, and Access to Resources in Highland Ecuador, *Economic Geography* 75, 4: 395–418.

Blunt, E. A. H. 1969 (1931). *The Caste System of Northern India with Special Reference to the United Provinces of Agra and Oudh*. New Delhi: S. Chand & Co.

Cohn, Bernard S. 1987. *An Anthropologist among the Historians and Other Essays*. New Delhi: Oxford University Press.

Crooke, William. 1888. *A Rural and Agricultural Glossary for the N.-W. Provinces and Oudh*. Calcutta: The Superintendent of Government Printing, India.

Dainik Jagran (08.10.01). *Reservation for Backward Classes*.

Dumont, Louis. 1998 (1970). *Homo Hierarchicus. The Caste System and its Implications*. New Delhi: Oxford University Press.

Gregory, Derek. 1994a. Social Theory and Human Geography. In: D. Gregory, G. Martin & R. Smith (eds.), *Human Geography. Society, Space and Social Science*. London: Macmillan, pp. 78–109.

Gregory, Derek. 1994b. *Geographical Imaginations*. Oxford: Blackwell.

Heitler, Richard. 1972. The Varanasi House Tax Hartal of 1810–11, *Indian Economic and Social History Review* 9, 3: 239–257.

Kumar, Nita. 1988. *The Artisans of Banaras: Popular Culture and Identity, 1880–1986*. New Delhi: Orient Longman.

Läpple, Dieter. 1993. Thesen zu einem Konzept gesellschaftlicher Räume. In: J. Mayer (ed.) *Die aufgeräumte Welt—Raumbilder und Raumkonzepte im Zeitalter globaler Marktwirtschaft*. Rehberg-Loccum: Evangelische Akademie Loccum, pp. 29–51.

Lynch, Owen M. 1969: *The Politics of Untouchability. Social Mobility and Social Change in a City of India*. New York/London: Columbia University Press.

Massey, Doreen. 1995. The Conceptualization of Place. In: D. Massey & P. Jess (eds.), *A Place in the World?* Oxford: Oxford University Press, pp. 45–85.

Mendelsohn, Oliver & Marika Vicziany. 2000. *The Untouchables. Subordination, Poverty and the State in Modern India*. New Delhi: Cambridge University Press.

Mohan, Giles & John Mohan. 2002. Placing Social Capital, *Progress in Human Geography* 26, 2: 191–210.

Rudolph, Lloyd I. & Susanne H. Rudolph. 1969. *The Modernity of Tradition: Political Development in India*. New Delhi: Orient Longman.

Schütte, Stefan. 2003. *Soziale Netzwerke als räumliche Orientierungssysteme. Konstruktionen von Raum und Lokalität der Wäscher von Banaras*. (Vol. 23 of "Studies in Development Geography", edited by H.G. Bohle and T. Krings). Saarbrücken: Verlag für Entwicklungspolitik.

Simonsen, Kirsten. 1996. What kind of space in what kind of social theory? *Progress in Human Geography* 20, 4: 494–512.

Soja, Edward. 1999. Thirdspace: Expanding the Scope of the Geographical Imagination. In: D. Massey, J. Allen & P. Sarre (eds.), *Human Geography Today*. Cambridge: Polity Press, pp. 260-278.

MARTIN GAENSZLE
(IN COLLABORATION WITH NUTANDHAR SHARMA)

Nepali Places: Appropriations of Space in Banaras

The image of Banaras as a sacred city of the Hindus—a "stronghold of Hinduism" (cf. Cape 1909)—has been rightly analysed as being a historical construction (see, e.g., Visuvalingam & Chalier-Visuvalingam and Freitag, this volume) and as such it is one-sided: several other religious traditions are also present and have left their mark on the city—especially Islam, but also Buddhism, Jainism, and more recently Christianity. But Banaras is not only a multi-religious place, it is in fact also a multi-ethnic, or multi-regional, and thus multi-cultural one, and has been so for long. As is well known, much of South Asia is represented in the city: there are Bengalis, Gujaratis, Maharashtrians, and many others (Singh 1996). It might even be argued that Kāśī was cosmopolitain long before the term became fashionable in the west: a cosmopolis in the literal sense of a "universal city".[1]

Though this paper also deals with the Hindu aspect of Banaras, we will focus on one particular community, the Nepalis, and try to show the ways they have appropriated the city and its spaces in their own distinct manner.[2] By appropriation we mean: the "making one's own" of spaces by giving them special meaning, mainly through cultural practices and ideas, thus turning them into domains which are marked as "familiar". In some cases, appropriation included control in terms of property rights, but this is not necessarily the case. As we will try to show, this appropriation has to be seen in a historical context, since the process has undergone significant change in the course of the last two hundred years, mainly due to the varying role which the formation of nation-states has played. Many of the acts of appropriation had to do with public spaces, temples, *ghāṭs* etc., but, as will become clear in the course of the argument, there was an increasing tendency to create what we will call "semi-public" spaces: spaces which are primarily meant for members of the Nepali community.

One might say that urban space in Banaras consists of three kinds of public places:

1. The *ghāṭs*, i.e. the riverfront with its unique architecture of embankment, stairs and temples. Here people meet for bathing, washing, worship, leisure and socialising.

2. The temples. Though there are many private temples, most are public and many are frequented on a daily basis. Some of these temples are also important meeting grounds and venues for collective celebrations, especially during festival days.

3. The lanes (*galīs*) in the quarters (*mohallās*). Much of public life takes place on the "streets", especially the small squares (*cauk*) which can be found in the centre of most *mohallās*.

Fig. 1: Location of the Nepali neighbourhood (drawing Niels Gutschow).

We will look at these three public spaces by focusing on three aspects of the Nepali presence in Banaras: 1. the tradition of pilgrimage; 2. the so-called "Nepali temple" at Lalitā Ghāṭ; and 3. the Nepali locality around Dūdh Vināyak (Fig. 1). As we go along we will provide further historical background.

The Tradition of Pilgrimage and Donations

The peoples in the Central Himālayan hills have long valued Kāśī as a place for pilgrimage. Most of the written historical evidence, of course, deals with kings and noblemen, but this may be taken to reflect a more wide-spread practice. It is likely that pilgrimage down to the plains became popular already during the first millennium, following the general trend in Hindu folk religion. In any case, in the second millennium we find more and more evidence that this practice was also cherished by the nobles in Nepal, who made generous donations to local temples and priests. In the following we will give some examples (see also Gaenszle 2002).

One inscription dating from 1576 states that Luṃgu Bhāro, a noble from Lalitpur (Patan), established an Umāmaheśvara icon at the Siddheśvarī temple in Banaras in the name of his son Deva Bhāro, who had died in a landslide. Before doing so he performed a fire sacrifice there. This icon can still be seen today in the premises of the temple in Siddheśvarī *mohallā*. After coming back to Nepal, the donor also built a temple for Gajendramokṣa Nārāyaṇ in Tuṃ Bahal, Lalitpur (Vajracarya 1999 [2056 V.S.]: 35–36). This temple still exists, though the icon has been stolen. The visit to Banaras, in this example, appears as a pious pilgrimage for the performance of death rites (*śrāddha*).

Newar Malla kings, too, travelled to Kāśī, among other places in Northern India, on pilgrimages. Again the major purpose for visiting these places was to commemorate the dead. As is written in the Bhāṣa Vaṃśāvalī, the king of Lalitpur (Patan) Siddhinarasiṃha Malla went on an extended pilgrimage in N.S. 772 [1652], prior to which he in fact relinquished his throne in favour of his son (Slusser 1982: 201). He further visited other sacred places, such as Badrinath, Prayag and Gaya (particularly known as sites for ancestral worship, *śrāddha*), and also Mathura, Puri and Rameshvaram in the south (Lamsal 1966 [2023 V.S.]: 64).

As is widely known, the Gorkha king Prithvi Nārāyaṇ Śāh went on a *kāśīyātrā* in the mid-18th century (probably in 1743) before starting his campaign to "unify" what was later to become Nepal. According to the account given in the Gorkhāvaṃśāvalī he performed *tīrthaśrāddha*, took ritual baths and made various donations.[3] He also is said to have performed all the usual pilgrimages (*pañcakrośī, antargṛha*, etc.). During his stay, we are told, a ritual called *khaḍgasiddhi* was performed for his benefit: it endowed his sword with special powers. Other sources recount that the king purchased guns of superior quality during this stay (Stiller 1989: 27).

By the late 18th century the king of Nepal had issued documents to certain Pandas (local Brahmans), authorizing them to act as pilgrimage priests for the Nepali pilgrims.[4] The successors of these priests still proudly call themselves Lālmohoriyā Paṇḍas (due to the royal red seal) and try to exert their ritual monopoly over the pilgrimage activities of Nepali pilgrims at Maṇikārnikā Ghāṭ, where they have posted a sign which advertises their services (Fig. 2). It reads: "*Kāśīstha tīrtha purohita—Nepālī Lālamohariyā Paṃḍājī ko Ghāṭa—Maṇikārṇikā Ghāṭa Kāśī—samasta Nepāliyo ke ekmātra tīrthapurohita*. [Kāśī pilgrimage priest—The Ghāṭ of the Nepali Lālmohoriyā Paṇḍa—Maṇikārṇikā Ghāṭ Kāśī—The only pilgrimage priest for all Nepalis]".

Another famous episode is the exile of Raṇa Bahādur Śāh (1775–1806), the grandson of Prithvi Nārāyaṇ Śāh, who resided in Banaras (from 1800 to 1804) after abdicating in favour of his son and becoming an ascetic called Svāmī Nirguṇānanda. Following the death of his beloved queen Kāntivatī, he had become a risk to the government, and fearing detention, he left for Kāśī. He lived there with his entourage, carefully watched by the British, with whose help he hoped to regain power. It seems that he was torn between leading the life of a renouncer and that of a king. According to local legends he stayed for some time at Lalitā Ghāṭ. In these legends he is depicted as a pious man who was generous in his donations and meditated facing the Gaṅgā.

Fig. 2: Signboard of Lālmohoriyā Paṇḍas at Maṇikārṇikā Ghāṭ (photo by Martin Gaenszle).

It seems that by the turn of the 19th century Kāśī was not only an important pilgrimage site but also a place to which the royal elites had a special relationship: they maintained permanent contacts (increasingly also with the British), and had envoys living there. For example, Paṇḍit Raṅganāth, *rājguru* at the royal court, stayed in Banaras for 30–40 years. During that time he gave a grand donation to the Ḍom *rājā* at Maṇikārṇikā Ghāṭ, and made an agreement which absolved the Nepalis who came for cremation rites from the obligation to pay the death tax—for all generations to come.[5] This may be seen as a further step in the Nepali appropriation of the riverfront.

The point to emphasize here is that during all these centuries the kings and other pilgrims from the hills regarded Kāśī as part of their own tradition. Of course, the place was controlled by local elites, local *mahārājās*, local Brahmans, and eventually the colonial rulers. But initially it was not seen as in a foreign country, in the modern sense of a foreign nation. Rather it was on the periphery of the realm, a true "centre out there" (in Victor Turner's sense: i.e. a respected high centre), and only gradually did it turn into a place on foreign, and increasingly hostile, territory.

So there is a longstanding tradition of visiting Kāśī: for *tīrthaśrāddha,* for *tīrthayātrā,* for *Kāśībās* and/or for engaging in other business. As a current proverb has it: "*Paśupati ko jātrā, sidre ko bepār* [When doing the Pashupati pilgrimage, one can also sell some fish]". This tradition continues until today: as we could observe during our fieldwork, every winter season several thousand Nepalis come down to Kāśī, visit the major temples and make particularly great use of the *ghāṭs* along the Gaṅgā. Our preliminary figures indicate that the majority of pilgrimage groups come for *tīrtha-*

śrāddha (though they do not represent the majority in terms of individuals). A large percentage are high caste (mainly Brahmans), many come from the Tarai region.

Fig. 3: Buddha with *dhāriṇī* verses at Lalitā Ghāṭ (courtesy Niels Gutschow).

But one of the more interesting findings is that this is not simply a Hindu tradition. There are still a considerable number of Newars from the Kathmandu Valley, and many of them Buddhists, who come for a pilgrimage to Banaras (approximately 14% of the groups were Newar, usually mixed, though one was entirely Buddhist). We even saw Buddhist Newars doing *śrāddha* at Maṇikārnikā Ghāṭ with their own family priest (Vajrācārya), and the local Paṇḍās do not object. From a historical perspective, this seems to be an old tradition which still continues to regard Kāśī first of all as a power place, and not so much as a "stronghold of Hinduism" in the nation called India. This aspect of a subliminal Buddhist tradition comes out clearly at the shrine of Lalitā Devī at Lalitā Ghāṭ: there is a small Buddha figure in a side chamber with verses in Newari script, which can be dated back at least to the 17th century (Fig. 3). Apparently Newar pilgrims donated it as a contribution to the sacred site.

The pilgrims thus have made use of the *ghāṭs* as public space in various ways. They visit them for bathing, performing rituals, making offerings. But of course this use of public space has its obligations. As the local pilgrimage priests are in firm control of the *ghāṭs* and have the right to demand fees and sizable gifts for each rite, they profit considerably from the visitors. The Pandas have a general reputation of being rather "predatory" and unscrupulous in their demands, which vary according to the wealth of the client (see Parry 1994: 119ff. for a discussion of the complex religious notions implied). Though most pilgrims accept this "bargaining over gifts" as part of the rite, there has been a tendency in recent decades to evade the services of the Pandas. Ever since a Nepali organisation founded in the 1940s provided hostel space for Nepali pilgrims, many have kept their distance from the Lalmohariya Pandas and simply performed their ritual at Pañcagaṅgā Ghāṭ. Thus the Nepalis in the Nepali locality eventually established a firm base to defend their interests, and could declare their independence from the dominant Pandas.

In any case, in recent years the number of pilgrims from Nepal seems to be decreasing continuously. Partly this may be due to the political instability in the country, which has made travel more difficult in general. But there is also an indication that this is a general long-term trend.[6] Now many prefer to visit places inside Nepal to do *śrāddha* or even *Kāśībās* ("residence in the sacred city until death"), such as Deoghat at the confluence of Gandaki and Trisuli. Increasingly, though transport has become easier, it is seen as an unnecessary exertion to travel to India—by many now perceived as a foreign country.

The Nepali Temple

There is evidence from the 12th century that it was a practice of kings of the Kathmandu Valley to build or renovate temples in Kāśī. One inscription of the medieval period, situated in the Kathmandu Valley and dating from 1143, mentions two different visits of a Saṅkarācārya from Kāśī to the Kathmandu Valley, which led to donations for temples in Kāśī.[7] Apparently ascetics from Kāśī maintained regular contacts with the courts in Nepal and even initiated princes into their order (ibid.). Having thus become part of a wider religious network, kings of Nepal came to worship and support gods outside their proper kingdom.

The best known appropriation of Kāśī's sacred topography is of course the so-called Nepali temple at Lalitā Ghāṭ (Fig. 4). The temple was built in 1843 by the reigning king Rajendra Bikram Śāh (1813–1881) and the crown prince Surendra Bikram Śāh (1829–1881) in the name of the senior queen Saṃrājya Lakṣmī Devī. Therefore the actual name of this Śiva temple is *Saṃrājyeśvar Mahādev Mandir*.[8] Nowadays the deity is also known as Paśupatināth, but it is not clear how long it has been identified with this deity. Our evidence suggests that the identification goes back to local Indian usage, since for most Banarsis Nepal is closely associated with Paśupatināth.

There is a story behind the dedication: Saṃrājya Lakṣmī Devī, the senior queen of Rajendra Bikram Śāh, had been in conflict with the junior queen for some time, and eventually decided to leave for Kāśī. This, by that time, had become a common way to avoid courtly intrigues. On the way, down in the Tarai, she fell ill and soon died of malaria, so she never reached Kāśī. But she expressed one wish: that a Śiva temple be built in her memory in the holy city. Thus, by building the temple, Surendra fulfilled the last wish of his mother.[9]

Fig. 4: Saṃrājyeśvar Mahādev Mandir at Lalitā Ghāṭ (photo by Martin Gaenszle).

As already mentioned, the construction of temples on the Indian subcontinent was a frequent pious act on the part of the hill kings. However, in the course of the 19th century this became increasingly difficult, once the British began to confiscate such land. It is known that Saṃrājya Lakṣmī Devī had strong anti-British sentiments, and so one wonders whether the wish to build the temple was perhaps an act of defiance, and thus had political overtones as well.

Another crucial question is: Why was the temple built at this particular site, which is in no way linked to any area of Nepali settlement? There are no Nepali residents in the area adjoining the temple, and though it is often assumed that many Nepalis live in Nepālī Khapra *mohallā*, which is a few blocks away, this is not the case at all. The traditional location of the Nepali community is around Dūdh Vināyak *mohallā*. Still the area around Lalitā Ghāṭ must have been associated with Nepal before the construction of the temple. As indicated above, the ex-king Raṇa Bahādur Śāh, who stayed in Banaras forty years before the temple was built (1800–1804), is locally linked to this place. Moreover, there is evidence that it was visited by pilgrims from Nepal even long before this.

In order to understand the Nepali appropriation of place, it is worthwhile to take a closer look at the sacred topography of the site. The Nepali temple is in fact situated on top of a densely packed sacred site comprising various important shrines built on different levels (see cross section, Fig. 5). It may be seen as a relatively recent layer in a long history of religious constructions.

On the lowest level, submerged in the river during the rainy season, is the shrine of Gaṅgā Devī. This shrine, housing a coloured *mūrti*, is part of the embankment architecture not much older than the Nepali temple.[10] Slightly above the path along the *ghāṭ* there is the much older shrine chamber built into the hillock on which the Nepali Mandir stands. It has two attached rooms. In the shrine chamber, which has a circular inner roof, is the main deity, Lalitā Devī, located in an open three-door compartment. To her left are the *mūrtis* of Bhairava and Kāśī Devī. This Lalitā Devī, it should be noted, is mentioned in the eulogic Sanskrit texts (*Kāśīkhaṇḍa*, *Kāśīrahasya*) and lies on the route of most major pilgrimage tours, such as the Pañcakrośīyātrā, Nagarapradakṣiṇāyātrā, Avimuktayātrā, Antargṛhayātrā (Singh 1993: 43–50). It is clearly an ancient and powerful religious place.

In the small room to the north of the main chamber is the small Buddha Sākyamuni with the verses in Newari script mentioned above. Such *dhāriṇī* verse indicates that it was established in the 17th century or even earlier (Fig 3). As Lalitā Devī, also known as Lalitātripurasundarī, is a central deity in the Śrīvidyā tradition of Tantrism, it is no surprise that Newar Buddhists were attracted to this sacred site.

On the same level as the Lalitā Devī shrine chamber (towards the south) is the *siṃhāsana* ("lion's seat") which, according to local legend, was used by Raṇa Bahādur Śāh during meditation. Thus the evidence taken together strongly indicates that this site was already revered by Nepalese devotees, including Buddhists and Tantrics, before Rājendra Bikram Śāh built the temple.

Between this *siṃhāsana* and Lalitā Devī's shrine chamber is the entrance to a narrow winding staircase which leads up to the third and fourth levels. The major deity on the third level is Gaṅgā Keśava, a form of Viṣṇu, located in a dark shrine chamber built into the hill. Like Lalitā Devī, this deity is mentioned in the Sanskrit eulogies (*Kāśīkhaṇḍa*) as being among those that should be visited during the main Yātrās of Kāśī. The shrine chamber also contains a large *liṅgaṃ* known as Gaṅgeśvar. Interestingly, both deities have traditionally been worshipped by the same family of priests as

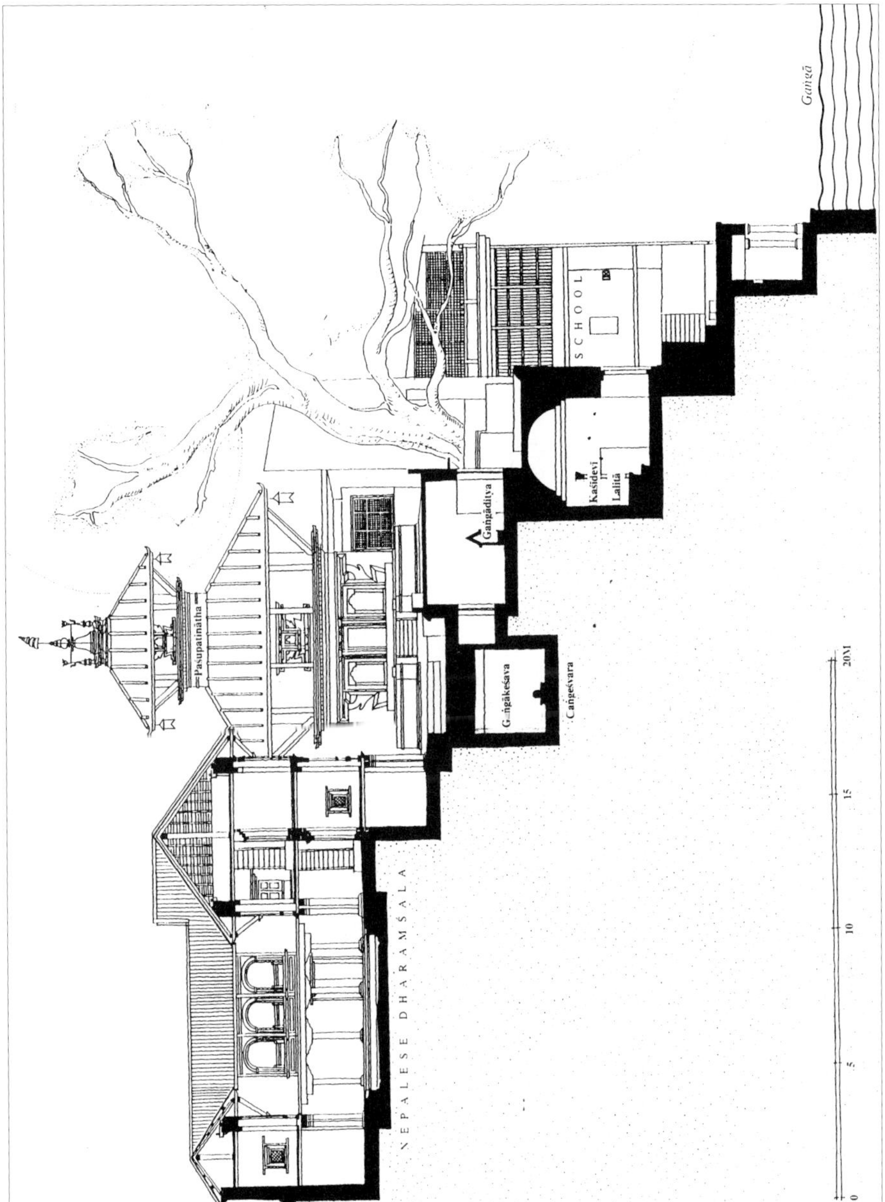

Fig. 5: Lalitā Ghāṭ (cross section, drawing Niels Guschow).

the one originally in charge of the Saṃrājyeśvar Mahādev Mandir. The pagoda temple built by the king of Nepal simply added further architectural and symbolic nuance to the site, and, as was the custom, ritual privileges were conferred on local priests.

But the horizontal plane, too, needs to be taken into account: to the north of Lalitā Ghāṭ is Rājārājeśvarī Ghāṭ, which is now "officially" (i.e. on the painted signs) labelled as Lalitā Ghāṭ as well. It used to be an important and (before the water tanks spoiled the view) picturesque location (Fig. 6). It is linked with Siddha Giri Maṭh, a Daśanāmī monastery, in which the Rājārājeśvarī temple is located.

Fig. 6: The Nepali temple seen from the north together with Siddha Giri Maṭh (photo by Martin Gaenszle).

As is well known, Rājārājeśvarī and Lalitā both are prominent in the Śrīvidyā tradition of Tantrism (where they are basically identical), and so it appears as no coincidence that these two deities are situated in close proximity to one another. Siddha Giri Maṭh once had a glorious reputation as a Tantric centre. But there are also other reasons for seeing a linkage between the two *ghāṭs*. Both sites are associated with Nepali kings in local legends. In the foundation myth of Rājārājeśvarī, it is said that a Nepali king was granted a boon by Siddha Giri, after which his queen gave birth to a girl, Rājārājeśvarī (Sinha & Saraswati 1978: 78f.). Also, a Nepali king, the one "who married a Maithili women" supposedly supported the monastery. This is further indication that Raṇa Bahādur Śāh may have had links with the place for some time.

In any case, the Nepali temple may be seen as "crowning" an ancient sacred site, which has been of Tantric importance for several centuries. It is an ornament, as it were, which frames the sacred sites below: on the one hand, it leaves the latter untouched, as what they are, but on the other it includes them into a larger "picture", thus making them part of a Nepali setting. It may thus be seen as a case of symbolic appropriation which preserves older layers of meaning while adding on new ones (cf. the notion of "inclusivism", Hacker 1983).

The temple is unique in being the only temple located outside Nepal still in possession of His Majesty's Government. Legally this is a somewhat awkward situation, and as the temple worship was under the control of a line of Indian priests until recently, for a long time the government in Nepal did not take much responsibility for it. But after the Maharaja of Kāśī pointed out the embarrassing situation to Nepal's prime minister Girija Prasad Koirala, reconstruction work was done by the Department of Archeology in the 1990s. In 1999 a trust was formed, in which there are both Nepali and Indian members, to take care of the compound.

Since the formation of the trust, there is a strong reappropriation taking place: the accompanying *dharmaśālā* (guest house) building is inhabited by Sanskrit students from Nepal who also conduct the temple worship. It is stipulated in the constitution of the trust that the priest should be a Nepali, so the traditional line of Indian priests has been discontinued. In 2001 the former incumbent received compensation for relinquishing all his rights. Today the temple (including the guest house) is seen as a symbol of Nepali culture, and increasingly is used as a forum and meeting point for the Nepali community. Though the Nepali temple is the only temple completely under the control of Nepalis, there are also others which are of special significance and which have been appropriated to a certain degree. All of these are located in the Nepali locality, which will be dealt with in the following section.

The Nepali Locality

In the course of history, and especially in the early 19th century, more and more Nepalis came to stay in Banaras on a permanent basis. They came to make use of and thus appropriate private space (i.e. houses, flats) as well as the third kind of public space—the *galīs* in the *mohallās*, the *cauks*, tea stalls etc. These expatriate Nepalis mainly settled down in one particular locality, the *mohallās* around Dūdh Vināyak. However, this appropriation was only partial, since many other castes and communities continued to reside there. In particular, there was a significant number of Marathi speakers. The 20th century witnessed an increasing tendency to create semi-public spaces by acquiring property for use by the Nepali community.

There is evidence that already in the early 19th century there was a concentration of Nepalis in the area around Rām Ghāṭ, Maṅgalā Gaurī and Dūdh Vināyak. According to the "Directory of Banaras", a by-product of James Prinsep's census of 1822, of seventy-nine major pandits in the city five were from the Himālayan hills—all living with their families in this area close to Pañcagaṅgā Ghāṭ.[11] The largest group at that time was that of Maharashtrian Brahmans, and Prinsep notes that most of the pandits were supported by the rajas of the Dekhan (Nair 1999: 250). The Nepali Brahmans, it is still recalled, were attracted by these numerous learned and respected pandits from Maharashtra and impressed by the generous support provided by their sponsors. All this clearly indicates that the first Nepali settlers in this area were Brahmans who came to study Sanskrit and eventually stayed on to do teaching themselves.

Fig. 7: The Nepali locality on an a section of the map by James Prinsep (1822).

In the second half of the 19th century this locality became the meeting ground for an illustrious mix of Nepali emigrants: intellectuals, poets, writers, businessmen and political refugees. The Ādikavī ("first poet") of Nepalese literature, Bhanubhakta, lived in Dūdh Vināyak as a student. He is known above all for his translation of the Rāmāyaṇa into Nepali. This translation was published in print in 1884–87 by Moṭi Rām Bhaṭṭa, a young poet and writer, who was also educated in Banaras. Being an energetic and enterprising man, he joined hands with Rām Kṛṣṇa Verma (the owner of Bharat Jivan Press) and others to set up the first non-state-controlled publishing venture catering for the Nepali language. Moṭi Rām Bhaṭṭa was also a member of the circle around Bhāratendu Hariścandra, who at the time was a leading activist in the Hindi language movement. Inspired by these ideas and encouraged by the possibilities of a fledgeling print capitalism, a nationalist movement promoting the Nepali language through books and journals was born.

Today the Nepali community in this locality is spread over about eleven *mohallās*, a densely populated area in what appears to be one of the oldest parts in the city. At the time of James Prinsep's survey, these *mohallās* were still distinct residential units, separated by gates which were closed at night. In fact, these gates are depicted in his famous map of 1822, which notes most of the *mohallā* names as we know them today (see Fig. 7). However, most of the gates were destroyed by the British in the late 19th century in order to have better control over the urban spaces. Nowadays the *mohallās* have only little significance as social entities, but there are still some cultural events organised on this level, and for many residents the *mohallā* is a central focus of identity.

This locality is also an area with a high concentration of temples and shrines, several of them not only of local importance. The map of major landmarks in the Nepali locality (Fig. 8) gives an overview of its religious topography: it lists the temples recognized as public sacred sites and worshipped on a regular basis by the local people. This does not mean that all Nepalis worship at these sites, since many of them are only important within the immediate neighbourhood. For example, most *mohallās* have a Gaṇeś shrine which is worshipped locally. However, among these temples, four are of special significance for the Nepalis in general.

1. *Maṅgalā Gaurī and Gabhastīśvara.* This is the most popular temple in the locality: many Nepalis go there daily to worship. The temple lies on the itinerary of the Pañcakrośīyātrā and once was within the compound of the Bindu Mādhav temple (according to Tavernier, as cited in Prinsep 1831 [1996]: 25). One of the *pūjārīs* (the main priest of Ādikeśava in a section of the temple), who practice on a rotating basis (*pāri*), has a Nepali family background: his grandmother was a Nepali woman, who was adopted by the heir to the priestly rights at the Nepali temple at Lalitā Ghāṭ. This reflects a link between the Maṅgalā Gaurī temple and the Saṃrājyeśvar (Nepali) temple, and indeed the traditional priestly line of the latter also held rights in the former. Moreover, there are many inscriptions, documenting donations by Nepali donors. One long inscription apparently goes back to the time of Surendra Bikram Śāh, though unfortunately it is in bad condition and can only be fragmentarily deciphered. This is the temple most frequently visited on a daily basis by the Nepalis in the locality. Close to this temple many Nepali Kāśībāsī widows have come to reside (most housing in rooms owned by *pūjārī* families), and they too visit the temple in the morning. However, the number has decreased significantly since 2002, when a home for these widows was opened near the Nepali temple (see below).

2. *Bindu Mādhav.* This temple, dedicated to Viṣṇu in his form as Kṛṣṇa, was once the biggest temple in Banaras. It has been destroyed several times since the 12th century. Though it was rebuilt in 1642 by the Maharaja Jai Singh I of Amber in Raja-

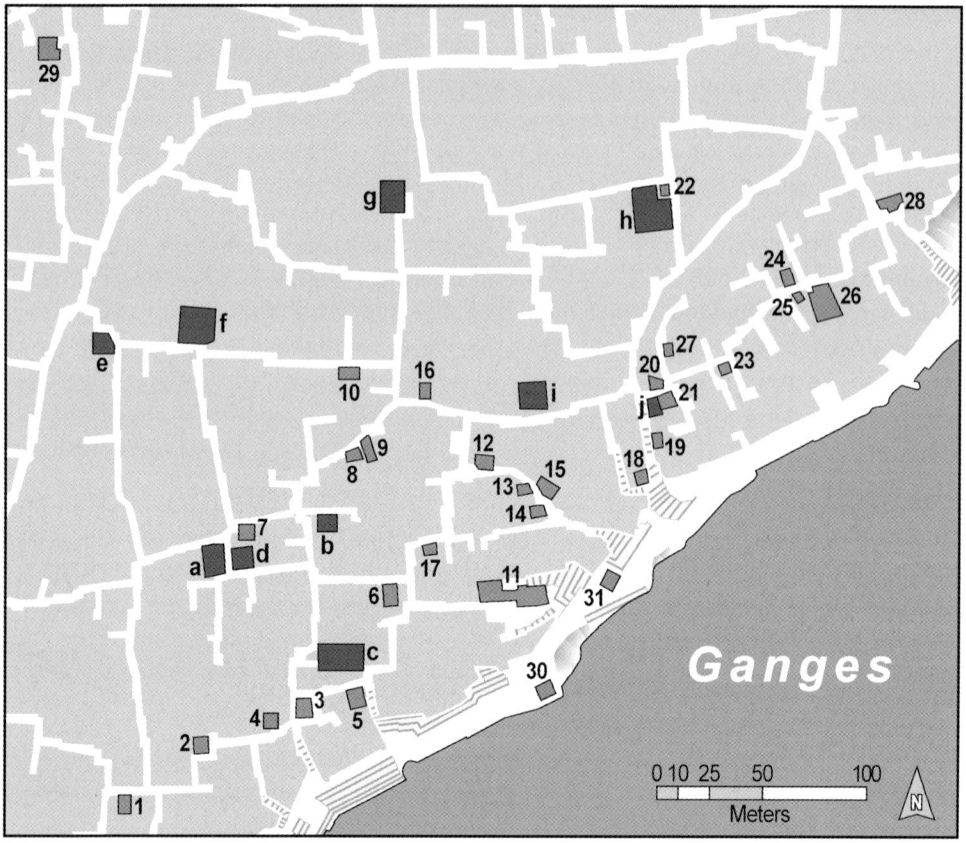

Fig. 8: Major landmarks in the Nepali locality.

Legend

Religious Topography

1. Ānanda Bhairava
2. Rām
3. Maṅgalā Gaurī, Gabhastīśvara
4. Gaṇeśa (Gaṇapatīśa)
5. Rām/Kaṅganvālī Havelī
6. Tailaṅga Svāmī
7. Dūdh Vināyak
8. Śiva
9. Śiva, Gaṇeśa
10. Kapileśvara
11. Bindu Mādhav
12. Brahmacāriṇī Durgā
13. Narasiṃha
14. Śiva
15. Gaṇeśa, 2 Śiva
16. Sunda Vināyak
17. Camatkārī Vināyak
18. Brahmeśvara
19. Gaṇeśa
20. Kṛṣṇa
21. Lāleśvara
22. Gupteśvara
23. Lakṣmī
24. Hanumān
25. Kālī
26. Viśvanāth + Annapūrṇā
27. Mātāmāī (Phulmātā)
29. Kāl Bhairava
30. Pañcagaṅgādevī
31. Pañcagaṅgeśvarī

Buildings of Significance for Nepalis

a. Jagadamba Bhavan (Nepālī Dharmaśālā)
b. Kamala Bhavan (Nepālī Dharmaśālā)
c. Jagannāth Bhavan + Kṛṣṇa Bhavan (Nepālī Dharmaśālā)
d. Former office of the Nepali Congress Party
e. Thāpā Koṭhī
f. Former residence of K.P. Bhattarai
g. Former residence of Pushpa Lal Shrestha
h. Nepālī Koṭhī
i. Nepālī Saṃskṛt Mahāvidyālaya
j. Office of the Nepali Congress Vidyārthī Saṃgha

sthan, it was again torn down by Aurangzeb in 1669, who built the famous Ālamgīr Mosque on its site (using some of the temple's stones). The present temple, housed in a simple building, is popular especially among the Nepali Kāśībāsīs who reside in the area. They meet there regularly in the morning to hear the pandit, who sits on a large *siṃhāsana* "seat" (much like the one in the sketch done by Prinsep, 1831 [1996]: 26), reciting and commenting on stories from sacred texts (see Fig. 9). One could say this has been (at least until recently) the spiritual meeting place of pious persons preparing themselves for their last days. That the Nepalis form a major section of this group is documented by the fact that 96 votive *śivaliṅgams* in an annex chamber have been given by Nepali women (as can be seen from the stone inscription). This was confirmed in conversations with the Kāśībāsī widows: they recounted that the donations were made together upon the suggestion of a local Indian pandit, with each of the women paying 600 Rs.

Fig. 9: Nepali widows at Bindu Mādhav Temple (photo by Martin Gaenszle).

3. *Durgā Brahmacāriṇī.* As the big Durgā temple (to the south of old Kāśī) is quite a distance away, this "Little Durgā", as it is also called, is of great local importance for regular worshippers of this goddess. The temple is located at the central square of Durgā Ghāṭ *mohallā*, which is brought alive in the mornings and evenings. The temple is not a free-standing building but integrated into an adjoining large mansion, which

currently belongs to a member of the Nepali community (the entrance sign says "Nepālī nivās"). There are a number of inscriptions on marble plates which testify to the contributions made by Nepali devotees. Though no dates are given (only devotional Sanskrit verses), these inscriptions are apparently of recent origin, not older than the reconstruction a decade ago.

4. *Ānanda Bhairava*. What makes this temple unique is the fact that it is the only temple in Banaras which is in the complete control of a Nepali lineage, and thus it may be seen as the clearest example of an "appropriation of sacred places" by the Nepali community. The temple does not have ancient roots, but according to its origin myth it is directly linked with the famous Kāl Bhairava temple. As the story narrated by the grandfather of the present temple priest goes, there was a *brahmacārin* (celibate) living in this house about 300 years ago–a Brahman from Kumaon. His name was Chandra Shekhar and he was a pious devotee of Kāl Bhairava. He went to that temple every day and performed 108 circumambulations. One day a quarrel ensued between him and the incumbent *pūjārī* of Kāl Bhairava. The following night Chandra Shekhar had a dream in which Kāl Bhairava appeared to him and said: "You do not have to go to Kāl Bhairava Mandir any longer in order to meet me. Instead I will come to your house". Since then Bhairava is worshipped in his form as Ānanda Bhairava. Thus the story identifies the deity as an equivalent of Kāl Bhairava. The ancestors of the present-day *pūjārī* hail from Nuwakot district in Nepal, but they are also Kumaon Brahmans. They are not descendants of the legendary Chandra Shekhar but acquired rights to the priesthood through a marriage link four generations ago. Thus the temple has remained in the hands of Kumaon Brahmans, who once boasted a sizeable number of households in Rām Ghāṭ.

As far as residential buildings are concerned, there are a number of houses which are or once were in the possession of Nepali families; some of these buildings are known to date back to the 19th century. The Thāpā Koṭhī, located in Govindaji Nāyak *mohallā*, seems to be the oldest Nepali-owned house in the locality. According to the present caretaker of the house, it was acquired by Hemdāl Thāpā in the 1840s and has since been the property of one line of his descendants. Hemdāl Thāpā was one of the colonels who accompanied Queen Rājya Lakṣmī Devī when she was expelled from Nepal after the Koṭ massacre (see Rana 1998: 44). He was a close associate of Jaṅg Bahādur Rāṇā, who sent him on this task. Hemdāl Thāpā's son Harka married a daughter of Jaṅg Bahādur, and since then the descendants have been intermarrying with Ranas and calling themselves Jang Thapa. The present owner is a Thapa from Kathmandu (Bansbari). Another example is the so-called Nepālī Koṭhī in Narāyaṇ Dikṣit *mohallā*, a fairly large compound housing several families. This building was originally owned by Dev Śamśer Rāṇā (approximately from 1901 to 1950), and at one time the young K.P. Bhattarai, one of the founders of the Nepali Congress Party, lived there. The wife of Dev Śamśer later gave the house away as dowry, and it came to be used by a Nepali doctor as a homoepathic clinic. In the early 1990s the house was sold and there are no more Nepalis living there. These examples show that even some powerful members of the ruling families had property in the Nepali locality. But generally they later preferred to buy mansions on the outskirts of Banaras.

According to a historical text written by a Nepali pandit in 1955, once (probably around the early 20th century) there were 50 "private houses" (*nijo ghar*) in the possession of Nepalis, but many of them were sold off at a cheap price later, apparently after independence. Today most Nepalis who live in the locality live in rented houses, even those who have resided there for several generations. Generally there is a high degree of fluctuation in the migration pattern. Students, for example, often come only for certain periods when exams are held. And as most Nepali residents have still strong links to family members in their home villages in Nepal, there is a considerable coming and going between the two places.

In terms of cultural practice, the Nepalis have simply joined the mainstream of local residents. They visit roughly the same temples and perform more or less the same Hindu rituals. Thus, there are few festivals celebrated in public spaces which have a distinctly Nepali character. One example is the unique way of celebrating Tihār (Dīpāvalī), when groups of singing youths go about from house to house asking for donations. An interesting case is the celebration of Dasaĩ, which in Nepal is accompanied by blood sacrifice. This, of course, is not possible in Kāśī, and so the Nepalis who insist on this tradition usually go to shrines outside the city, e.g. Vindhyāvāsinī. In any case, the majority of religious festivals are celebrated at home, i.e. in a private setting, and so does not have the character of a public performance.

But in the 1930s there was a movement launched by a Nepali pandit which eventually led to the creation of a semi-public space. Jagannāth Guragaĩ, a Brahman from Taplejung who later settled in Kalleya (a little kingdom linked to the Nepali Shah dynasty at the border in north Bihar), had witnessed the problems which Nepali pilgrims experienced when staying in Banaras. Many pilgrims fell prey to greedy *paṇḍas* or their *dalāls* (helpers). The Lālmohoriya Paṇḍas put up the pilgrims in their own house, but many Nepalis, already pressured for ritual gifts (*dān*), felt uneasy about this dependancy and obligation. Though there was the Nepali *dharmaśālā* at Lalitā Ghāṭ, it was only a small building, and at that time was mainly used by ascetics (*sādhus*). Students, too, were constantly on the lookout for accommodations. So Jagannāth Guragaĩ began to collect money in order to buy a house to provide a safe haven, as it were, for Nepalis. Guragaĩ was a sometime poet and composed the "*Guṇaratna Mālā*", which was printed in a little booklet and sold. He travelled throughout North India and Assam and raised funds for his project.

Eventually in 1942 Guragaĩ and other leading Nepali pandits founded an organisation called *Śrī Nepālī Chatrabās Dharmaśālā Vidyādharmapracāriṇī Samitī* in Dūdh Vināyak. This organisation bought a house in Maṅgalā Gaurī (in 1943), which was later extended, and until today it has served as a *dharmaśālā* and general meeting ground for Nepalis. It is known, in memory of the founder, as Jagannāth Bhavan. The student hostel, however, was closed in 1992 after some internal quarrels. The *samiti* soon developed into a kind of expatriate organisation for the Nepali community. It was even used by the Nepali government in order to channel funds to the Nepali temple at Lalitā Ghāṭ and the Kāśībāsīs who receive a small pension for their living from the Guṭhi Saṃsthān.

In the 1960s two more houses, both in Dūdh Vināyak, were acquired with the help of donations from members of the Rāṇā family: one from Jagadamba, wife of Madan Śamśer Rāṇā, in 1960, and one from Rāni Kamalā, the wife of Padma Śamśer Rāṇā in 1963. These are known as Jagadambā Bhavan (Fig. 10) and Kamalā Bhavan respectively. In a way these donations repeated the pattern of "royal gifts for the public", which we already saw forthcoming in the foundation of the Nepali temple. But there are differences: here the acquisition of property was a joint endeavour, initiated by individuals and merely assisted by aristocrats. A crucial point is that the beneficiaries of the gift were to a certain extent the donors themselves. And, as is made clear in the constitution as well as the name, the property is meant primarily for use by Nepalis.

Fig. 10: Drawing of Śrī Jagadambā Nepālī Dharmaśālā in Dūdh Vināyak (undated, ca. 1960).

Up to the present the Nepali *dharmaśālās* have remained an important focus for the Nepali community. Their premises not only provide space for putting up pilgrims and other visitors from Nepal; they are also used for the celebration of festivals, such as Dasāī̃, Rām Navamī and Sarasvatī pūjā, and also weddings and other life-cycle rituals. Sometimes also more secular events take place (e.g. meetings of the Nepali student union, the anniversary celebrations for the co-founder of the Nepali Congress Party, B. P. Koirala). Through the *dharmaśālā* network the Nepalis in Banaras have created a "nationalized" space for their community, and though it is not restricted to Nepalis, it is primarily used by them.

More recently, however, the focus has shifted somewhat to the newly revived Nepali temple complex. Whereas previously the annual Sarasvatī pūjā, which is celebrated by the Sanskrit students and their teachers, was done in the Nepali locality's Jagadambā Bhavan, in 2002 it was held in the *dharmaśālā* compound at Lalitā Ghāṭ. And, most recently a house was bought by the Trust (with money from the Ministry of

Culture and Tourism) to house widows (Kāśībāsīs) from Nepal (*vṛddhāśram*). It was inaugurated by the incumbent minister on the first death anniversary of king Birendra (June 1, 2002). Since then around twenty Nepali widows have shifted to this new residency and worship regularly at the Nepali temple. Though the Nepali locality remains the centre of Nepal's presence, ritual activities increasingly take place at the temple which is *the* Nepali temple in Banaras.

Conclusion

In summary, one can say that there is a long tradition of appropriating space in Kāśī by peoples from the central Himālayan hills: pilgrims have left their imprint on the riverfront, devotees have built shrines and temples, and migrants have settled down, constructing their own neighbourhood with its own infrastructure. But this appropriation has taken different forms in the course of history. In earlier times it occurred primarily in the form of donations to deities, following a pattern which is widespread in all of South Asia: donations are made (often to translocated deities brought along) for the public's benefit, usually in memory of deceased persons (see Burghart 1987). During the last century, at a time of rising diaspora nationalism, more and more ordinary Nepalis have acquired property in Kāśī, not only for private use but also with similar religious motives, i.e. to obtain merit. But the property thus acquired for the "public" was primarily meant for use by the Nepali community. Thus a "nationalization" of space was achieved, which created little islands of Nepali semi-public space in the city of Kāśī. The concept of nation has affected the quality of places in different ways for different actors: a religious site that before was simply Hindu may now also be categorized in national terms. That this is an ongoing process of negotiation is most clearly seen in the case of the "Nepali temple": ever since the Indian temple priests were removed from their hereditary position recently, the place has more and more beem taken over and "revived" by the Nepali community. Thus the temple, managed by a new temple trust, is becoming a new focus of Nepali self-representation in Banaras, and this can be seen as an expression of ongoing change in the appropriation of space.

Notes

1 Sheldon Pollock, for example, coined the term "Sanskrit cosmopolis" to characterize the period (roughly between 300 and 1300 CE) in which Sanskrit was a kind of translocal idiom throughout South Asia. After 1000 CE this period was gradually superseded by the "Vernacular Millennium", in which local languages increasingly gained importance (Pollock 1998, 2000). From this perspective Banaras, a centre of Sanskrit learning, may be seen as a descendant of such a pre-modern, alternative cosmopolitanism (also see Nandy 2001).

2 Fieldwork in Banaras was made possible through funding by the Deutsche Forschungsgemeinschaft (German Research Foundation). We are grateful for its support.

3 See the text published by Yogi Naraharinath in his *Itihāsprakāśmā sandhi-patra-saṃgraha* (V.S. 2055)

4 This emerges from documents in the National Archives, Kathmandu, reel # E2426\3, # E2430\2.
5 The Ḍom are an "untouchable" caste of funeral attendants who supervise the process of cremation and collect a tax for every corpse (see Parry 1994: 90f.)
6 In the late 1970s Parry reports that in the month of Pus about 50,000 Nepalis came to do their ancestral rites with the Lalmohariya priest (Parry 1994: 98). This figure is well above the present one of less than 10,000 during this peak month.
7 Nepal 1986 (V.S. 2043): 46-50.
8 This emerges from the inscription in the pinnacle of the temple (see Shrestha 2001: 49).
9 Contradicting this version, there were rumours that the senior queen was poisoned by her enemies. Apparently this rumour was published in an English-language Indian newspaper, enraging the king in the presence of the British Resident (Shaha 1990 I: 175).
10 The riverbank is still without a built embankment in the painted panorama published by Rai Ananda Krishna (2003), which he dates to the early 19th century.
11 The directory has only recently been published, in Nair 1999: 226-260.

References

Burghart, Richard. 1987. Gifts to the Gods. Power, Property, and Ceremonial in Nepal. In: S. Price & D. Cannadine (eds.), *Rituals of Royalty: Power and Ceremonial in Traditional Societies*. Cambridge: Cambridge University Press, pp. 237–270.

Cape, C. Phillips. 1909. *Benares: The Stronghold of Hinduism*. London: Charles H. Kelly Publ.

Gaenszle, Martin (in collaboration with Nutan Dhar Sharma). 2002. Nepali Kings and Kāśī: On the Changing Significance of a Sacred Centre, *Studies in Nepali History and Society* 7, 1: 1–33.

Hacker, Paul. 1983. Inklusivismus. In: G. Oberhammer (ed.), *Inklusivismus. Eine indische Denkform*. Wien: Institut für Indologie der Universität Wien, pp. 11–28.

Krishna, Ananda. 2003. *Banaras in the Early 19th Century: Riverfront Panorama*. Varanasi: Indica Books.

Lamsal, Devi Prasad (ed.) 2023 V.S. *Bhāṣā Vaṃśāvalī*. Vol. 2, Kathmandu: Royal Nepal Academy.

Nair, P. Thankappan. 1999. *James Prinsep: Life and Work*. Vol. I. (Background and Benares Period). Calcutta: Firma KLM Private LTD.

Nepal, Jnanamani. 1986 (2043). *Paśupatināthako darśan sparśan pūjana saṃvandhamā samikṣa*. Kathmandu: Maheshraj Panta.

Nandy, Ashis. 2001. Time Travel to a Possible Self: Searching for the Alternative Cosmopolitanism of Cochin. In: *Time Warps: The Insistent Politics of Silent and Evasive Pasts*. New Delhi: Permanent Black, pp. 157–209.

Naraharinath. 2055 v.s. *Itihāsprakāśamā sandhipatrasaṃgraha*. Lalitpur: Balaram Gharti Magar.

Parry, Jonathan. 1994. *Death in Banaras*. Cambridge: Cambridge University Press.

Pollock, Sheldon. 1998. India in the Vernacular Millennium: Literary Culture and Polity, 1000–1500, *Daedalus* 127, 3: 41–74 (Special Issue "Early Modernities", ed. by Shmuel Eisenstadt, Wolfgang Schluchter, Björn Wittrock).

— 2000. Cosmopolitan and Vernacular in History, *Public Culture* 12, 3: 591–625.

Prinsep, James. 1831–33. *Benares, Illustrated in a Series of Drawings*. Calcutta: Baptist Mission Press (Reprinted 1996, Varanasi: Vishvavidyalaya Prakashan).

Rana, Purushottam Shamshere J. B. 1998. *Jung Bahadur Rana: The Story of his Rise and Glory*. Delhi: Book Faith India.
Shaha, Rishikesh. 1990. *Modern Nepal: A Political History 1769–1955*. 2 vols. New Delhi: Manohar.
Shrestha, Shobha. 2001. Sāmrājyaśvara Mahādeva mandira tathā dharmaśālāko saṃrakṣaṇa tathā saṃvardhana eka jānakārī, *Ancient Nepal* 147: 49–54.
Singh, Rana P. B. (ed.) 1993. *Banaras (Varanasi): Cosmic Order, Sacred City, Hindu Traditions*. Varanasi: Tara Book Agency.
Singh, Shashi Bala. 1996. Settling Process and Spatial Pattern of Linguo-Cultural Groups in Varanasi City. *National Geographical Journal of India* 42, 1–2: 116–132.
Sinha, Surajit & Baidyanath Sarasvati. 1978. *Ascetics of Kashi: An Anthropological Exploration*. Varanasi: NK Bose Memorial Foundation.
Slusser, Mary S. 1982. *Nepal Mandala: A Cultural Study of the Kathmandu Valley. Vol. I: Text; Vol. II: Plates*. Princeton: Princeton University Press.
Stiller, Ludwig F. 1989 (1968). *Prithvinarayan Shah in the Light of Dibya Upadesh*. Kathmandu: Himalayan Book Centre.
Vajracarya, Dhanavajra. 1999 (2056). *Mādhyakalakā abhilekha*, Kathmandu: CNAS.

VASUDHA DALMIA

Visions of a New Banaras and the New Novel in Hindi

Premchand (1880–1936), the best-known Hindi author of the twentieth century, first rose to fame with the publication of *Sevasadan* (1918), a novel which focussed on Varanasi,[1] the holiest of all cities in India. But there was little that was holy about the city, as Premchand saw it at the turn of the century. It was, much more, a city caught in the throes of modernizing and refashioning its image, even as it made a bid to drive away its once famed courtesans[2] from the vicinity of Chauk or the central square at its heart.

Some suggestion of the many layered social texture of the city is contained in the very title *Sevasadan*[3] or the House of Service, the name of the educational institution which was erected to house the daughters of former courtesans. *Seva* and *sadan*, however, also carried other meanings. Seva or service could also be of the kind which the courtesans of the city supplied to their clientele, and Sadan or abode was also the name of one of the central male protagonists of the novel, an erring and largely unfulfilled young man. Sadan had been in love initially with the one sister, the beautiful and enticing Suman, whom he had first met in the Bazaar of Beauty. But eventually he married the other sister, the more pacific Shanta, to whom he had been betrothed all along and who had suffered the misery of her lot with virtuous resignation. Though the various connotations of the title must also have existed for contemporary readers, it was surely its most apparent meaning, with its obvious allusion to social reform, the puritanical rather than that which suggested the lurid, which accounted for the initial appeal of the novel. It was the era of high nationalism, Hindi was seeking to set up its own respectable literary canon, and if nation, woman, social reform, and the reordering of the city landscape could be brought together thus educatively, the novel could do none other than win public approval. But the novel's enduring popularity may have had more to do with the interplay of the various meanings, and of voices other than the calm, analytical auctorial, which make possible more than one reading of how *seva* and *sadan* came variously to be conjoined but also queried. This essay is an effort to trace and recover this interplay of meaning in order to gain some understanding of the enduring power and popularity of the novel and the image in which it chose to recast the old city of Kashi.

In addition to its vividly dramatic dialogues, its deep psychological insight into persons and processes, the narrative is also girded at all times by a keen sense of social history. It registers crucial societal transitions, even as they take place, from known patterns of life to new, untried ones. One transition is that from the country to the city; many of the central characters undertake or undergo this traumatic passage. After her marriage, Suman is transported from her well-to-do home in a rural district to the stench of the narrow lanes in the inner city and the claustrophobia of the small dark rooms in which she houses with her elderly husband. Sadan in running away from his landowning village home finds himself filled with fear and foreboding as he dashes across field and wood to catch his night train to Benares. Two miles outside the village there is a pipal tree, its dense foliage filled with ghosts which appear in all shapes and forms to hinder the wary traveller. Sadan freezes into inaction and can cross its dark presence only when he hits upon the idea of singing verses from Tulsidas's *Ramayana* at the top of his lungs to fend off the evil spirits which lurk there.

The city itself is changing rapidly. In the second decade of the new century, it continues to be dominated by the old time ra'is or gentry, the great landowners of the district, residing in their town homes, and the equally powerful merchant aristocracy with its widely spread kinship networks. But alongside the gentry, the colonial government has allotted some public space to the new professionals, whom it also nominates to the municipal board, which has newly been granted some autonomy in civic government. The rhetoric of social reform, as yet thinly represented by these new professionals, is beginning to gain ground and provide unexpected twists to matters considered long settled. But even social reform is a two edged sword, which can be used to further various ends, especially when it focuses on women. At the centre of the novel is yet another crucial transition: the courtesans of the city are no longer to practice their art publicly, for the social reformers have taken up the cause of fallen women. This touches the heart of the relationship Suman and Sadan could have to each other; Suman's private problems become linked to the city's problems and its changing face. Private and public are thus inextricably intertwined.

The Courtesans of the City and the New Novel in Hindi

The courtesans of the city had long occupied central space in public life. Their reputation went back many centuries; there are references to them in Jataka tales and in Sanskrit works of different genres and periods.[4] As legend had it, the gentry had ever flocked to the best known amongst them, in search not only of the more obvious pleasures but also to enjoy the best in the arts: music, dance, poetry and to luxuriate in an atmosphere radiating refinement in matters of aesthetic taste and judgment. Once a courtesan had acquired a reputation for beauty and the mastery of the arts, she could attain high social standing. She would be asked to appear at the court of the Benares Maharajas, in the mansions of the rich and the powerful, but also in the most prominent temples and on the bank of the holy river on important religious occasions. The number of these women was small, though their power was often staggering, as the sober voice of James Prinsep, the first British census-taker of the city, informs us in the second quarter of the nineteenth century:

> If, however, beauty and high vocal skill are comparatively rare, the witching influence of the arts and graces of these women is as much acknowledged and as powerful as ever. Examples are not wanting of large accumulations of wealth from the successful

> exercise of the skill and accomplishment of the profession: some of the best houses and the handsomest temples of the city, have been erected by ladies of dancing notoriety.
> According to the census taken in 1827, there were 264 Hindooee and 500 Mosulmanee professional *Nach* girls in the town, not a very large proportion to the population. There were, however four times the number of *khanehgee-kusbees*, who do not profess the accomplishments of singing and dancing. (Prinsep 1833: 33)

Fig. 1: ULEE BUNDEE. Hindoo Nâch Girl—Drawn on Stone by W. Sharp from an original Sketch by James Prinsep Esqr. Printed by C. Hullmandel.

Prinsep spoke prudishly of "dancing notoriety", dancing fame would have been the more adequate expression. The city was proud of the beauty of its leading courtesans and of their mastery of the arts.

Bharatendu Harishchandra (1850–1885), acknowledged father of modern Hindi literature, old style merchant aristocrat and modern style publicist, who seldom spent a night at home, had confessed that he went to these women in order to acquire cultivation and knowledge of the arts.[5] He was to note their all-pervasive presence in Kashi and the power they exercised on men, and to eulogize them in one of his short plays, *Prem jogini*, the yogini of love (1874–5), which provided vignettes of the city at its most pious but also at its most raucous.[6] He captioned the scenes, when they were published in book form, as *Kashi ke chayachitra ya do bhale bure fotograf*, "reflections of Kashi or a couple of good and bad photographs". In choosing the word "photograph", he both indicated the modern character of his undertaking—he was considered a good amateur photographer—, as well as the reality of the individual in social situations that the photograph as a medium sought to capture and hold fast. In seeking to ground the reality of city life, he placed the value of these public women at the top of the social spiral and thus of Kashi's claims to fame.

The city as a whole clearly had two sides, the exalted and the seamy. In fact, with the exception of Babu Ramchandra, the central protagonist and his *musahib* or agent Sudhakar, the characters in the play consisted almost in their entirety of the rogues and charlatans of the city. At least half of the city, at least as the rogues and charlatans saw it, was run over by prostitutes of all hues and classes:

> *Adhi kashi bhand bhanderia bahman aur sanyasi*
> *adhi kashi randi mundi rand khangi khasi* (333)
> Half of Kashi is run over by bards and bhanderias, brahmans and ascetics
> The other half with whores, widows with shorn heads, and very special harlots.

But even on the grander map, which offered the more representational face of the city, and which Sudhakar was to go on to offer as the corrective to the seamier version, the courtesans were placed at the pinnacle of the social order. Sudhakar's depiction of the city was mythical and modern at the same time. He waxed eloquent on the beauties of riverside landscape, the well laid out ghats and the high mansions, five to seven storied, towering on the one side of the river, interspersed with the temples of Gopallal and Bindu Madhav, of Vishvanath and Annapurna. It was a city graced by the presence of other godly beings as well. He eulogized the various rungs of the city's hierarchical order, beginning with the Maharaja, who was the religious and cultural head of the city, backed by the civic and military authority of the British commissioner, the collector and the magistrates. There were many wealthy merchants, there were scholars, the learned in Sanskrit and the servants of Hindi, there were the able craftsmen. There were educational institutes and a number of public libraries and monuments. There were the famous ancient ruins of Sarnath. There were people from all parts of the country, and ascetics of the most diverse orders, who in their diversity encapsulated and encompassed all parts of the incipient nation. There was wealth, trade, textiles, handicrafts. The city blossomed and glowed. And finally, as a climax almost:

> The youthful *barabilasini*, courtesans, exotically dressed and ornamented, with fragrant betel leaves in their mouths, sauntering on the royal paths on the flimsiest pretext, whether it is going to the temples, to sing or to roam in parks, in order to dupe and ensnare men, poor, dim-witted men.
> What more can I say? Kashi is Kashi. There is not another city like Kashi in the three worlds. You'll believe it when you see it. It is useless to say more. (342)

Kashi was the pinnacle of all cities in the subcontinent, wherein all regions and crafts were represented, and atop this pinnacle were perched the courtesans of the city. This exalted reputation of the women of the city was to continue unabated into the first half of the twentieth century.

When E. B. Havell (1864–1937), the great apologist for the high aesthetic value of Indian fine arts, came to write his two travelogues to highlight the tradition of art and religion in the subcontinent, he not only naturally turned to Benares, but in doing so, could not refrain from mentioning the central role of the courtesans in the culture of the city.[7] He compared the musicians and dancing girls, whom he saw as "an indispensable institution" in the Hindu social and religious system, to "the Gandharvas, the mythical musicians of Indra's heaven, who attend the feasts of the gods, and the Apsarases, the voluptuous charmers". Passing lightly over the matter of the social origin of the dancing girls—he saw them as the unmarried daughters of the Kathaks—he went on to describe their mode of habitation and reputation in the city:

> They live in the quarter known as Dal-ki-mandi, a long street with houses of several stories, some of them resplendent with silver furniture and crystal chandeliers. Unlike the dancing girls of southern India, they are not attached to any particular temple, or "married with the god", but at special festivals and religious ceremonies they are engaged to chant the praises of Rama, or to sing Sita's love, in the classic songs of Tulsi Das, or the more voluptuous odes which tell of Krishna and his amours. Of secular songs for pleasure parties they have an extensive repertoire, both old and modern. They are often very generous with the wealth they acquire, and in old age, when virtue has become a necessity, spend it freely in works of charity and religion.
>
> Benares from very ancient times has been famed for these sirens, whose amorous glances, alluring mimic, and pretty shuffling feet have troubled many a Hindu sage. (Havell 1905: 86–87)

The best-known courtesans clearly exercised great social power; they played a significant role on public occasions, as even the most puritanical retrospective accounts are constrained to admit:

> Nevertheless, there is no doubt these women musicians and dancers made a substantial contribution to the music centre of Varanasi. Furthermore, on certain occasions, such as the worship of the Ganga, the songs of these women formed an essential part of the rituals.
>
> Fees were taken by tawaifs for performing, excepting in certain religious places and on certain occasions that were attended by commoners. They sang free in the akhada of Sadhu Kinaram, in the Vagishwari Temple on the occasion of Saraswati Puja, in front of the Shitala Temple in the month of Chaitra, near Panchaganga Ghat in the month of Kartik, in the akhada of Naga Baba near Gai Ghat, near Panchaganga Ghat, near the Kali Temple on the Dashashvamedh Ghat; and also on the various Ghats of the Ganga during worship of the river. (Medhasananda 2001: 668)

When the well known Hindi writer Amritlal Nagar set out to document the power and presence of the courtesans of North Indian cities after their profession was banned by law in independent India, Ramchandra Varma, one of the many *rasikas* or connoisseurs he interviewed in Benares, spoke once more of their musical talent and of the importance of their role in the ritual life of the city:

> The assembly that takes place on the Lolark sixth of Kartik is quite famous here. At Baba Kinaram's place, the entire city's courtesans would perform. 'Adi Vishveshvar,' the older Vishvanath, was visited by courtesans every Monday. Each year, they held a *mela* there on the day of Gopashtami. That's why the trendy youth of the day called Adi Visheshvar 'randhibaz Mahadev' or courtesan-addict Shiva. (Amritlal Nagar 1994: 154)

New poetic compositions gained wide circulation if they were taken up and performed by courtesans. A vital nerve thus connected the performing arts with literary composition and its informed appreciation and circulation. The connection would continue to be operative well beyond the era of early moderns such as Bharatendu Harishchandra. It is explicitly evoked in Premchand's novel by Kunwar Aniruddh Singh, powerful Hindu Zamindar, involved in the controversy surrounding the eviction of the courtesans from their quarters at the centre of the city. Aniruddh Singh saw the paucity of literary works in modern Hindi as being directly linked to the decline of Indian musical culture. *Navin* or new Hindi literature could at most lay claim to a couple of plays by Harishchandra, to the popular novel *Chandrakanta*,[8] and for the rest, there were translations from the English, made not even directly but via Bengali and Marathi. There was a great dearth of novels. This meagre production was the direct result of the decline of the musical tradition and of the artistic sensibility which was coupled with it:

> Yes, it's true, I like the sitar. I start feeling queasy when I hear the harmonium or the piano. These English instruments have destroyed our own music, which is barely discussed today. And if there were something left to be destroyed, theatre has come forward to complete the task. Whoever you see harps on and on about the *ghazal* and *quawali*. In a while, our music will disappear altogether, like the art of archery. It is music which gives birth to pure thoughts. With the disappearance of music, we have become devoid of these and the worst impact of this loss has been on literature. How sad is it that in the country where an invaluable work like the *Ramayana* was created, where a work as delightful as the *Sursagar* was created, we have to take recourse to translation, even for the simplest novels.

But the indigenous music which was now being invoked as essential to the survival of the arts, performative as well as literary, had begun to change its contours and context. Aniruddh Singh spoke of Tulsidas's Awadhi *Ramayana* and the Brajbhasha devotional verse of Surdas in connection with the musical tradition he wished to see regenerated. Clearly, in being projected as national heritage, music was being entirely severed from its original performers in the bazaar of beauty. It was Hindu, middle class and respectable, as indeed the new novel in Hindi and it was as such that it was sought to be situated within the city, which was also losing some of its mythical status. The new voices in the city were seeking to create an image of Kashi which was far removed from the one which was socially and ritually centred around the courtesans. In the public disputes about the location of the courtesans, the city's ill reputation was seen as reflected in the sad fate of the nation, of its music and also that of the Hindu woman. For, provocatively enough, Premchand had chosen to cast Suman as a Brahman woman who had deserted her husband, rather than as Muslim and at least nominally single.

Suman's move to the bazaar of beauty challenged the status of Hindu marriage as an inviolate social institution and in portraying the psychological motivation of her move with more than a measure of sympathy, the author seemed, at the least, to be also querying its self evident claim to sanctity.

Wife and Courtesan in the Changing City

Suman's father had been a district Sub-Inspector of Police, who had brought up his two daughters to expect the best in life. He had indulged them by buying them all manner of beautiful clothes and expensive baubles and arranged for them to receive some education from a Christian lady who tutored them at home. But he had laid no money aside for the considerable sum he would be expected to expend as their dowry and various disasters overtook the family when it came to finding a suitable match for them. Suman's mother was finally able to marry her off with little expenditure to an elderly bridegroom, relatively poor but with high standing within her own caste group. Confined within the narrow walls of her new home in the city and initially humoured and pampered by her husband Gajadhar, Suman managed to lead an exemplary life as a dutiful housewife for the first two years of her married life. But she was bored and frustrated at the same time and just opposite her lived the plump Bholi, luxuriating in the comfort of her well-lit and well-aired home and visited nightly by the town gentry. Bholi would beckon to her to come over, though Gajadhar reacted with disapproval and censor, when she finally succumbed to the temptation to go over to her house. Money did not make the men who visited Bholi great by any estimation, he explained. In the arrogance of their being, the rich chose to ignore religion but Bholi was not purified by their coming, and he quickly forbade Suman to visit her. Suman then tried to become pious in her ways. She bathed daily in the holy river, she recited the Tulsi *Ramayana* to the women of the neighbourhood. If this did not bring peace to her soul, it gave her the satisfaction of being morally more elevated than the likes of Bholi. On the occasion of *Ramnavami*, the celebration of the birthday of Rama, Suman visited a great temple in the company of her friends. The temple was beautifully decorated; it glittered with electric lighting and so crowded was it, that there was no space in its overfilled courtyard even for the proverbially tiny sesame seed to be accommodated in it. The sound of sweet music wafted in the air and her neighbour Bholi sat at the centre of all this festivity, each gesture she made observed eagerly by the most prominent citizens of the city, the rich, the learned and the pious, many of whom she had seen at the river front. Suman stood there as if struck by a thunderbolt, her pride shattered to bits. Not only did wealth bow at Bholi's feet, even religion sought her favour:

> The very prostitute whom I wanted to humble with my pious ways is the recipient of esteem and honour in this assembly of great men, in this pure abode of the lord, while I find not even standing space.

Suman abandons religion thereafter. Gajadhar's earnest arguments regarding the ultimate worthlessness of the men who crowd around Bholi fail to convince her. The city's public spaces are reserved for public women, not for poorly dressed creatures like her, as a second incident brings home to her with painful clarity. She has accompanied her friends to Beni Bagh, which houses a small zoo and an aviary. Her peace is rudely shattered when she attempts to sit for a moment on a bench. The park watchman immediately unseats her. Just then, two finely dressed women arrive, one of them her neighbour Bholi. The watchman not only seats them on the bench, he offers them a hastily plucked bouquet of flowers. Once they depart, Suman upbraids him for his unseemly behaviour. His words sting her to the quick: "Do you aspire to be their equal?" He manhandles her when she tries to sit on the bench again. She is rescued by a well to do couple passing by in their carriage. The house of her benefactors, the lawyer Pandit Padam Singh and Subhadra, his cultivated wife, becomes a refuge for her.

But not for long. Padam Singh is elected as a member of the powerful Municipal Board of the city and to celebrate the event, he allows himself to be persuaded to hold a *mujra* one evening. It is here that the second key event in Suman's city life takes place. The focalization of the scene is entirely through Suman. From behind the curtain, she and Subhadra watch Bholi Bai perform. Bholi is neither as pretty as Suman, nor does she sing particularly well. Suman has a good ear and a quick memory. She would not need more than a month to learn to sing better than Bholi. She can also cast sidelong glances and she also knows how to smile demurely:

> Suman sat there for a long time, trying to disentangle cause from effect. Finally she came to the conclusion: Bholi is independent while my own feet are in shackles. Her warehouse (*dukan*) is open and there is a rush of clients, while my warehouse is shut, so no one stands and waits. She has no thought for barking dogs, I am scared of social censure. She is outside purdah while I am inside it. She chirps freely as she hops from branch to branch, while I cling to one branch. It is shame, the fear of ridicule, which makes me the servant of others.

Suman's deeply reflexive and dry conclusions, her estimation of marriage as prostitution though with a single client, can be regarded as no other than revolutionary for her times. Wives are easily acquired and easily disregarded. Social esteem is reserved for the very public women each married man would be quick to condemn at home. Men are fickle, easily swayed, and when it comes to taking quick, bold action to retrieve a situation, they are usually found to be missing. If they come to the rescue at all, it is usually after the event.

Suman is beautiful and talented; she aspires to be free, to be admired and respected, to live in luxury, to be rid of her tyrant. The mujra at Padam Singh's house gets over very late, by the time she gets home, it is well past midnight. Gajadhar refuses to let her in, he tells her that she has been afflicted by the water of the city (*shahar ka pani*). His insane jealousy, her own extreme dissatisfaction and her pride lead to a violent exchange between the two, which bars her further entry into her own house. Just as Nora in *The Doll's House*, Suman picks up her suitcase and departs, into an unknown and uncertain future.

She seeks refuge with Padam Singh and his wife but her stay at their house is painfully brief. Padam Singh fears for his reputation once word about her putative relationship to him begins to spread in the bazaar. Only Bholi welcomes her warmly and takes her in. Bathed with perfumed soap, expensively clad, Suman views herself in the mirror and is struck by the radiance and beauty of her own person. It does not take her long to learn how to sing, for as Bholi tells her, there is no need to labour at dhrupads and tillanas any more, nowadays a few catchy ghazals, some thumris and a couple of popular theatre tunes can do the trick. Suman is an instant hit. Word of her beauty and her talent spreads quickly, for Dalmandi, the bazaar of beauty, thrives on scandal and sensation. Sadan Singh, young nephew of none other than Padam Singh, seduced by the wares of Dalmandi, is just one of the many ardent suitors who linger at her door. And Suman cannot help being gratified by his wooing.

But the winds of change are also blowing in the holy city: theosophy, Arya Samaj, new education, and public lectures have all made an impact. If the narrative sets out to explicitly affirm the need for social reform, it also makes clear that at times it has a very blunt edge, if indeed it is not a two-edged weapon. For once again, the author works with dual perspective. We are introduced to Vitthaldas, intrepid social reformer and zealous missionary in the cause of public weal. Vitthaldas' endeavours are viewed with some scepticism and not a little irony by the author. We find, as so often in Premchand's work, that the speech surrounding the characters creates "highly particularized character zones. These zones are formed from the fragments of character speech, from various forms for hidden transmission of someone else's word, from scattered sayings belonging to someone else's speech, from those invasions into authorial speech of other's expressive indicators (ellipses, questions, exclamations). Such a character zone is the field of action for a character's voice, encroaching in one way or another upon the author's voice" (Bakhtin 1981: 316). We find reflected in the author's words of introduction not only Vitthaldas's own estimation of his worth, but also a suggestion of the response of his sorely tried wife and the gratitude almost forced out of the recipients of his relentless service.

Babu Vitthaldas was the life and soul of most civic-minded institutions in the city. No task could be accomplished without his help. This personification of manly vigour (*purushartha*) carried his heavy burden gladly. Even if now and then he was pressed down by it, he never lost heart. He had no time for meals, found no leisure to sit at home, his wife complained of his affectionless behaviour. But in his obsession to serve the nation (*jati*), Vitthaldas had forgotten to look out for his own happiness and self-interest. He roamed around with single-minded intent, raising funds for orphanages or arranging tuition fees for poor students. And if some calamity befell the nation, he brimmed over with his love for it. In times of famine, he could be found going from village to village, bearing a load of flour on his head. The extent of his selfless service during cholera and plague epidemics simply amazed people. When just recently the Ganga was in flood, he did not look into his house for days at end; he had donated his entire wealth to the cause of the nation, but he showed not a trace of pride in this deed. He had not received higher education. His powers of oratory were also not above the ordinary. Often his thinking lacked maturity and far sight. He was neither particularly skilled in maneuver, nor clever or wise. But his patriotism was his virtue; this won him the respect of all and sundry in the city.

Vitthaldas inspires grudging respect, for he rushes in where angels fear to tread. He finds simple remedies, he eradicates social ills, even if the people he set out to help could themselves be eradicated in the process. He is not one to wallow in doubt. He makes up his mind to rescue Suman, though it is he who had spread word in the bazaar about Padam Singh's putative relationship with her in the brief period she could stay at his house. He knocks at her door and takes her to task immediately. The dialogue which ensues is a classic case of two people talking well past each other, the one hell bent on social reform, the other equally relentless in her dry and caustic analysis of the social processes of which she herself has been the victim:

> Vitthaldas: Now that our much revered Brahman ladies are treading this disgraceful path, there will be no stopping our decline. Suman, you are causing the head of the Hindu jati to be lowered.
>
> Suman answered very earnestly: *You* may think so, no one else does. A few gentlemen have just left after listening to the mujra, all were Hindus, but none of them seemed to have lowered heads. They are very glad that I have come here. And then, I am not the only Brahmani in the Mandi; straight away I could name a couple of others of very high families to you, whose biradaris were not prepared to take care of them; they were left feeling so helpless, that they came here. If the Hindu jati is not ashamed of itself, how long can frail women like us continue to protect it?

Vitthaldas responds by admitting that the Hindu jati had indeed fallen very low. But its moral repute (*maryada*) had always been held aloft by Hindu women. It was their truth and their stainless reputation which saved it from further decline. It was to protect Hindus that lakhs of women burnt to ashes in crackling flames. They chose to bear pain themselves in order to protect others. And if ordinary women did not shrink from doing so, the expectations of Brahman women could be higher still. It was only the lower caste woman who refused to bear up to suffering of any kind, who ran to her natal home if she took offence at her husband's behaviour. And if she did not get along even there, she ran off to the brothel. He ends with the ringing words:

> Suman you have lowered the head not only of the Brahman jati but of the entire Hindu jati.

Suman has to correct him repeatedly when he insists that people don't respect her. The highest in the city bow at her feet. There is not one in the city who does not go wild with delight when she deigns to cast a gracious look at him. But Suman is also on the defensive, Vitthaldas's words have stung her to the quick. She had resolved, she tells him, that she will dance and sing but have no sexual contact with the men who surround her.[9] Vitthaldas, never at a loss for remedies, recommends work at home, cottage economy, to her. But even he can respond only evasively to Suman's query: "Does a single lover of the Hindu jati exist anywhere, who would be willing to spend fifty rupees a month for my upkeep?"

Vitthaldas shows little understanding for the predicament in which Suman finds herself; he is more occupied with her position as a symbol of respectable Hindu womanhood. As a once-married Brahman woman she has to be extracted from the moral filth in which she finds herself presently mired. When he offers to find a place for her in a home for widows, she is extremely sceptical about the welcome she would receive there. She asks: is she expected to give up her life of relative security and comfort only to be regarded as a barely tolerated object of charity in a pleasureless abode? But Vitthaldas has yet larger goals in mind. He wants to remove prostitution from public places and do away entirely with *mujra*, the social custom of song and dance on festive occasions.

In focalizing primarily through Suman, on the one hand, and Sadan Singh and Padam Singh on the other, the author seems to be propelling the reader to view the action through at least three perspectives which seem diametrically opposed to one another and often at odds with the professed auctorial stance. Sadan seeks to situate himself in the city as a man of some affluence and influence. He needs financial resources in order to woo Suman and he maneuvers to come by them. Later he goes through a spell of enthusiasm for social reform, and we share his reflections. But he finds short cuts and takes morally dubious decisions, finding a measure of stability only at the end. Padam Singh sees through the falsity of many social situations but seldom finds the courage to act boldly when directly challenged to do so. Once again we are privy to his thought processes. His marriage, troubled as it is, comes nearest to representing marital bliss, and his social reform endeavour comes nearest to being well-reasoned and reflected. But we see him falter in his actions and seek shelter behind self-justification. If the author approves of marriage as a social institution, if he expressly supports social reform, in the action of its proponents, we see these aims questioned if not contradicted. Of the three, the most empathetically portrayed character is that of Suman, and it is her speech that is psychologically more persuasive as also more convincing in its analysis of societal ill and of the resolutions she herself is compelled to seek. Suman's reasoned reflection, her spirited exchange with Gajadhar and her decision to accept Bholi's offer rather than seek reconciliation with her husband through the mediation of Subhadra, seem entirely comprehensible, given her particular situation. It is then the speech which surrounds Suman which most consistently refracts professed authorial intention.[10]

As Suman's reflections bring home to the reader, marriage is a commercial transaction, particularly when negotiated on the basis of dowry. It is not unlike prostitution, which deals more blatantly with the financial exchange involved in buying sexual and other services. Society officially sanctions marriage while at the same time virtually imprisoning the woman concerned in cheerless domestic servitude. It does not explicitly sanction prostitution but at the same time it elevates the woman concerned, if she possesses the required talent and beauty, to a position of glamour and relative autonomy, celebrating her charms in lavish public display. Suman makes a radical statement when speaking of both arrangements, the ones in which she and Bholi respectively find themselves, in terms of *dukan*, or warehouse. In what ways is the one institution morally more elevated than the other? The social conduct of the men who patronize the warehouse would contradict their public stand on the issue.

Then there is the vexed issue of social reform which sets out to change the older social order, elitist and decadent, while attempting to put into its place a more middle class morality, itself often flawed and feelingless. Those who advocate social reform, in self-promotion or bouts of self remorse, seem more concerned with their own public and private ends than with the fate of individuals such as Suman, who are caught in these winds of change and buffeted about, compelled to believe in a public cause rather than in their own hard-won comfort and sense of well being. The Hindu jati Vitthaldas speaks of seems to be a caricature of its own projected image and the social reform being called for no more than one more measure of social hypocrisy. Out and out social reformers such as Vitthaldas are often callous and unfeeling or those such as Padam Singh are narcissistic and noble in turns. The author seems to regard them with some aversion and suspicion, though his ostensible sympathies are reserved for the cause they represent.

The rhetoric of social reform, as the author proceeds to show us, seems particularly amenable to misuse for communal purposes. Though Suman is Hindu, many of the courtesans are Muslim and the leading members of Hindu society are not slow to attribute the moral responsibility for the evils of this profession on Muslim culture. The clientele is mixed, but most shops around the Chauk are Hindu owned, and it is the Hindu merchants who stand to suffer the greatest financial loss if the women are made to shift to another part of town. Could this be a ploy of the Muslim faction in the Municipal Board to use this opportunity to harm Hindus? The communalization of the debate seems inevitable.

Social Reform and the Communalization of the Dalmandi Debate in the Municipal Board

Most North Indian cities underwent radical structural transformations in the years after the 1857 uprising. Not only was there a major reshuffling of social hierarchies, space itself was differently allocated and distributed. At the centre of most cities of any size or splendour, was situated the Chauk or town square and somewhere near it if not located right there, were the courtesans of the city, the *tawa'ifs* or *ba'is*. In the cleansing operations which followed in cities such as Lucknow and Delhi, whole mohallas were uprooted, though in spite of widespread damage and destruction, the famous Chauk in Lucknow was saved, and with it, its equally renowned courtesans, sustaining and surviving the loss of their old clientele.[11] In most cities, they would be driven out finally only after 1956.[12]

The civic administration of the city also underwent rapid change after 1857. The Acts of the preceding decades regarding city government had called upon the inhabitants to take the initiative to restore civic order and administer it, for the system of mohalla or neighbourhood control and administration had begun to dissolve in the early decades of the nineteenth century. But these pre-1857 measures to put a civic government in place had been desultory and half-hearted. However, once the Army Sanitary Commission (1861) painted a dismal picture of the hygiene observed in Indian towns, more directives were issued to look after water supply, lighting and sanitation under a more comprehensive policy laid down by the Government of India. Between 1864–68 three further acts were passed, for Lucknow (1864), Punjab (1867) and the

North West Provinces (1868). The municipal administrative body was no longer to be voluntary, it was henceforth to be a nominated body, though still expected to cater primarily to the needs and interests of the British. If the nominated members largely consisted of those prominent citizens, the bankers, merchants and members of the aristocracy who had remained loyal in 1857, willy-nilly it also ended up including the new professionals.[13] However, the state government and the district magistrate continued to control, dissolve or supersede a municipal board. A power tussle was almost preprogrammed. "The attraction for chairmanship to the intelligentsia was its executive power and the bureaucracy was unwilling to be divested of the influence of the board chairmanship".[14] Once the Municipalities Act of 1916 came into force, the non-official members jostled for power amongst themselves. The municipal boards were chided in the *East India Progress and Condition of India During 1917–1918* for the prevalence of fissures, though these fissures had almost certainly also been fostered by the imperial government. It complained of the "spirit of faction" which prevailed on several boards, it admitted, surely not without a smirk, that the "maintenance of a double record of proceedings was not unknown". It is necessary to recall this long history of power constitution within the city and tussle it involved, in order to get some idea of the privilege it meant to be nominated or elected as a Board member. The stakes could not but be high.

Benares in the second decade of the twentieth century was also undergoing these changes, which in turn was restructuring the very social fabric of the city. Hindus and Muslims saw themselves as having to band together in their religion-based groups in order to defend what they saw as their common interest. But the views held by the individual members of the respective groups within each faction were anything but uniform. The merchants differed from the landed gentry in the interests they represented, while the new professionals, who came into more contact with modernizing trends, were likely to be the ones who were most earnest about social reform. However, even they could ill afford to ignore group loyalties and were thus automatically sucked into the respective Hindu and Muslim factions. What kind of changes did the different social strata desire and with what rationale? As Alok Rai puts it, "(t)he early 1900s were a time of considerable ideological ferment in India. A sort of damaged modernity seemed available under colonial aegis, a modernity at once embryonic and addled" (Rai 1991: 197).

Premchand followed these developments with much interest. In a remarkable fictional representation of these debates and the vested interests which sought to control city spaces for financial and social ends, he showed in precise detail, how in the matter of the displacement of the courtesans, communalism itself was a construct, a public face which masked commercial and political agendas and sought validation through it.[15] The professed concern for the welfare of the courtesans was an issue which was bandied about in a power battle between the communalized Hindu and Muslim factions in the Municipal Board in a manner which made the courtesans seem more like victims of social reform than its beneficiaries.

If Vitthaldas's indefatigable efforts did not galvanize people into action, once Padam Singh took up the challenge, the movement to save the courtesans while banishing them from the city centre gathered unexpected momentum. Padam Singh was the son of a landowning family and a lawyer of repute; he was part of the old gentry, but also one of the new professionals, who was liberal in his views, in spite of the mujra he had permitted to be held in his house. Propelled ever forward by his guilty conscience, for he realized that he had allowed Suman to be cast out of his house too hastily, he brought the matter to the attention of the Municipal Board.

Premchand records the Municipal Board debates on the issue of the courtesans and the cleansing of the city centre with some earnestness and some outright satire. Once again, the individual members of the Board, introduced individually within their own factions, the Hindu and the Muslim, enter the stage under the spot light of slanted speech which forms veritable character zones, integrating the speech of the characters themselves in invisible quotation marks, as also the public appraisal of them. The Board has eighteen members, eight Muslims and ten Hindus, the majority well educated. Within each camp there are social reform enthusiasts but also dissidents who take shelter behind communal verbiage, for entirely different ends, as we shall see.

The interest groups represented in the Muslim and Hindu camps are merchants and shopkeepers, religious leaders, landowners (*zamindars*), retired civil servants and the new professionals. By and large, both the Hindu and the Muslim members are equally unwilling to lose the income the courtesans generate for them and at any rate do not wish to seal their own source of pleasure and enjoyment. However, they choose their words carefully, for the new professionals give the tone, and all subscribe to it publicly.

The deliberations of the Muslim camp are presented first. Haji Hashim Sahib exercises a great deal of influence on the populace and is generally considered the leader of the Muslims. He is against the removal of the prostitutes. Maulana Teg Ali is the Vali of the Imam Bara. Munshi Abulwafa is the owner of a perfume and oil factory. He owns a chain of oil and perfume stores in several big cities across North India. Munshi Abdul Latif is a big landowner, an absentee landlord who resides in the city. He loves poetry and is considered a good poet himself. Shakir Beg and Sharif Hasan are lawyers. They have very progressive social views. Sayyid Shafakat Ali is a pensioned Deputy Collector and Khan Sahib Shohrat Khan is a famous Hakim. These last two stay away from meetings and assemblies, but they do not lack generosity and deeper reflection. Both are religiously inclined. They command great respect in society.

Haji Hashim makes of the matter a communal challenge. He asks: are you aware of this new ruse employed by our brothers of the same native land?[16] And Abulwafa, the oil and perfume factory owner, pointing out that ninety percent of the tawa'ifs are Muslim, proclaims that this is an attempt by the Hindus to reduce the number of Muslims in the city. The Hindus are out to claim as large a number of the populace as they can, including the Doms.

The two gentlemen who generally stay away from meetings are the most sweeping in their condemnation. Sayyid Shafakat Ali, the pensioned Deputy Collector, wants the prostitutes to be removed altogether from the city. They are contagious, they spread immorality around them. Hakim Shohrat Khan goes a step further and wants to remove them altogether from Hindustan. They are black serpents with poison in their eyes. He views it as a matter of great misfortune that most *tawa'ifs* call themselves Muslims.

The new professionals are more charitable. Sharaf Hasan, the lawyer, says that he sees no harm in the fact that they call themselves Muslims. The problem is that no one tries to help them rectify their ways. Once a woman falls, Islam becomes indifferent to her. Surely these women also repent of their ways, but they have to continue in their profession because they know no better and because they have to consider the future of their children. Now if grooms could be found for their daughters, the situation would improve considerably. He does not want to reject the suggestion on a political basis, that is, merely because it comes from the Hindus. When Teg Ali, the Vali of the Imam Bara, taunts him, Shakir Beg, the other lawyer, asks people to calm down and offers instead a more moderate suggestion. The *tawa'ifs* should be removed from the centre to some corner of the city.

The landowning gentry are not unlike the merchants in their aversion to this change, though their reasons are very different. Thus Abdul Latif, the Zamindar, does not want to let go of the pleasure of the courtesan's cultivated company. His views will be echoed by his Hindu counterpart. He speaks in jest but means it seriously: there are all kinds of shops in the Chauk, soaps, leather, oil, textiles, pots and pans—why not the warehouse of beauty? This kind of levity provokes sharp responses. Haji Hashim persists in regarding it as a Hindu-Muslim matter and has no hesitation in proclaiming that he will resist it precisely because it comes from the Hindus. Abulwafa, the factory owner, agrees, for the merchants on both sides are in any case against the move. But they present their arguments in communal terms.

However Sayyid Shafakat Ali, the pensioned Deputy Collector is a genuine social reformer. He says that if the Hindus are doing something which is advantageous for all, he has no hesitation in joining them. He is against opposition for opposition's sake. There is no conclusive agreement on the matter and the assembly disperses.

Though the Hindus and Muslims regard themselves and each other as homogeneous groups, they are actually split into different interest groups, which have almost parallel interests, which cut across the religious divide. The Hindu camp in its agreements and disagreements presents thus almost a mirror image of its Muslim counterpart. The financially and socially most powerful are against the move. The financial losses the Hindu merchants incur will be regarded as political victory for the Muslim group. Seth Balbhadra Das, wealthy merchant, is the chairman of the Board and Lala Chimanlal and Dinanath Tiwari are also important leaders of the merchant community while Lala Bhagat Ram is a contractor. Seth Balbhadra Das and Chimanlal own most of the stores in the Chauk, Dinanath most of the real estate in Dalmandi. Lala Bhagat Ram's transactions are almost entirely dependent on Dinanath. These four men are decidedly against the removal of the courtesans from the centre of the city. Only the four professionals are clearly for the removal of the prostitutes from Dalmandi: the two lawyers, Pandit Padam Singh and Rustam Bhai and two others in new professions: Ramesh Datt, College Professor and Theosophist and Prabhakar Rao, editor of the daily *Jagat*. Vitthaldas is not a member of the Municipal Board.

The Vice Chairman of the Board is Dr. Shyamacharan, medical practitioner, and ostensibly the most Anglicized of the group. He fears official disapproval. He wavers in his decision, preferring to shift the responsibility on the larger Municipal Council. And finally, there is Kunwar Aniruddha Bahadur Singh, the biggest Zamindar in the district. He is liberal in his views, he knows the best of both cultures, Surdas and Mozart, but he is reluctant to give up the old style pleasures of the cultivated. Both sides hope to win over Dr. Shyamacharan and Kunwar Sahib, for victory or defeat seems to depend on them.

Since Padam Singh, the prime mover in the game is still away at the wedding of Shanta and Sadan Singh (which will not take place, once it is discovered that Shanta is Suman's sister), Seth Balbhadra Das calls a meeting to his *susajjit baradari*, ornamental reception rooms. Prabhakar Rao was known to be a fanatic opponent of the Muslims, so Balbhadra Das hopes to win him over by giving the debate a communal twist.

Dinanath Tiwari, owner of most of the real estate in Dalmandi, begins the debate in words similar to Haji Hashim's: "Our Muslim brothers have shown great generosity in this matter. They are operating on the principle of killing two birds with one stone. On the one hand they hope to acquire a good reputation through social reform, on the other, they have found a pretext to damage the Hindus. How can they lose such a good opportunity to gain a good name?"

Chimanlal who owns most of the stores in Chauk, says with a straight face:

> I have no dealings with politics, nor do I go anywhere near them. But I have not the least hesitation in saying that our Muslim brothers have a vice-like grip on our necks. Most of the real estate in Dalmandi and the Chauk belongs to Hindus. If the Municipal Board accepts this, the Hindus will be wiped out. If one wants to learn how to hurt on the sly, one need only look at the Muslims ... Sadly, some of our Hindu brothers have become puppets in their hands. They don't know what harm they will cause their own jati by this ill enthusiasm.

Prabhakar Rao knows he is being trapped by this logic.

> He looked helplessly towards Rustam Bhai, asking to be saved. Rustam Bhai was a fearless and clear speaking man. He stood up to answer Chiman Lal. "It grieves me to see you people twist a social matter to give it the appearance of Hindu-Muslim conflict (...). It can bring profit to some Hindu merchants to riddle it with conflict, but it is difficult to estimate the harm it does to the cause of the community. There is little doubt that if the proposal is accepted, Hindu merchants will suffer the greater loss, but surely there will be repercussions for the Muslims also. There is no small number of Muslim shops in Chauk and Dalmandi. We should not, moved by the spirit of dispute and opposition, suspect the integrity of our Muslim brothers. They are acting in the interest of common weal; it is another matter if the Hindus are suffering greater losses thereby. I firmly believe that even if the Muslims were hit by the greater loss, they would take a similar decision. If you agree truly that the proposal has been made with view to the reform of an evil social custom, there should be no hesitation in accepting it, whatever the financial loss. Money should play no part when it comes to a matter of social conduct.

Prabhakar Rao is much reassured by this clear analysis of the matter. It transcends the communal. The issue is to be decided on moral and social grounds. One takes losses in one's stride, if one is to act for the welfare of the populace at large. He gives the example of the loss the government agreed to face in view of the moral rectitude of giving up the opium trade with China. This seems to be explicit social satire. Kunwar Aniruddha Singh, the leading Zamindar of the region, points out dryly:

> Sir, you remain engrossed in the editorial work of your journal. When do you find the time for the pleasures of life? But we, who are free of such anxieties, need to have some amusement (...). Today, you bring the proposal of ousting the prostitutes of the city, tomorrow you will say that no one within the municipal district should dance, sing, perform mujra without permission: it will become impossible for us to live.

When Prabhakar Rao suggests that he take up reading as a pastime, Kunwar Sahib answers with obvious pleasure: "We are forbidden to read". He knows the dance steps of France and Spain, he can play so well on the piano as to put Mozart himself to shame, he has full knowledge of English social custom. He knows when to don a solar hat and when the turban. He also reads books, he has cupboards full of them, but has no need to remain glued to them. He also seeks other refined pleasures of the senses. "We will be wiped out, if this proposal of yours comes through". Dr. Shyamacharan, the other undecided member, remains evasive in his response. He first wants to put these questions in the Legislative Council.

Now follows the delicious social comedy. Seth Balbhadra Das is convinced of victory, he expects the proposal to be rejected. "Now he could afford to remain impartial, which was the prerogative of the Chairman. He analyzed the proposal in a weighty speech. He said:

> I do not believe in social upheaval. It is my belief that the society which needs change takes care of it naturally, in the course of things. Foreign travel, caste difference, the useless restrictions on commensality, all bow their heads, when faced with the changes wrought by time. I want to let society operate freely in this regard. Once there is full consensus that people do not want to see prostitutes in the Chauk, what power in the world would be able to remain deaf to it?

In conclusion, Sethji spoke in very moving tones:

> We are very proud of our music. Even those who are acquainted with the music of France and Italy acknowledge the *bhava*, *rasa* and blissful peace of Indian song. But such is the wheel of time, that the very institution which our reformers are straining to rip out by its roots, holds the proprietorship of this divine store of riches. Do you really want to destroy this institution and cruelly turn into dust this priceless gift of our ancestors? Did you know that all the national and religious feelings that remain to us are due entirely to our music, otherwise, no one today would even know the names of Rama, Krishna and Shiva? (...) No ill custom is eradicated by neglect or cruelty. It can only be destroyed by education, knowledge and compassion. There is no direct path to heaven. We will have to cross the *Vaitarani*. Those who think they can spring directly into heaven by the good graces of some Mahatma are not more laughable than those who think that by banishing prostitution, all the poverty and sorrow of India will be eradicated and a new sun will rise in the East.

In order to appreciate the irony of these words, we need to recall that in the second decade of the twentieth century, Indian music had come increasingly to be projected as Hindu music with an ancient, Sanskritic past, its Muslim component, in production and performance, had been almost entirely sidelined. This classicizing process had been set in motion in the last decades of the nineteenth century on both sides of the subcontinent. In Calcutta, Sourindra Mohan Tagore had begun to write treatises on Hindu music, in the Bombay Presidency, the Parsis, followed by like-minded citizens in Pune, had begun to found musical societies which made this heritage available to middle class boys and eventually also to girls. The dissemination of this music through the towering figures of Vishnu Narayan Bhatkhande (1860–1936) and Pandit Vishnu Digambar Paluskar (1872–1931) completed the process in the following decades. Bhatkhande wrote its history, set up a standard notational system and codified the vast storehouse of ragas and raginis for teaching use. Paluskar evolved new modes of music pedagogy and set up a chain of schools, the hugely successful Gandharva Mahavidyalayas. A series of national conferences, convened severally by both men, further spread the word about the ancient genealogy of Hindu music as also its contemporary respectability.[17] However, public singing by females continued to be the domain of courtesans till well into the 1920s. It was only in 1924 and thanks entirely to the efforts of Paluskar, that it became possible for a female performer to *sit* on stage while performing. "Till then, female performers, who were mostly prostitutes, were supposed to stand on the stage when they sang. It is significant that Paluskar's school was the first to break this custom" (Bhagwat 1987: 112).

Seth Balbhadradas, to serve his own ends, uses precisely this newly won status of music as national heritage to eulogize the courtesans and the monopoly of music that they still hold. In order to save the national heritage, as he pleads so eloquently, these women must be allowed to practise their art publicly, till such future date when others can come forward to relieve them of their burden. To be effective, social change has to be gradual. It cannot be brought about by putting ill-conceived notions of social reform into operation.

"*Unfortunate Suman*"

However, in the final run, it is the social reformers who prevail. Padam Singh returns to the scene and the matter is taken up again with renewed energy. The Municipal Board meets, all its members are present on the occasion, and Padam Singh presents the proposition. It has three sections. 1) The prostitutes should be removed from the central part of the city to some settlement far away from it. 2) They should be forbidden to move about in the central spots and parks of the city. 3) Those who arrange for their performances should be heavily taxed and all such festivities should be banned from public places. Kunwar Aniruddha Bahadur Singh had feared just this.

After some debate, whereby the same kinds of opinions are aired once again, Seth Balbhadra Das calls for a vote. The first section of the proposition is passed with the provision that those who marry within nine months or learn some craft to earn their livelihood be allowed to remain where they are. When Padam Singh accepts the provision, the reformist members who had thus far supported him, desert him. Nine members vote for it, and eight against it. Thus it is passed with the majority of one vote, that of Seth Balbhadra Das. The social reformers, in spite of being in actual minority, and in spite of their last minute split, have won a narrow victory by virtue of their moral stand, which not all feel free to oppose publicly. There is much acrimony after this is brought to pass, for most are unhappy with the outcome and Prabhakar Rao blocks all further activity. Once the dust settles, the other two sections are also passed. The prostitutes leave to be resettled on the outskirts of the town, repenting, almost all of them, of their lives. They now find themselves in Alaipur, the first station after Benares Cantonment on the Choti or narrow gauge railway line. Have all social problems connected to prostitution been resolved? One needs only recall Seth Balbhadradas's passionate words appealing for the retention of the women at the centre of the city: "Those who think they can spring directly into heaven by the good graces of some Mahatma are not more laughable than those who think that by banishing prostitution, all the poverty and sorrow of India will be eradicated and a new sun will rise in the East".

Meanwhile, pushed by events in the public sphere, matters have also moved on in the private sphere. When Suman and Shanta were cast out of the widow's home where they had initially found refuge, Sadan had happened to meet the two sisters. Through a series of coincidences and chance encounters, it becomes possible for Shanta and Sadan Singh to be united as a married couple. A now pregnant Shanta is reminded incessantly of Sadan's former attachment to Suman and she watches each move Suman makes with suspicion and dislike. Engrossed in their vision of the future, Shanta and Sadan, now backed by his family, pay little positive heed to *abhagini Suman*, unfortunate Suman. She finds herself increasingly unwelcome in their house.

On her way to drown herself in the river, Suman meets Gajanand Swami, who is none other than Gajadhar, turned secular Sadhu and social reformer. He is at the centre of a movement to found a house for the education of the daughters of former prostitutes and he invites her to head that ascetic institution, set up somewhere on the outskirts of the city. The little girls will learn how to perform domestic chores effectively, sew, tend the vegetable patch and sing patriotic songs. Suman is a wan, wry figure at the end of the novel, educator of prostitutes' daughters, who have no future in respectable society as wives or mothers. It was this pious end, which seems to have made not only for the respectability of the novel but also its contemporary popularity.

In this puritanical period at the beginning of the twentieth century, a novel which dealt with vice without sinking into it, which made a prostitute, who did not prostitute herself, into its central protagonist, had found thus a way to legitimate its intent, while capturing the reader's attention with its sensational subject matter and deep psychological insight. And this, in fact, was Premchand's great achievement as a novelist. He conveyed several messages at once, making for surprising turns and twists in the plot, for people behaved unexpectedly, motivated by complex needs and desires. On the one hand, beautiful, passionate, headstrong Suman had been tamed and this could be seen as the triumph of virtue. But on the other hand, she was also a tragic figure, the victim of a social system that first failed to accord her the respect she claimed as a virtuous wife, then made much of her as a beautiful courtesan, but the moment she stepped out of that role, proceeded to regard her as little more than a social outcaste who could only be accommodated as one who looked after others of her kind. In the final run, very few were left to support her in the virtuous role allotted to her. Padam Singh did not even get down from the carriage, out of a mixture of shame, embarrassment and unwillingness to get involved, when in the last pages of the novel Subhadra, his wife, visits Suman at Sevasadan, the exemplary institution located somewhere on the outskirts of the town. The dialogue between the two women, with which the book closes, is provided without any further narratological intervention. And it is in citing this conversation that I too should like to close this analysis of a remarkable novel. It is a dialogically particularly poignant moment, saying one thing, while also indicating several others.

For one, the author shows us that the usefulness of Sevasadan, the institution, is itself contestable, for it continues to be regarded as a house of ill repute. The little girls have an uncertain future, it is unlikely that any man will come forward to marry them, as Suman tells Subhadra. For another, even its benefactors shun it. It is clear that Shanta does not consider visiting her sister there. Suman fears also that Padam Singh will not let his wife come for another visit, though Subhadra assures her that, in fact, she intends to drag Shamaji along to visit with her next weekend. But Subhadra herself is suspicious. Does Sadan still visit Suman? Suman's denial and closing words of gratitude speak more of her loneliness and isolation, and her need to be connected to the world she once knew, than of any real satisfaction derived from being there.

> When Subhadra was leaving, Suman said in a piteous voice—I will be waiting for you this Sunday.
> Subhadra—I shall definitely come.
> Suman—Shanta is well, isn't she?
> Subhadra—I just received a letter from them. Doesn't Sadan come here?
> Suman—No, but he sends a two-rupee donation each month.
> Subhadra—It's time for me to leave, but you don't need to see me off.
> Suman—Your coming here fills me with content. Your devotion, your affection, your extraordinary efforts, which shall I praise first? You are a true jewel in the community of women. (With moist eyes) I shall consider myself your servant. As long as I live, I will remain grateful to you. You held my arm and saved me from drowning. May you both receive God's grace always.

It is this level of realism, I would suggest, which asks to be read against the grain, which accounts for the novel's enduring claim to fame. The story of Suman, of her decline, rise and decline—the questions it raises about the social institutions fostered by the modernizing middle classes operating under the benign/malignant aegis of British imperial rule, about the moral claims of the self-cleansing undertaken by the city and of the marginalization thereby not only of the courtesan but also of the wife—has lost neither its poignancy nor its significance in today's India.

Notes

1 I refer to the city as "Kashi" when speaking of it as a pilgrimage site with all the mythical connotations that this conjures up. Otherwise, in order to remain near Premchand's usage—he calls the city "Banaras" in Hindi—I use the British-Indian term "Benares", since that was the usage in English till very lately, when the name of the city was officially changed to "Varanasi".

2 The Urdu term for courtesans is "*tawai'f*", the one more current in Hindi now is "*Bai*". For the sake of uniformity, I have used the term "courtesan" throughout this essay.

3 The text used for the translation of passages from the novel is the twenty-volume edition of Premchand's collected works, *Premchand Rachnavali*, published by Janavani Prakashan in Delhi in 1996; *Sevasadan* is contained in the second volume.

4 See Motichandra [1962] 1985: 46, 89, 145.

5 Harishchandra is reported to have told a friend, when he was moved to write a poem after the visit to a courtesan: "It is for this that I cultivate intercourse with them. Tell me, how else could this true motif have been obtained?" For further details see Dalmia (1997: 128).

6 The play was published, *Premyogini* subsequently as *Premjogini,* in three installments in *Harishchandra chandrika*, the literary journal Harishchandra edited, between August 1874 and April 1875. The text is also accessible in the first volume of his Complete Works, *Granthavali* I (319–354), whence the following quotations.

7 See Guha-Thakurta (1992: 149–59 and 175–87) for an account of the vast influence exercised by E.B. Havell (Superintendent of the Madras School of Arts in 1884 and later of the Calcutta School of Art), along with Ananda Coomaraswamy, in turning the tide of public opinion in Britain in the cause of Indian "fine arts". Havell was to write *Benares the Sacred City. Sketches of Hindu Life and Religion* in 1905.

8 Written by Devakinandan Khatri (1861–1913), the novel (1888), a modified version of the Urdu *tilism* or thriller, proved to be an all time best-seller, inspiring Khatri to write several follow-ups, amongst them, the 11-part *Chandrakanta santiti* or the progeny of Chandrakanta.

9 Premchand thus finds a route to preserve a measure of respectability for Suman; she is the courtesan who manages to remain physically unscathed by the profession she practices. This is the escape route also taken by Bhagavaticharan Varma (born 1913) in his evergreen novel *Chitralekha* (1934) which is set in the Gupta period and also focuses on a Brahman woman, this time a young widow, who rises to fame as a dancer in the city of Pataliputra. This sexual abstinence allows for a reading of Suman's character as an "angelic victim" and makes for a peculiar kind of relationship to the society which has victimized her. Alok Rai has spoken of "the cult of the angelic victim. This fecund genre, particularly in writings dealing with the question of women, may be considered a sub-set of the cult of the ennobled poor. Here also the game consists of representing the woman-victim as being simultaneously damaged and undamaged, wronged but essentially unharmed, both needing salvation and deserving of it. The fact that such representations are necessarily ambiguous in the fashion just indicated, interacts curiously with the confusion regarding the proper relation between wrongs and rights. The consequence is a muddled valorization of victimhood which, curiously, works to soften and make manageable the critique of victimizing societies and institutions" (1991: 201).

10 As Bakhtin has shown: "The language used by characters in the novel, how they speak, is verbally and semantically autonomous; each character's speech possesses it own belief system, since each is the speech of another in another's language; thus it may also refract authorial intentions and consequently may to a certain degree constitute a second language for the author. Moreover the character speech almost always influences authorial character speech (and sometimes powerfully

so), sprinkling it with another's words (that is, the speech of a character perceived as the concealed speech of another) and in this way introducing into it stratification and speech diversity" (1981: 3–4).

11 See Oldenburg (1984) for an account of the changes in the city landscape of Lucknow, following the violence perpetrated by the British in the wake of the 1857 uprising, particularly the havoc wrought in the lives of the courtesans.

12 Amritlal Nagar has recorded the reactions of the courtesans of Lucknow to this final ousting from the city and their profession (1994).

13 See the pioneering study by Narayani Gupta (1981) for the changes brought about by the British in the imperial city of Delhi, once they took charge. On the politics of the municipal boards they created, see pp. 115–122 and 213–216.

14 This account of the evolution of municipal government of cities is based on the first chapter, "Historical Perspective of Urban Local Government in India", of Debidas Nanda's book (1998). The citation is from page 15. This Municipalities Act of 1916 continues to govern Uttar Pradesh even today.

15 The move to oust the courtesans from the centre of the city was apparently a real consideration at the time, though it was not ever effected. Premchand's is therefore a fictional account, which Professor Anand Krishna attributes to his Arya Samaj indoctrination (personal interview, January 7, 2003). The famous courtesans of Benares continued to operate in Dalmandi till well into the early 1950s. Professor Anand Krishna recalls that a *majlis* or assembly used to be held in Girija Bai's *kotha* which was situated just opposite Chitra cinema hall near the Chauk.

16 As Harish Trivedi points out in his review of the English translation of the Urdu version of the novel (2004), Haji Hashim uses the term *biradaran-e-vatan*, brothers of the same native land, when speaking of the Hindus and the more intimate *biradaran-e-millat*, brothers in religious faith, when addressing his Muslim compatriots.

17 See Bakhle's forthcoming monograph on the nationalization of Indian music, whence also the above information.

References

Bakhle, Janaki. forthc. *Two Men and Music: Nationalism and the Making of an Indian Classical Music*. New York: Oxford University Press.

Bakhtin, M. M. 1981. *The Dialogic Imagination: Four Essays*. Austin: University of Texas Press.

Bhagwat, Neela. 1987. Vishnu Digambar Paluskar, *Journal of Arts and Ideas*. 14/15, July–December.

Hariścandra. *Bhāratendu granthāvalī. Pahlā khaṇḍ* (cited as *Granthāvalī* I), ed. Śivprasād Miśra "Rudra" Kāśikeya. 2nd rev. ed. Banaras: Nāgarī Pracāriṇī Sabhā, 1975 (first edited by Brajratnadās 1950).

Dalmia, Vasudha. 1997. *The Nationalization of Hindu Traditions: Bhāratendu Hariśchandra and Nineteenth-Century Banaras*. Delhi: Oxford University Press.

East India Progress and Condition of India During 1917–1918. London, 1919.

Guha-Thakurta, Tapati. 1992. *The Making of a New 'Indian' Art: Artists, aesthetics and nationalism in Bengal, c.1850–1920*. Cambridge: Cambridge University Press.

Gupta, Narayani. 1981. *Delhi Between Two Empires 1803–1931: Society, Government and Urban Growth*. Delhi: Oxford University Press.

Havell, E. B. 1905. *Benares the Sacred City. Sketches of Hindu Life and Religion*. London. (Reprinted by Vishwavidyalaya Prakashan, Varanasi, 1990).
Medhasananda, Swami. 2001. *Varanasi at the Crossroads: A Panoramic View of Early Modern Varanasi and the Story of its Transition*. Kolkata: The Ramakrishna Mission Institute of Culture.
Motīcandra. [1962] 1985. *Kāśī kā Itihās*. Banaras: Viśvavidyālaya Prakāśan.
Nāgar, Amṛtlāl. 1994. *Ye Koṭhevāliyāṃ*. Allahabad: Lokbhāratī Prakāśan.
Nanda, Debidas. 1998. *Municipal Administration in India*. Varanasi: Ganga Kaveri Publishing House.
Oldenburg, Veena Talwar. 1984. *The Making of Colonial Lucknow, 1856–1877*. Princeton: Princeton University Press.
Premchand. *Nirmala*. [1927] 1999. Translated with an Afterword by Alok Rai. Delhi: Oxford University Press.
Premchand. 1996. *Premcand Racnāvalī*. 20 Volumes. Delhi: Janavāṇī Prakāśan.
Prinsep, James. 1831–33. *Benares Illustrated. A Series of Drawings*. Third Series. Calcutta: Baptist Missionary Press London: Smith, Elder and Company. (Reprinted by Viśvavidyālaya Prakāśan, Varanasi, 1996).
Trivedi, Harish. 2004. The Power of Premchand, *The Hindu Literary Review*, May 2.
Varmā, Bhagvatīcaraṇ. [1934] 1993. *Citralekhā*. Delhi: Rājkamal Prakāśan.

Contributors

Hans Bakker is Gonda Professor of the History of Hinduism in the Sanskrit Tradition and Indian Philosophy at the University of Groningen. He is editor of the *Indo-Iranian Journal* and the *Groningen Oriental Studies*. His research focuses on the history and religions of classical and early medieval India. Among his major publications are books on Ayodhya, the Vakatakas, and a critical edition of the *Skandapurāṇa*.

Joachim K. Bautze is an Indian art historian who has conducted several years of field-research in India and Myanmar. He has both overseen and participated in a good number of exhibitions on Indian art in Germany and abroad. In addition, he has been lecturing on the arts of India at the Free University of Berlin, the South Asia Institute of the University of Heidelberg, and Wako and Chuo universities, both in Tokyo. Several of his publications deal with the interaction of Indian and Western painting.

Elizabeth Chalier-Visuvalingam is an independent researcher who holds a state doctorate in the anthropology of Indian religions from the University of Paris-X. Her interests are Indian philosophy, especially Kashmir Śaivism, Tantric rituals in India and Nepal, and the Jewish minorities of India. She currently teaches in Chicago (under the French Ministry of Education). A number of her publications focus on Bhairava in Banaras and Nepal.

Vasudha Dalmia is Professor of Hindi at the University of California, Berkeley. She has researched and published widely on Hinduism, colonial and post-colonial Hindi literature and medieval Indian religiosity. She has authored books on Bharatendu Harischandra in Banaras and on the politics of modern Indian theatre.

Sandria B. Freitag is Associate Director of the Center for South Asian Studies in the Research Triangle area of North Carolina, and also teaches art history and history at the three member universities (University of North Carolina, Duke University, and North Carolina State University). Her research links popular culture to larger political and socio-cultural developments in the subcontinent in the 19th and 20th centuries. Her publications include a volume on the emergence of communalism in North India.

Martin Gaenszle teaches anthropology at the South Asia Institute, University of Heidelberg. Presently he is affiliated with Leipzig University and works in a project documenting two endangered language traditions in Nepal. He has carried out fieldwork in various parts of east Nepal as well as in Banaras and published on local history, ethnic identity, and performative traditions.

Jörg Gengnagel is a research fellow of the Collaborative Research Centre "Dynamics of Ritual" at the South Asia Institute, University of Heidelberg. His fields of research include court ritual in Rajasthan, the sacred topography of Banaras and dualistic Śaivism. He has co-edited a book on ritual in South Asia and currently is preparing a monograph on the religious cartography of Banaras for publication.

Niels Gutschow was born in Hamburg and lives in Abtsteinach and Bhaktapur. He is an architect, architectural historian and architectural anthropologist who has worked since 1970 in Nepal and India, His research has resulted in numerous books, the most recent being a comprehensive monograph on the sacred landscape of Banaras.

Nita Kumar is a historian and anthropologist of urbanism, modernity, education, family, and culture. She is a co-editor of the journal *Gender and History* and the Honorary Director of the NGO *Nirman* that runs the Centre for Postcolonial Education and Vidyashram—the Southpoint School. In addition, she holds the Brown Family Chair in Modern South Asian History at Claremont McKenna College, California. She has authored books on the artisans of Banaras, urban fieldwork, and the history of education in Banaras.

Axel Michaels is head of the Department of Classical Indology at the South Asia Institute and speaker of the Collaborative Research Centre "Dynamics of Ritual" at the University of Heidelberg. Among his fields of specialization are Nepalese culture and religion, ritual studies and Dharmaśāstra. His publications include volumes on Hinduism, asceticism and the Paśupati temple in the Kathmandu Valley.

Sumathi Ramaswamy is Associate Professor of History at the University of Michigan, Ann Arbor. Her previous publications and research interests focused on language and gender in Tamil Nadu, and on the politics of place and identity. Her current research interests centre on the history of cartography in modern India and visual studies.

Stefan Schütte is a social geographer who has carried out extensive research into rural vulnerability and social networks in South India. His work includes a study of the impact of tea-growing on the rural livelihoods and food security of tribal groups in Tamil Nadu. He currently works as an urban livelihoods researcher for the Afghanistan Research and Evaluation Unit (AREU).

Nutandhar Sharma is a freelance researcher of culture and history in Nepal. At present he is working as a research collaborator for the Collaborative Research Centre "Dynamics of Ritual" of Heidelberg University at the Department of Classical Indology, South Asia Institute. His fields of research and publication are concerned with the cultural practices of Newars and other ethnic groups of Nepal.

Rana P. B. Singh is Professor of Cultural Geography at Banaras Hindu University, Varanasi. He has done extensive research in the Varanasi region and has been involved in studying and promoting heritage planning and spiritual tourism for the last two decades. He has lectured at various centres in America, Europe, East Asia and Australia and published numerous books on pilgrimage and the sacred geography of Banaras.

Ravi S. Singh is Reader (Associate Professor), Department of Geography, Banaras Hindu University, Varanasi. His doctoral dissertation deals with the sacred geography of goddesses in India, with special reference to the Varanasi region. His recent book is a study of paths of development in Arunachal Pradesh.

Sunthar Visuvalingam is currently spearheading a multi-volume collaborative project on "Abhinavagupta and the Synthesis of Indian Culture", of which the first collective volume on aesthetics was published in September 2005. He also maintains the *svābhinava* website (*www.svabhinava.org*) and associated public forums for comparative culture and interreligious dialogue.

Annette Wilke is Professor of the Study of Religion and Head of the Seminar of the Study of Religion ("Allgemeine Religionswissenschaft") at the University of Muenster, Germany. Besides systematic studies in religion, her major fields of research are Hindu traditions (devotional, Tantric and Vedāntic), goddess worship and the Tamil Hindu diaspora in Germany. She has carried out fieldwork and published on goddess worship in India, Nepal, Sri Lanka and Germany.

Index

Ethno-Indology

Heidelberg Studies in South Asian Rituals
Edited by Axel Michaels

Volume 1: Jörg Gengnagel, Ute Hüsken, Srilata Raman (Ed.)

Words and Deeds Hindu and Buddhist Rituals in South Asia

2005. 299 pages, 10 ill., pb
ISBN 3-447-05152-3. € 48,– (D) / sFr 83,–

Words and Deeds is a collection of articles on rituals in South Asia with a special focus on their texts and context. The volume presupposes that a comprehensive definition of "ritual" does not exist. Instead, the papers in it avoid essentialist definitions, allowing for a possible polythetic definition of the concept to emerge. Papers in this volume include those on Initiation, Pre-Natal Rites, Religious Processions, Royal Consecration, Rituals which mark the commencement of ritual, Rituals of devotion and Vedic sacrifice as well as contributions which address the broader theoretical issues of engaging in the study of ritual texts and ritual practice, both from the etic and the emic perspective. These studies show that any study of the relationship between the text and the context of rituals must also allow for the possibility that different categories of performers can and do subjectively constitute the relationship between their ritual knowledge and ritual practice, between text and context in differing and nuanced ways.

Volume 3: Niels Gutschow, Axel Michaels

Handling Death

The Dynamics of Death Rituals and Ancestor Rituals Among the Newars of Bhaktapur, Nepal
With Contributions by Johanna Buss and Nutan Sharma and a Film on DVD by Christian Bau
2005. 216 Seiten, 140 Abb., gb
ISBN 3-447-05160-4. € 68,– (D) / sFr 116,–

In a rare combination of competence, an architectural historian (Niels Gutschow) and an indologist (Axel Michaels) have documented death rituals of the ethnic community of Newars in the Kathmandu Valley, Nepal. The first part of the book focusses to a specific setting, the ancient city of Bhaktapur and its calendric rituals of death and renewal. An introduction to the urban fabric with its cremation places, routes of death processions, places of spirits and ancestor deities is followed by a presentation of specialists involved in the death and ancestor rituals – illustrated by 28 maps. The second part presents a detailed description of the union of the deceased with his forefathers, a ritual which is also documented on a DVD. In addition, local handbooks and manuals used by the Brahmin priest during this ritual are edited and translated. This ethno-indological method of combination of textual and contextual approaches aims at understanding both the agency in rituals and the function of the text in contexts. Formalized rituals turn out to be by no means strict, stereotypical and unchangeable. The uniqueness of the actors, places and time has prompted the authors to name places and actors and to date time.
The study of death rituals represents the first part of a trilogy of studies of life-cycle rituals in Nepal.

HARRASSOWITZ VERLAG · WIESBADEN
www.harrassowitz-verlag.de · verlag@harrassowitz.de